FRACTALS IN GEOGRAPHY

Nina Siu-Ngan Lam

Lee De Cola

PTR Prentice Hall

Englewood Cliffs, New Jersey 07632

Library of Congress Cataloging-in-Publication Data

Lam, Nina Siu Ngan and De Cola, Lee
Fractals in Geography/edited by Nina Siu Ngan Lam and Lee De Cola.
p. cm,
Includes bibliographical references and index.
ISBN 0-13-105867-3
1. Geography--Mathematics. 2. Fractals. I. Lam, Nina Siu Ngan. II De C
G70.23.F7 1993 92-37383
910'.01'51--dc20 CIP

Editorial/production supervision
and interior design: *Dit and Dominick Mosco, Ann Sullivan*
Cover design: *Jerry Votta*
Manufacturing buyer: *Mary E. McCartney*
Acquisitions editor: *Michael Hays*

The publisher offers discounts on this book when ordered in bulk quantities. For more information, contact:

Corporate Sales Department
PTR Prentice Hall
113 Sylvan Avenue
Englewood Cliffs, NJ 07632

Phone: 201-592-2863
Fax: 201-592-2249

Printed in the United States of America
10 9 8 7 6 5 4 3 2 1

ISBN 0-13-105867-3

Prentice-Hall International (UK) Limited, *London*
Prentice-Hall of Australia Pty. Limited, *Sydney*
Prentice-Hall Canada Inc., *Toronto*
Prentice-Hall Hispanoamericana, S.A., *Mexico*
Prentice-Hall of India Private Limited, *New Delhi*
Prentice-Hall of Japan, Inc., *Tokyo*
Simon & Schuster Asia Pte. Ltd., *Singapore*
Editora Prentice-Hall do Brasil, Ltda., *Rio de Janeiro*

CONTENTS

FOREWORD

I am delighted to have this opportunity to write a foreword to this very interesting collection of articles on fractals in geography. Although many of the principles of fractal mathematics were worded out in the first half of this Century, and some even earlier, it took computer graphics to reveal the potential value of what were often viewed as bizarre mathematical curiosities. And when drawn, many fractals looked like topography, cities, or other geographic features. Fractals received widespread attention in the geographical sciences after the appearance of the English-language edition of Benoit Mandelbrot's first book on fractals in 1977. Since then, fractals have influenced many areas of geographic research. So far, fractals have been applied to geomorphology, meteorology, and urban growth and form. Fractals also have had a major impact on computer graphics, especially terrain visualization. And perhaps most importantly, fractals have drawn unprecedented attention to the intimate interconnection between scale, resolution, and form when continuous phenomena are represented in the discrete mathematics of digital computers.

The present volume is really two books for the price of one. Part I, written by the editors, provides an excellent tutorial on fractals in general and on methods in fractal analysis. I especially like Chapter 2, which provides a well-written catalogue of methods for the empirical estimation of fractal dimension. I will not comment specifically on the contents of the 10 excellent research chapters that make up Parts II - IV of the book, since the editors have provided an excellent overview in their preface. In any case, these articles speak for themselves.

When new techniques or theories come into a discipline, they are often met with skepticism, distrust, or even hostility. But some techniques and methods exceed some unwritten threshold, and become popular, perhaps even "trendy". At this stage, the new approaches may be used incautiously, or even incorrectly, often by people who do not understand their assumptions and interpretations. This bandwagon phase is often in turn slowed by criticisms of over-use and misuse, and if an approach makes it through the criticism phase, it joins the spectrum of proven methods that are part of normal practice. This book may signal that fractal analysis has matured to take its place as an essential approach to the analysis of spatial data.

David M. Mark
National Center for Geographic Information and Analysis and Professor of Geography,
State University of New York at Buffalo
October 1992

PREFACE

This book is about the use of fractals in geography and the fractal characterization of geographic phenomena. The book was written and compiled for those who are interested in fractals, particularly for geographers and especially for students in advanced undergraduate and graduate geography programs. In this book you will learn what a fractal is (both noun and adjective), how to describe fractal phenomena, and how to make fractal patterns. Of course, there are by now quite a few books on fractals and they can be found on the shelves of bookstores in the science, art, nature, computer, and even philosophy sections.

So why do we need another book on fractals? To begin with, the fractal paradigm is fundamentally important not only to the sciences but also to the other fields mentioned above. No educated person can afford to be ignorant at least of the basic ideas of complexity, self-similarity, and recursion. But fractal concepts are complex, and the subject itself is changing rapidly. A systematic introduction to the uses of fractals in geography and geosciences as presented in this book will be very useful.

The second reason is that there is no work that provides a methodical introduction to the use and analysis of fractals in the geosciences and especially geography. Indeed, a common comment from many colleagues, usually after reading another article on fractals in a popular magazine or newsletter is "so what?" How can geographers use it? This book is aimed at answering these "what" and "how" questions. We have read about and worked with fractals firsthand and have brought to this treatment a geographic sense of the importance, value, and applicability of this new paradigm to our field. After reading this book, you may not look at the world in quite the same way as before.

The third reason is that the book is an organized compendium of research in the field. We have invited 18 scientists to contribute 10 specialized research chapters that demonstrate in detail the way they use fractal concepts in their research in physical and human geography, as well as in cartography, remote sensing, and geographic information analysis. In leafing through the volume, you may have noticed that it is not just another pretty book. Its appearance reflects the fact that we aimed to present working papers that capture the evolving spirit of scientific research, rather than finished products that, however beautiful, leave the reader mystified about "how that was done."

We hope from this book to bring to you the fascinating subject of fractals and its relevance and potential in geography; it is by no means definitive, and many important problems still remain untackled. Ahead is an avenue of more exciting and fruitful research.

The book is divided into an introductory Part I of 4 chapters and three research Parts consisting of 10 chapters. Part I was written by the editors for readers with little formal background in fractals or quantitative geography. The examples we use to illustrate the various principles of fractals were taken deliberately from conventional geography, yet one can see that these "old" examples are tied closely with "new" fractals, as well as with different fields of science. The section should give you a firm understanding at least of certain kinds of fractals and a good notion of "how it is done". Parts II to IV, which were written by a broad group of scientists, cover wider ground, and are intended as surveys of state-of-the-art research.

The 10 research chapters fall into three parts. Chapters 5 through 9 in Part II center on research and applications in physical geography, where the focus is mainly on natural phenomena. This subfield is the earliest to use fractals and is still dominating the literature. More and more applications, however, are found in human geography, as in Chapters 10 through 12 in Part III, and in mapping sciences (cartography, GIS, and remote sensing) as in Chapters 13 and 14 in Part IV. What follows is a brief review of the research chapters.

Borrough in Chapter 5 gives a comprehensive review of most of the significant uses of fractals in landscape studies. While the fractal model has proved its usefulness in simulation and description, more analytical uses of fractals, especially toward the understanding of processes and multiscales, are urged.

Goodchild and Klinkenberg in Chapter 6 argue that the fractal model is the null process for river networks, thus setting an example of the use of the fractal model in hypothesis testing. The fractal paradigm is maturing, as evidenced in this chapter, by its transition from being the alternative to the null hypothesis.

Chapter 7 by **Phillips** provides a critical review of the misuse of the fractal model in interpreting river networks in previous research, and demonstrates with examples that existing methods for estimating the fractal dimension for river networks based on traditional morphometric parameters are unreliable.

Lavallée, Lovejoy, Schertzer, and Ladoy in Chapter 8 stress that monofractal models are inadequate to characterize anything but ideal fractal sets. They extend the monofractal analysis of sets to the multifractal analysis of fields and demonstrate a technique for estimating the multifractal parameters. Their research represents one of the latest challenges to classical fractals and has broad implications in geophysics and geography.

Plotnick and Prestegaard in Chapter 9 provide both a concise review of the leading techniques for the analysis of time series data and an illustration of how these techniques lead to consistent results using the authors' sediment transport data. Here we have an application of the fractal model to a spatial phenomenon with a time series component.

Arlinghaus in Chapter 10 demonstrates clearly and graphically the venerable central place theory as an ideal fractal recursive set, which has implications for partitioning of space for GIS developments. Potential linkages between fractals and other classical geographical concepts, such as the Hagerstrand diffusion model, are also suggested.

Chapter 11 by **Batty, Fotheringham, and Longley** focuses on the use of fractal simulation in understanding urban growth and compares a number of cluster growth models. It is a concise overview of fractal applications in urban morphology studies, with several major issues addressed, including the role of urban morphology studies in urban geography, the argument between process and form, and the rationale and potential of using fractals.

Lam, Qiu, Zhao, and Jiang in Chapter 12 explore the applicability of the fractal model in analyzing and characterizing cancer mortality patterns in China. A comparison of fractal measurements from different methods is also provided to illustrate the issue of measurement with socioeconomic data based on irregular political boundaries.

Clarke in Chapter 13 presents new ways of simulating fractal terrain. With the use of the Mt. Everest elevation data, the chapter provides a striking example of how models, images, and the real world are linked through fractal analysis. It is interesting to learn that the author is able to create artificial Mt. Everests that are more convincing than the real thing!

In the last chapter, Chapter 14, **De Cola** demonstrates the use of multifractal analysis in visualization and image processing. Using a satellite image, he shows how deviations from smooth multifractality may reveal features in landscapes, thereby allowing the generation of hypotheses about the underlying processes.

Many individuals have helped to make this book possible. We would like to thank, first and foremost, the contributors for their encouragement, time, and willingness to modify, sometimes extensively, the manuscripts to suit this book. Keith Clarke, Dale Quattrochi, David Mark, Norman Bliss, Pat Chavez, Jr., and Nicholas Matalas kindly read and commented on the editors' chapters, with David Mark also providing a foreword. Louisiana State University graduate students Stephanie Weigel, Changyong Cao, and Hong-Lie Qiu assisted in proofreading, data preparation, and programming. Thanks are also extended to Clifford Duplechin, Mary Lee Eggart, and Kerry Lyle of Louisiana State University for their professional cartographic and photographic work; to the faculty and staff of the Geography and Geology Department and Gar-On Yeh, both of the University of Hong Kong, for providing facilities to finish this book while Lam was on sabbatical leave; and to Mike Hays of Prentice Hall for his insight and support. Finally, the support from our families is gratefully acknowledged.

Part I

BASIC FRACTALS

CHAPTER 1

INTRODUCTION TO FRACTALS IN GEOGRAPHY

Lee De Cola
U. S. Geological Survey, Reston, VA

Nina Siu-Ngan Lam
Department of Geography and Anthropology, Louisiana State University, Baton Rouge

From the *Random House Dictionary of the English Language* (1987), we have:

> geography, from greek: *geo* (earth) + *graph* (*graphein*: to mark or carve).
>
> geometry: the branch of mathematics that deals with the deduction of the proportions, measurements, and relationships of...figures in space, from *geo* (earth) + *meter* (measure)
>
> fractal: a geometrical or physical structure having an irregular or fragmented shape at all scales of measurement....

The objective of this introductory chapter is to use fractal concepts to present the key ideas of classical and fractal geometry as applied to geographic science. We begin by presenting a few examples of complex spatial phenomena in the context of a simple typology of objects. We then explore the problem of describing the forms of three ideal fractals, offering the fractal dimension as a simple measurement. This concept is next applied to three real phenomena manifesting different varieties of complexity. These exercises should give the reader a basic understanding of key fractal ideas.

WHY FRACTALS?

In an article written almost a decade before he coined the term "fractal", Mandelbrot (1967) posed a question that geographers began asking much earlier: how long is the coastline of Britain? The idea that decreasing the divider step used to measure a mapped coastline generally increases the number of steps more than proportionally is a fact that has puzzled many researchers. This "Steinhaus paradox" (1960) is related to a number of questions that have long preoccupied cartographers and geoscientists, including:

- What is the appropriate scale and generalization level for a map?
- How does scale relate to the nature of such phenomenon as the curvature of the coastline and the importance of individual bays and promontories?
- What sampling interval or resolution (that is, step size) should be used?

These issues have played a key role in the development and definition of fractals.

Real-world spatial phenomena typically reveal more detail with increasing resolution. If anything, this "paradox" resides in our insistence that all objects have some scale-independent measure of extent. We will discuss later that complete scale-independent measure, however, is rather unrealistic. In this chapter we introduce a simple approach to fractal description of curves (and other things) with an emphasis on geographical applications.

Fractal geometry arose in part out of frustration that classical geometry is not capable of characterizing or measuring many natural forms. A coastline is not straight or circular; nor can a landscape be representable by any analytical function. Similarly, it is cumbersome and sometimes impossible to describe snowflakes, terrain, urban patterns, and river networks through such traditional means as a set of complicated polynomials. Fractal dimension can therefore be used primarily as an index for measuring the complexity of curves and surfaces.

Why should researchers want to learn about fractals in the first place? One answer is that measurement is a key initial step in scientific understanding. The measurement of forms and patterns, and particularly of distance, curvature, and shape, provides information about the underlying structures and processes.

But perhaps the greatest reason for learning about fractals is the sheer volume of research produced from these new ideas. Fractal techniques for scale parameterization are being used in numerous studies as an index for describing the complexity of curves and surfaces. For example, fractals have been used to examine coastlines and islands (Goodchild, 1982; Lam, and Qiu, 1992), shoreline erosion (Phillips, 1986), topography and terrain (Mark and Aronson, 1984; Roy et al., 1987), soil properties (Burrough, 1983), rainfall and clouds (Lovejoy and Mandelbrot, 1985), coral reefs (Bradbury and Reichelt, 1983; Mark, 1984), ocean bottom relief (Barenblatt et al., 1984), underside of sea ice (Rothrock and Thorndike, 1980), central place theory (Arlinghaus, 1985), urban structure (Batty and Longley, 1986), point patterns (De Cola, 1991), and remote sensing images (De Cola, 1989; Lam, 1990).

Moreover, Goodchild (1980) has argued that the fractal dimension can be used to predict the effects of cartographic generalization and spatial sampling, a result that may help in determining the resolution of pixels and polygons used in studies related to geographic information systems and remote sensing. Muller (1986) proposed the use of the fractal dimension as a guiding principle for implementing generalization algorithms in automated cartography. In this book (Chapter 5) and in 1981, Burrough shows that many environmental variables are fractal and that the examination of the fractal dimension is useful for separating scales of variation that might be the result of natural processes (Burrough, 1981).

Over the years, then, applications of the fractal dimension and its associated scale ranges have evolved from mere description and measurement of phenomena to hypothesis formulation and investigation of scaling properties (Lam and Quattrochi, 1992). The research chapters in Parts II to IV describe in detail the various applications of fractals in studying geographical phenomena. One can see that the potential of fractal applications is boundless.

INTRODUCTION TO BASIC CONCEPTS

It is relaxing to read a new book while having something to drink. Please take a minute to do the following experiment. First, pour yourself a cup of black coffee or strong dark tea, stir the coffee, and immediately pour in about a teaspoon of cream (better than milk for this experiment). What kinds of turbulence patterns do you see? What is their shape? Are they smooth, regular, or uniform? Do they have shapes at all? How big are these patterns? Do they have a typical size? What do we mean by pattern and shape in the first place, and how did such forms arise? Now look at Figure 1.1. Do you see any similarities? What do

Figure 1.1 A digital image of the surface of Jupiter. (Provided by Dr. Bradford A. Smith, Voyager Experiment Team, National Space Science Data Center, NASA.)

we mean by similar, and does this correspondence mean that analogous processes are at work? What do we mean by process? How should something that is apparent in a cup of coffee at a scale of centimeters (10^{-2} meters) correspond in any way to something on the planet Jupiter at a scale of a 100,000 kilometers (10^8 meters)? Do these systems have anything to do with one another? Could a cup of coffee on your desk influence or be influenced by turbulence on Jupiter? Finally, what does all this have to do with geography and with fractals? After you have thought about these problems, you can drink your coffee!

Now consider Figure 1.2, which represents a phenomenon at a scale roughly of kilometers (10^3 meters), midway between your coffee cup and Jupiter. Can you guess what this object represents? As you puzzle over this question, ask how could the object be described, characterized, and analyzed? Many words come to mind (complicated, irregular, bloblike, and so on), but what rigorous language could we use to explore the characteristics of this thing? Now hold the figure upside down in front of a mirror and see if you can figure (!) out what it is. You should be able to recognize "BosWash", the vast U.S. northeastern megalopolis.

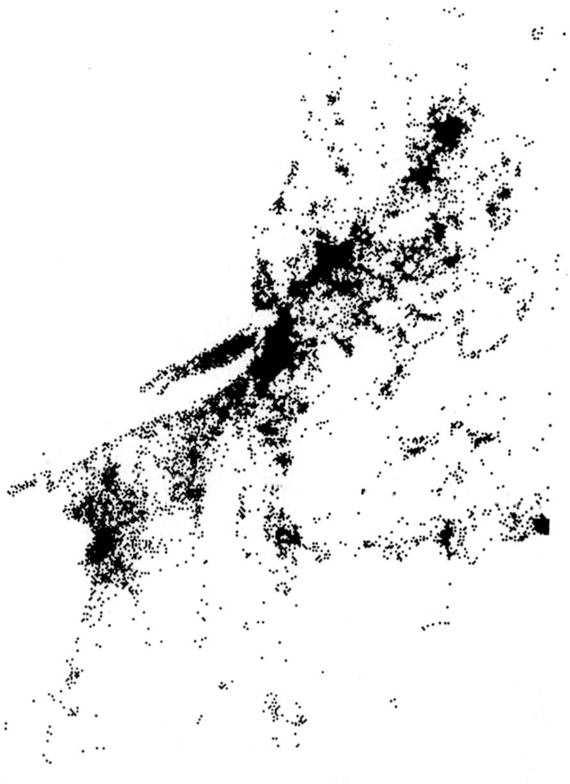

Figure 1.2 An upside down image of population concentration in the "BosWash" northeastern U.S. megalopolis, created from the 1980 U.S. Census. (Source: Argonaut Press, Glasglow Road, Fairfield, Iowa 52556.)

STEP 1: Begin with a line segment AB:

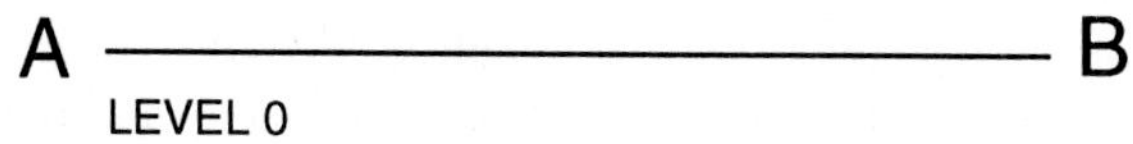

STEP 2: Replace the segment with the Koch generator:

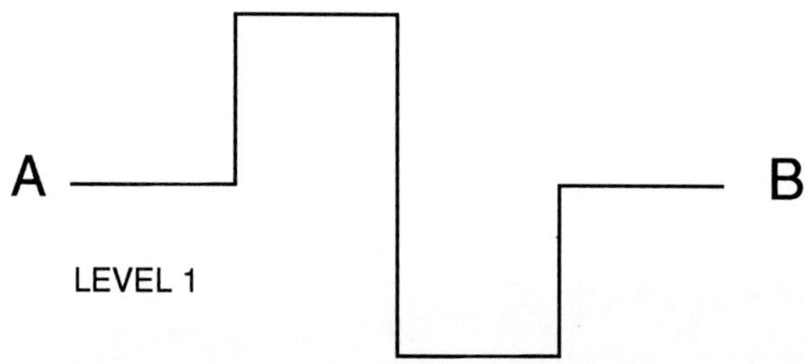

STEP 3: Now do two things with the generator. First, scale it down by 1/4:

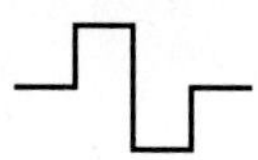

STEP 4: Next, replace each line segment in the generator by the scaled-down version:

STEP 5: Scale down the generator of STEP 2 by 1/16 and replace each of the line segments of STEP 4 with the result.

STEP 6 ... ∞: Repeat STEP 5 endlessly, scaling the generator by $(1/4)^n$ and replacing each of the line segments in the new figure with the scaled-down version of the generator.

Figure 1.3 Steps in the construction of a Koch quadratic curve.

What geographic techniques would you use to describe, characterize, and analyze the objects found in coffee cups, on the earth, and in outer space? To help answer these questions, consider the definitions at the beginning of this chapter. The first definition implies that geography writes about, graphs, and even carves up the earth to help us understand the world and its objects. In fact, geography studies many kinds of complex and dynamic objects that lie on or near the surface of the earth in a space of three physical dimensions and one temporal dimension. Within space–time geographers have abstracted many kinds of things (for example, people, rivers, cities, and mountains) whose relationships are important to human welfare. The discipline remains true to its origins by carving up this space in various ways by creating abstractions of real objects through physical and mathematical models, as well as by performing rescalings, intersections, and projections on maps. In fact, geography deals with real objects that often look similar to pieces of themselves when examined at different cartographic scales and that have forms that manifest significant randomness.

To make these ideas clear, we begin with an ideal fractal object, the Koch quadratic curve (Mandelbrot, 1983). Figure 1.3 shows the first few steps in the construction of this curve, which will be used below as part of a discussion of fractal measurement. The Koch curve can be constructed by taking a LEVEL-0 straight line segment AB and replacing it

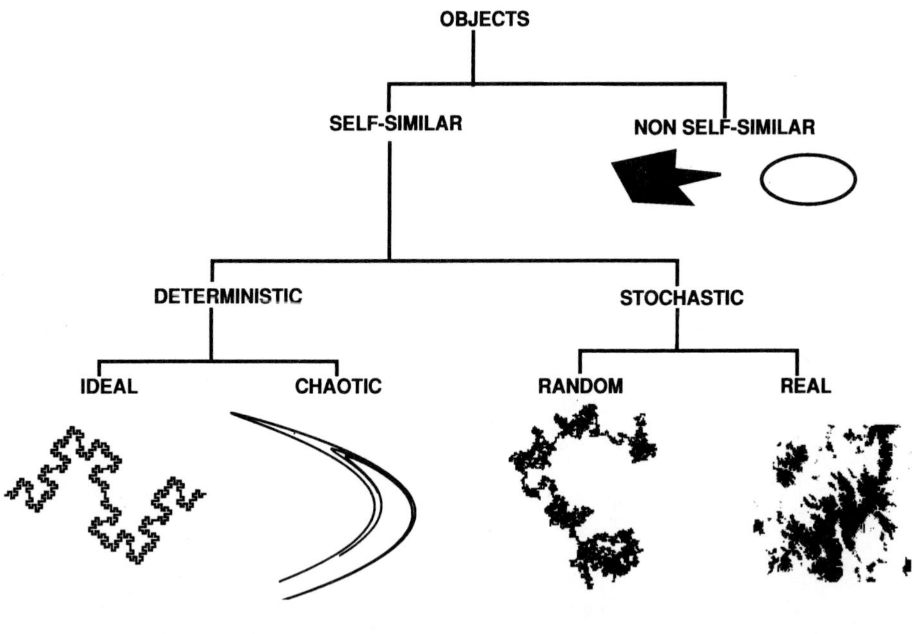

Figure 1.4 A simple typology of geometric objects.

with the squiggly eight-segment (LEVEL-1) Koch generator that also spans *AB* but is much less direct. We then reduce the generator to one-quarter of its original size and replace each of its segments with this new smaller Koch generator, giving us the LEVEL-2 construction. When each of the little segments of LEVEL-2 is replaced by another one-quarter scale reduction of the generator, we get the LEVEL-3 construction. Obviously, we can continue this process indefinitely, adding more and more squiggliness but still only spanning *AB*.

Three things stand out about the Koch quadratic curve. First, it is irregular, rough, squiggly, complicated, and so on. Second, it manifests self-similarity in that pieces of it are a scaled-down version of the whole of it. Third, it is ideal in that we can easily construct levels of it by following simple rules.

Figure 1.4 is a rough taxonomy of spatial objects. Classical geometry deals mainly with objects that, unlike the Koch curve, cannot be recursively subdivided into self-similar components, while fractal geometry provides an approach to the analysis of self-similarity. Much of the early work on fractals focused on ideal objects that are constructed to be completely self-similar, such as the Koch quadratic curve. Although we have shown only its fourth level, an ideal fractal would bifurcate infinitely. Another kind of fractal may be represented by the chaotic trajectory of a nested function. In the same figure we have represented chaos as the recursive values of a so-called strange attractor.

Self-similar objects whose evolution is not deterministic, such as the random-walk trajectory shown in Figure 1.4, manifest a degree of stochasticity that makes forecasting their behavior impossible. The real-world objects of geography, represented in the figure as a black–white image of an area of the Rocky Mountains, are also often self-similar, and they are certainly stochastic. We shall return to these issues in Chapter 4. The rest of this chapter presents several examples of basic fractal objects and how they can be characterized by use of the concept of dimension, an approach to understanding geometric form through the attributes of number and size.

EXAMPLES OF IDEAL FRACTALS

Size of a Line Segment

The simplest way to introduce fractal analysis is by analyzing a basic geometric object, so we begin with a line segment. We shall use the fundamental notions of size and number referred to above to arrive at an understanding of the form of this object. Then we shall apply this argument to the analysis of other more complex entities.

Consider the line segment connecting points *A* and *B*:

A ____ *B*

The discussion that follows refers to the particular line segment above and not some ideal line. Several things stand out about this object: it is bounded, it exists at a physical scale of centimeters (any smaller and it would become invisible; any larger and it would not fit on the page), it is straight (or as nearly straight as printing can make it), and (as a consequence of these characteristics) it has a definite size. To measure this size we need a definition and a standard, so let us assume we have a standard line segment of size STEP = 1/4 (decimeter). We realize we have begged the question of how to define the size of our measuring step, but it is necessary to begin at a level lower than the almost completely abstract level of formal geometry, and if we are going to measure, we need something to measure with!

Now let us define the DISTANCE from *AB* as the product of COUNT, the number of times that STEP marks off *AB*, and STEP, the size of the measure. Suppose we find that COUNT = 4, so that the product COUNT × STEP = 1. Does this mean the size of *AB* is 1 decimeter? We are inclined to think so, but what happens for other STEP values? Let's generalize this process. Try the following experiment. Begin with a divider STEP = 1/4 (decimeter) and count off the number of COUNTs that STEP fits into *AB*. The answer is 4, as we found above as well as show in the second row of Table 1.1

TABLE 1.1 Relationship between COUNT and STEP for a Line Segment

LEVEL	STEP	COUNT	DISTANCE	Log (STEP)	Log (COUNT)
0	1	1	1	0	0
1	1/4	4	1	-2	2
2	1/16	16	1	-4	4
3	1/64	64	1	-6	6
4	1/256	256	1	-8	8
5	1/1024	1024	1	-10	10
LEVEL	$(1/4)^{\text{LEVEL}}$	4^{LEVEL}	1	-2LEVEL	2LEVEL

If we try this value for various powers of STEP = $(1/4)^{\text{LEVEL}}$, we always get the results as shown: the product COUNT × STEP = 1, which seems to indicate that the size of *AB* = 1. A plot of these results gives Figure 1.5a, which is not particularly informative, because COUNT increases very rapidly as STEP decreases, as Table 1.1 also shows. But if we examine the logarithms of the data (where logs are taken to base 2 for convenience), we get the results shown in the last two columns of the table as well as in Figure 1.5b, which is a straight line. These data conform to the following linear equation:

$$\log(\text{COUNT}) = a + b\log(\text{STEP}) \tag{1.1}$$

where the intercept $a = 0$ and slope $b = -1$, so that log(COUNT) = 0 - log(STEP) or COUNT = STEP^{-1}. Thus, a formula that yields a way of computing the size or length of the line segment becomes DISTANCE = COUNT × STEP = 1, which is a constant.

Two points can be made. First, at least in the case of the straight line AB, the product of COUNT × STEP = constant, which would have to be if the concept of length as size is to have any general meaning (we shall see that it often does not). Second, this product equals 1, which motivates us to assign the number 1 (decimeter) to the size (length) of the line. But these observations should not obscure what is central to the argument here: that Equation 1.1 allows us simply to characterize the line AB as having a form specified by the relationship

$$\text{COUNT} \propto \text{STEP}^b \tag{1.2}$$

That is, the COUNTs are inversely proportional to STEP size, and in this particular case $b = -1$. Now let's see how this approach to COUNT and STEP analysis applies to other objects.

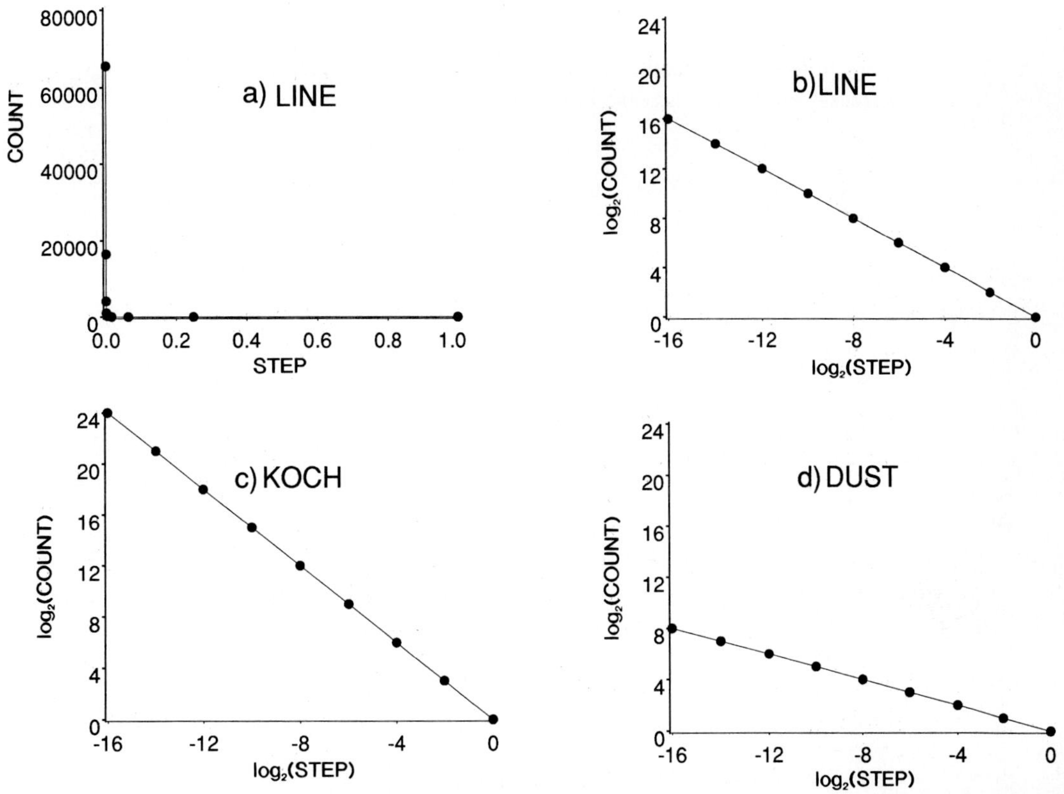

Figure 1.5 Plot of COUNT versus STEP for three ideal fractals.

Length of a Fractal Curve

Look at the Koch quadratic curve of Figure 1.6, which also connects the points *A* and *B*. The curve is constructed at different resolutions for each LEVEL value as shown in Figure 1.3. (More complex examples of this kind of self-similar construction are presented in Chapter 3.) At LEVEL = 0, STEP = $(1/4)^0 = 1$ and COUNT = 1 (because 1 of these STEPs fits the curve). At LEVEL = 1, STEP = $(1/4)^1 = 1/4$, but COUNT = 8 (as opposed to the case of the line segment, where COUNT = 4). At LEVEL = 2, STEP = $(1/4)^2 = 1/16$ and COUNT = 64, and so forth, as shown in Table 1.2.

The value of COUNT increases even faster than in the case of the line *AB*, yet a graph of the unlogged values does not look much different from Figure 1.5a. When the logged values are plotted, as in Figure 1.5b, we again get a straight line, but in this case Equation 1.1 yields log(COUNT) = 0 - 3/2 log(STEP), while Equation 1.2 becomes COUNT = $\text{STEP}^{-3/2}$, where $b = -3/2$ is a *scaling relationship* relating the number of COUNTs for STEP at each resolution LEVEL. In general, when the logs of two variables have a linear relationship, we speak of the dependent variable as *scaling* with respect to the independent variable.

Note that, in contrast to the line segment, DISTANCE is not a constant, but increases with each increase in resolution, and DISTANCE = COUNT × STEP = 2^{LEVEL}

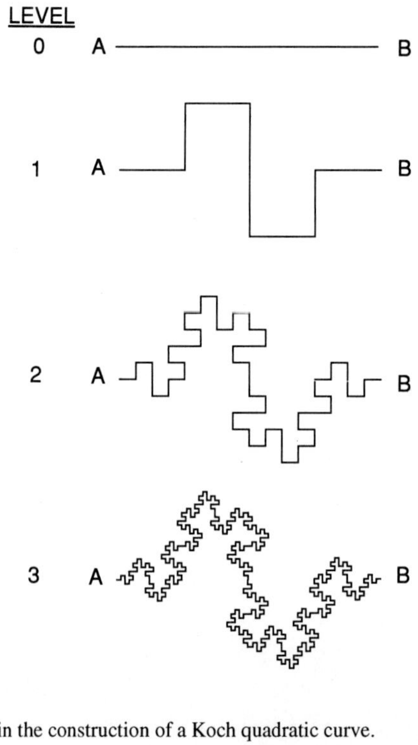

Figure 1.6 Levels in the construction of a Koch quadratic curve.

as shown in the table, so examining (or constructing) the Koch curve at larger cartographic scales reveals a longer distance between points A and B. As we examine the Koch curve at higher and higher resolutions (decreasing STEP size), we see ever more (and apparently endless) detail, so that distance increases. In more rigorous language, we say that "COUNT scales as the -3/2 power of STEP," just as an engineer might say that "the volume of a sphere scales as the 3/2 power of its area: VOLUME $\propto$ AREA$^{3/2}$."

How do we characterize this intriguing behavior? Certainly, DISTANCE is not a meaningful parameter, at least in the usual sense of the term. Yet the scaling relationship between COUNTs and STEPs is regular and unambiguous, parameterized by the scaling exponent $b = -3/2$. Now recall that for the line segment, $b = -1$, representing the balance between COUNT and STEP that makes the measurement of DISTANCE for an ideal line possible. This is one reason why we call line segments one-dimensional and assign lines the fractal dimension $D = -b = 1$. In the case of the Koch curve, whose irregularity and complexity increase in a consistent fashion with each resolution level, we use the same logic and give this curve a fractal dimension $D = -b = 3/2$, that is, as the length of STEPs decreases by the second power of 2 (or 1/4), the number of steps increases by the third power of 2 (or 8), so that the slope of the line in Figure 1.5c is -3/2.

TABLE 1.2 Relationship between COUNT and STEP for a Koch Quadratic Curve

LEVEL	STEP	COUNT	DISTANCE	Log (STEP)	Log (COUNT)
0	1	1	1	0	0
1	1/4	8	2	-2	3
2	1/16	64	4	-4	6
3	1/64	512	8	-6	9
4	1/256	4096	16	-8	12
5	1/1024	32768	32	-10	15
LEVEL	$(1/4)^{LEVEL}$	8^{LEVEL}	2^{LEVEL}	-2LEVEL	3LEVEL

The slope of Equation 1.1 can therefore be used as an index of the irregularity of a curve, and indeed a useful definition of fractal dimension is $D = -b$. We assign a fractal dimension of $D = 1$ to the line (as well as to all smooth curves that essentially become linear at high resolution), for which COUNTs and STEPs exactly balance, and $D = 1.5$ to the Koch curve, for which COUNTs increase faster than STEPs, so that length is a function of step length (or, more generally, of resolution), a phenomenon familiar to cartographers, as we shall demonstrate in Chapter 2.

To recapitulate, the fundamental equation of fractal scaling analysis is therefore

$$\text{COUNT} \propto \text{STEP}^{-D} \tag{1.3}$$

What do we then say of a curve for which $D > 1$, that is, a curve that reveals more length at higher resolution or cartographic scale? This behavior is quite common in the real world in that closer examination of something almost always reveals more detail. You might be tempted to say, "Of course the Koch curve is *constructed* to look different at each scale level, and so it is not 'natural' like a straight line." But, on the contrary, it is a common characteristic of "natural" (that is, real world) phenomena that, unlike straight lines, they look different at different scales, as we shall see when we look at real data.

Length of a Fractal "Dust"

A straight line segment is a fractal set that can be recursively subdivided into smaller line segments whose number (COUNT) scales exactly inversely with their size (STEP), but for the Koch curve the number of sublines increases faster than size decreases—the closer you look, the more you see. What of a set that reveals *less* length at greater resolution? Consider an example of the Cantor "dust" shown in Figure 1.7 (Mandelbrot, 1983). At LEVEL = 0, as before, STEP = 1, COUNT = 1, and LENGTH = 1, but higher resolution reveals less distance, as shown in Table 1.3.

Again, this behavior is strange and unlike that of a straight line segment, but the relationship between COUNTs and STEPs is still scaling, as shown in Figure 1.5d, as well as by a fit to Equation 1.1:

$$\log(\text{COUNT}) = 0 - 1/2 \log(\text{STEP})$$

so that $D = -b = 1/2$, a generalization of the Koch curve but in the other direction—the closer you look, the less you see!

In general, therefore, to estimate the dimension (D) of some curve, take a STEP of some size < 1 and COUNT the number of times that STEP marks off the curve. Then do this for various LEVELs, COUNTing the number of times that $\text{STEP}^{\text{LEVEL}}$ marks off the curve. Finally, fit Equation 1.1 to your data and $D = -b$. Generally, the rapidity with which COUNT increases is a measure of complicatedness or irregularity, and the goodness of fit of the estimate is a measure of self-similarity (the tendency of an object to retain its form with changes in scale). Furthermore, DISTANCE, while a determinable number, may not

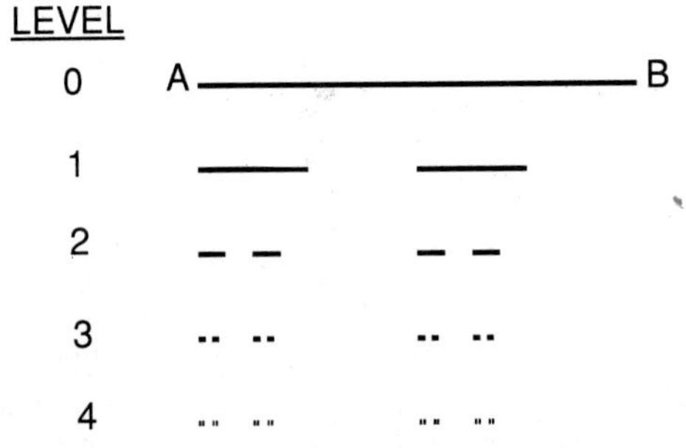

Figure 1.7 Levels in the construction of a Cantor dust.

be a useful concept when dealing with complexity, either in the real world or in the world of fractal constructions.

TABLE 1.3 Relationship between COUNT and STEP for a Cantor Dust

LEVEL	STEP	COUNT	DISTANCE	Log (STEP)	Log (COUNT)
0	1	1	1	0	0
1	1/4	2	1/2	2	1
2	1/16	4	1/4	-4	2
3	1/64	8	1/8	-6	3
4	1/256	16	1/16	-8	4
5	1/1024	32	1/32	-10	5
LEVEL	$(1/4)^{LEVEL}$	2^{LEVEL}	2^{-LEVEL}	-2LEVEL	LEVEL

EXAMPLES OF REAL FRACTALS

So far we have been analyzing the complexity of curves embedded in a space of Euclidean dimension $E = 2$. In the next chapter we shall present the analysis of a cartographic line using a technique based on the COUNT/STEP approach used above. But what of the estimation of size for objects that exist in $E = 2$-space but for which line segment STEP-counting may not be an appropriate measure of size? Some objects "occupy" space in a way that the topologically one-dimensional objects we have considered up to now do not. For a simple object, such as a rectangle, we would just count the number of unit squares needed to fill it. In an exactly analogous fashion to the discussion of a line segment above, we would find that the COUNT of squares of size STEP $(1/4)^2$ would be

$$\text{COUNT} \propto \text{STEP}^{-2} \tag{1.4}$$

so that Equation 1.1 gives the following fit:

$$\log(\text{COUNT}) = A - 2\log(\text{STEP}) \tag{1.5}$$

and in the same way that we assigned a fractal dimension of $D = -b = 1$ to the line we can assign $D = -b = 2$ to the square. This logic conforms to the common notion of a square as being an object of fractal dimension $D = 2$ (any other of whose "dimensions" should also equal 2 as befits the defining object for an $E = 2$-space). For the simple and ideal figures treated in classical geometry, we know that, like the square, the AREA measurement is constant, and the product of COUNT $\times$ STEP^2 = AREA is a constant, so the scaling relationship COUNT $\propto$ STEP^{-2} holds, which gives a fractal dimension $D = -b = 2$.

Area of Remotely Sensed Feature

But now consider a space-filling object that is not square, nor even a rectangle, but that, as with the Koch curve, changes with resolution level. In the analysis of satellite images, scientists seek to understand the phenomenon of land cover. Consider in the present case the problem of characterizing dry land versus wet land (we shall not explore the problem of what is or might be a "wetland," but our analysis does bear on this complex and quite political issue). A data structure for this problem is a square window subdivided into 256^2 identical, adjacent, square cells each of size $(10\text{ m})^2$. Let us decide that any cell completely filled with liquid water shall be called a WATER cell, while any other kind of cell shall be called OTHER. These data can then be represented as a picture of $256^2 = 65{,}536$ cells, with the WATER cells shown as black (as in Figure 1.8). So we have reduced the very complex problem of characterizing land cover to the simpler (but still quite sophisticated) problem of analyzing a binary or monochrome image.

Let's see how the AREA of WATER changes at different resolutions. Again, we can COUNT the number of squares of size STEP that are contained in the set. We have therefore classified the image at three other resolution levels, using cells of size 10^2, 20^2, 40^2,

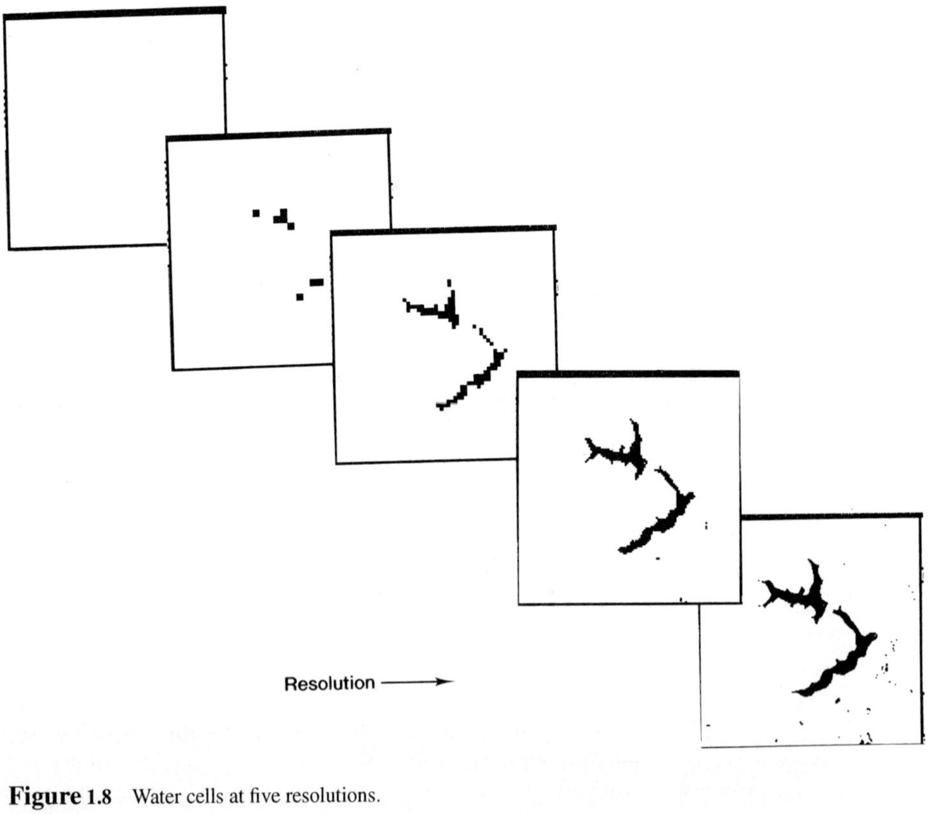

Figure 1.8 Water cells at five resolutions.

and 80^2 square meters. Our results (now using logs to base 10) are shown in Table 1.4. It is clear that the COUNT of WATER cells increases faster than the area of the cells decreases, so that the total area of the window devoted to WATER rapidly increases as cell size decreases. We determined above that the way to characterize scale-dependent behavior is to examine log(COUNT) at different log(STEP) resolutions, and we can do the same thing here. Figure 1.9a shows what happens to the logs. As expected, COUNTs increase with increasing resolution, although the points are not so perfectly linear as they were for the previous artificial data. Yet we can fit Equation 1.1 to the above data, yielding

$$\log(\text{COUNT}) = 6.35 - 2.76 \log(\text{STEP}), \text{ where } R^2 = 0.99.$$

The WATER cells are much like the Koch curve that they proliferate faster with increasing resolution, so a strict application of the fractal argument used above requires that we give this object a dimension of $D = -b = 2.76$, which is greater than that of a simple $D = 2$ polygon.

TABLE 1.4 Relationship between COUNT and STEP for WATER Cells in a Satellite Image

LEVEL	STEP	COUNT	AREA	Log(STEP)	Log(COUNT)
0	80	10	64,000	1.90	1.00
1	40	109	174,000	1.60	2.04
2	20	658	263,200	1.30	2.82
3	10	3241	324,100	1.00	3.51

Are the WATER regions fractal? Yes and no. They certainly exhibit multiresolution behavior inconsistent with a $D = 2$ set, say a square. The fact that more detail is revealed as we increase resolution is characteristic of fractal behavior, and this effect is quite striking. Nevertheless, Figure 1.9a reveals that the relationship between resolution and measurement is not perfectly linear, as required by ideally fractal behavior. Moreover, an extrapolation of the scaling equation with a STEP size of 0.24 meter would predict that the whole image would be WATER. Although this is not realistic (and the curve is not linear anyway), we can imagine that, as STEP approaches millimeter size (LEVEL = 34), we would have to be counting individual drops of water, which in a humid environment exist almost everywhere on the ground, so that the "wetland" would be nearly universal. We can at least conclude that the WATER phenomenon does exhibit fractal characteristics, but also has some scale dependency, as expected of a real-world pattern.

Sizes of Cities

For our penultimate example, we examine a population distribution, which is a phenomenon that is as much demographic as spatial. The index we used for lines and squares was a linear STEP, and for settlements viewed as regions an obvious index would also be some areal measure (see Chapter 11). But for settlements as collections of people their populations are an appropriate index. Many spatial phenomena, as we have seen, have no typical scale or size. We tend not to speak of an average-sized country, island, mountain, lake, city, whether by size we mean area, height, population, or any other measure of magnitude. Rather, we often rank a list of objects from the biggest, longest, most populous member of a set (see any sample of "Global Pursuit" questions). In fact, fractal phenomena that are scaling tend to have a single and often quite large extreme value. Table 1.5 shows the size distribution of the populations of the 1615 largest cities in the world in 1970 (De Cola, 1985).

TABLE 1.5 Relationship between POPULATION and COUNT (number of cities larger than a given POPULATION) for the world, 1970

POPULATION (Greater Than)	COUNT	Log (POP)	Log (COUNT)
4,000,000	24	6.60	1.38
2,000,000	63	6.30	1.80
1,000,000	159	6.00	2.20
500,000	345	5.70	2.53
250,000	726	5.40	2.86
100,000	1615	5.00	3.21

From Table 1.5 we can see that there is certainly no typical size here; cities of all sizes from over 4 million to perhaps 1 (if we count hermits!) can be found throughout the world. In some cases the largest element of a geographical set is very much bigger than the next largest—often twice as big—and about three times as big as the third element, four times larger than the fourth, and so forth. One aspect of spatial phenomena, as we have seen, is the tendency of entities to be scaling or to have no characteristic size.

Once again we can fit Equation 1.1 to these data (see Figure 1.9b), yielding

$$\log(\text{COUNT}) = 9.36 - 1.20 \log(\text{POPULATION}), \text{ where } R^2 = 0.997.$$

Interpretation of the scaling coefficient $D_{\text{CITY}} = 1.20$ is not so straightforward as in the strictly spatial cases discussed above, but the human geographic consequences of this analysis are clear. Imagine that world urbanization were much more uniform; the slope of the regression line characterizing the data in Figure 1.9b would be much flatter and the scaling coefficient D_{CITY} would be smaller; the world urban system would have a lower fractal dimension. In fact, we can regard settlements as scaling and therefore fractal (see Chapter 10): regional sets of settlements manifest rank-size regularity, metropolitan areas can be disaggregated into constituent cities and towns, and big cities can be regarded as aggregations of neighborhoods, which are like settlements in their socioeconomic organization.

Berry (1964) used one of the most descriptive phrases in geography to title his paper "Cities as Systems Within Systems of Cities."

Indeed, this situation is analogous to the scaling self-similarity of turbulence in your coffee cup and on Jupiter (and even in BosWash?) that was illustrated above. Sets of cities are scaling, in the sense that any set, regardless of its extent, is similar to any other in the interrelationship of the sizes of its elements. In fact, we could go beyond this and claim that any set is merely made up of smaller subsets, which are made up of even smaller sub-subsets, and so on. If the subsets are in some way similar to the supersets, then the phenomenon is probably fractal.

Numbers of Map Features

For our last example we compare the kinds of features on maps. The U.S. Geological Survey has established policies for the numbers and kinds of features it represents on

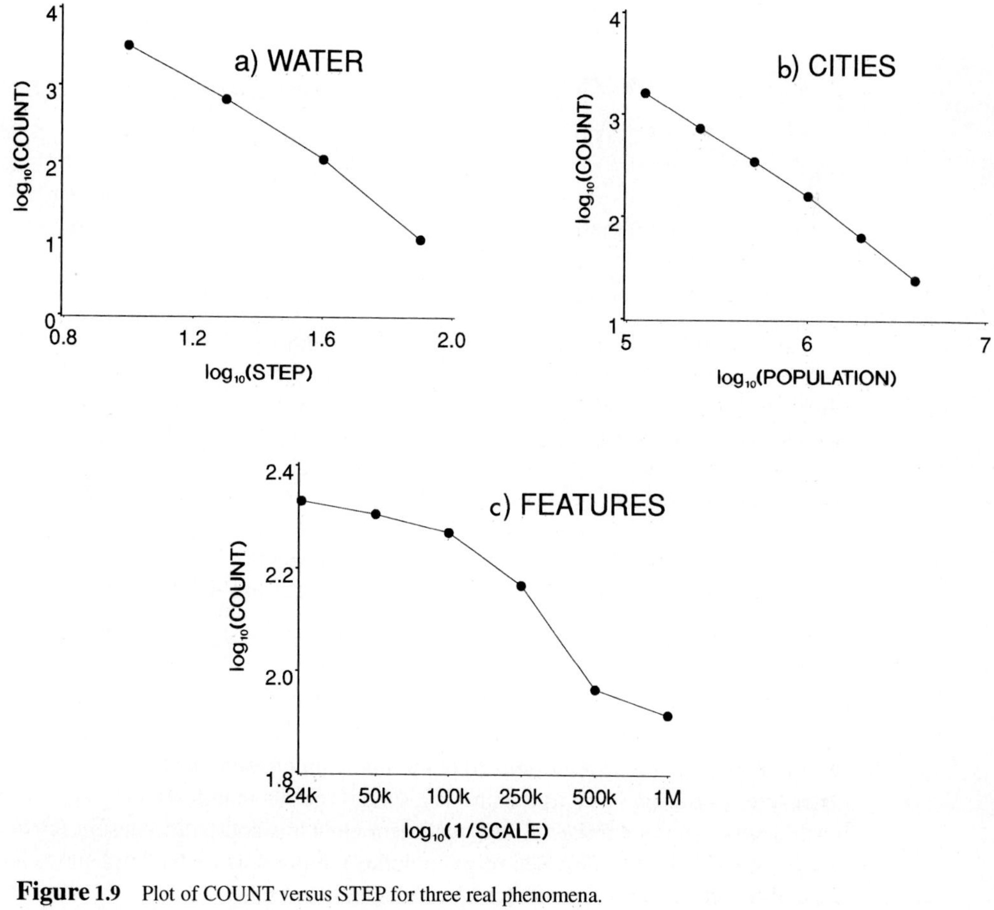

Figure 1.9 Plot of COUNT versus STEP for three real phenomena.

its various map series. An interesting and quite detailed table in Thompson (1981) specifies the features shown in six of these map series. For example, on the 1:24,000-scale topographic maps, 54 kinds of miscellaneous cultural features (airports, archeological sites, boardwalks, and others) are depicted, while on the 1:250,000 maps only 22 kinds of such features may be depicted. On any given map, of course you will find various instances of any given feature, but policies dictate what kinds of objects are supposed to be represented. Table 1.6 is a summary of Thompson's data.

TABLE 1.6 Relationship between FEATURE COUNT and SCALE RATIO for USGS Maps

LEVEL	SCALE RATIO	FEATURE COUNT	Log(1/SCALE)	Log(COUNT)
0	1:1,000,000	82	6.00	1.91
1	1:500,000	92	5.70	1.96
2	1:250,000	147	5.40	2.17
3	1:100,000	186	5.00	2.27
4	1:50,000	202	4.70	2.31
5	1:24,000	214	4.38	2.33

When we graph these numbers (Figure 1.9c), we see the telltale increases in detail with cartographic scale: increasing resolution results in a change in the "measure" of a phenomenon. Again, the standard approach is applying Equation 1.1, yielding

$$\log(\text{COUNT}) = 3.58 - 0.275 \log(1/\text{SCALE}), \text{ with } R^2 = 0.91.$$

But we are not strictly justified in saying that these features have a $D_{\text{FEATURE}} = -b = 0.275$, because, as in the above case of cities, we are not examining a correspondence between linear steps and linear or even areal counts. Although SCALE and COUNT are clearly conceptually related and are (weakly) scaling, this is not a geometric correspondence. Nevertheless, the analysis does reveal that an increase in scale has an important effect on feature representation. Moreover, an examination of Figures 1.9c suggests that large-scale maps show less information than an exactly scaling relationship would predict.

CONCLUSION

We conclude with a few definitions of our own to complement those with which we began this chapter. A *fractal* (used as a noun) is a spatial set that manifests a regular scaling relationship between the number of its constituent elements and their measure (size, density, intensity, and so forth). This definition includes temporal or dynamic phenomena, which can be spatially represented and therefore fractally measured.

Then we can define *fractal dimension* as the absolute value of the scaling exponent relating the number of these elements and their size. Note that this definition is not unambiguous; changes in the specification of a basic element may change the scaling relationship: we used STEP, as well as POPULATION and SCALE above, as we shall see, other elements are possible for other kinds of real and more complex topological sets.

We then can define *self-similarity* as the extent to which the relationship between the number of constitulent elements and their measure is truly linear, and complexity can be defined as the quality of having a fractal dimension that diverges significantly from a non-integral value.

Finally, *fractal* (used as an adjective) is the quality of manifesting such scaling regularity. We see that this characteristic is quite general—we have already shown that lines are fractal! Anything at all can be subjected to fractal analysis: language, history, pictures, prices—the list is endless—but only those things that manifest scaling can properly be called fractal.

REFERENCES

Arlinghaus, S. L. 1985. Fractals take a central place. *Geografiska Annaler* 67:83–88.

Barenblatt, G. I., et al. 1984. The fractal dimension: a quantitative characteristic of ocean-bottom relief. *Oceanology* (USSR) 24(6):695–697.

Batty, M., and Longley, P. 1986. The fractal simulation of urban structure. *Environment and Planning A* 18:1143–1179.

Berry, B. J. L. 1964. Cities as systems within systems of cities. In *Regional Development and Planning: A Reader*, J. Friedmann and W. Alonso, eds. Cambridge, MA: MIT Press.

Bradbury, R. H., and Reichelt, R. E. 1983. Fractal dimension of a coral reef at ecological scales. *Marine Geology Progress Series* 10:169–171.

Burrough, P. A. 1981. Fractal dimensions of landscapes and other environmental data. *Nature* 294(19):240–242.

———. 1983. Multiscale sources of spatial variation in soil. I. The application of fractal concepts to nested levels of soil variation. *Journal of Soil Science* 34:577-597.

De Cola, L. 1985. Lognormal estimates of macroregional city size distributions, 1950–1980. *Environment and Planning A* 17:1637–1652.

———. 1989. Fractal analysis of a classified Landsat scene. *Photogrammetric Engineering and Remote Sensing* 55(5):601–610.

———. 1991. Fractal analysis of multiscale spatial autocorrelation among point data. *Environment and Planning A* 23:545–556.

Goodchild, M. F. 1980. Fractals and the accuracy of geographical measures. *Mathematical Geology* 12:85–98.

———. 1982. The fractional Brownian process as a terrain simulation model. *Modeling and Simulation* 13:1133–1137.

Lam, N. S.-N. 1990. Description and measurement of Landsat TM images using fractals. *Photogrammetric Engineering and Remote Sensing* 56(2):187–195.

———, and Qiu, H. L. 1992. The fractal nature of the Louisiana coastline. In *Geographical Snapshots of North America*, D. G. Janelle, ed. New York: Guilford Press.

———, and Quattrochi, D. A. 1992. On the issues of scale, resolution, and fractal analysis in the mapping sciences. *The Professional Geographer* 44(1):89–99.

Lovejoy, S., and Mandelbrot, B. B. 1985. Fractal properties of rain, and a fractal model. *Tellus* 37A:209–232.

Mandelbrot, B. B. 1967. How long is the coast of Britain? Statistical self-similarity and fractional dimension. *Science* 156:636–638

———. 1983. *The Fractal Geometry of Nature*. San Francisco: W.H. Freeman & Co.

Mark, D. M. 1984. Fractal dimension of a coral reef at ecological scales: A discussion. *Marine Ecology Progress Series* 14:293–294.

———, and Aronson, P. B. 1984. Scale-dependent fractal dimensions of topographic surfaces: An empirical investigation, with applications in geomorphology and computer mapping. *Mathematical Geology* 16:671–683.

Muller, J. C. 1986. Fractal dimensions and inconsistencies in cartographic line representations. *The Cartographic Journal* 23:123–30.

Phillips, J. D. 1986. Spatial analysis of shoreline erosion, Delaware Bay, New Jersey. *Annals of the Association of American Geographers* 76(1):50–62.

Random House Dictionary of the English Language, 1987, 2nd ed. New York: Random House.

Rothrock, D. A., and Thorndike, A. S. 1980. Geometric properties of the underside of sea ice. *Journal of Geophysical Research* 85(C7):3955–3963.

Roy, A., Gravel, G., and Gauthier C. 1987. Measuring the dimension of surfaces: A review and appraisal of different methods. *Proceedings, Eighth International Symposium on Computer-Assisted Cartography (Auto-carto 8)*, Baltimore, pp. 68–77.

Steinhaus, H. 1960. *Mathematical Snapshots*. New York: Oxford University Press.

Thompson, M. M. 1981. *Maps for America*. Washington, DC: U.S. Geological Survey.

CHAPTER 2

FRACTAL MEASUREMENT

Nina Siu-Ngan Lam
Department of Geography and Anthropology, Louisiana State University, Baton Rouge

Lee De Cola
U. S. Geological Survey, Reston, VA

The measurement of the fractal dimension D of geographic data is a first step in the understanding of spatial complexity. In this chapter we present approaches to estimating D for curves based on the COUNTing of STEPs. This technique is then generalized for the measurement of D for surfaces, and then an approach based on the variogram is presented.

MEASUREMENT METHODS FOR CURVES

In Chapter 1 we used a line and the Koch quadratic curve to illustrate that measuring the fractal dimension of a curve is relatively straightforward. We showed that Equation 1.3 is the fundamental relationship:

$$\text{COUNT} \propto \text{STEP}^{-D}$$

where the scaling exponent D represents the degree to which COUNT increases with increasing resolution. If we let scale-dependent

$$\text{DISTANCE} = \text{COUNT} \times \text{STEP} \tag{2.1}$$

then substituting Equation 1.3 to Equation 2.1, we obtain

$$\text{DISTANCE} = A \times \text{STEP}^{1-D}$$

which can be shown as a linear relationship, using logs, as:

$$\log(\text{DISTANCE}) = \log(A) + (1\text{-}D)\log(\text{STEP}) \qquad (2.2)$$

so that fractal dimension $D = 1 - b$, where b is the slope of the line estimating the relationship between log(DISTANCE) and log(STEP).

Walking-divider Method

A technique that implements this approach is the walking-divider method, which computes D for a chain of lines represented by a series of x, y coordinates. A FORTRAN subroutine, modified extensively from the program by Shelberg et al. (1982) is included in Appendix 2.1. The program requires as input a curve defined by a series of x, y coordinate pairs, plus two parameters specifying initial divider STEP and the number of divider STEPs (number of walks) required for the regression. The output of the program includes the fractal dimension value, the R^2 value for assessing the goodness of fit of the regression, a listing of divider STEPs versus line STEPs in both logarithmic and unlogged forms, and a plot of the regression.

The algorithm works as follows. An initial divider STEP (that is, chord length) must first be given to begin walking the dividers along the curve. To provide a guideline for selecting the initial step size, the program calculates the average distance between every pair of adjacent points as digitized; one-half of the average distance is suggested as an appropriate initial value (Shelberg et al. 1982; Clarke and Schweizer, 1991). The algorithm next walks through the curve and counts the number of steps it takes. Intermediate points are interpolated using ordinary trigonometric functions and Euclidean distance calculations. (For a complete list of equations used in the algorithm, see Shelberg et al., 1982.) Each chord step size doubles the step of the preceding level so that, if the initial STEP were 2, then the subsequent STEPs would be 4, 8, 16, and so on. When plotted on a double log form, these geometric series provide equal spacing between the chord sizes.

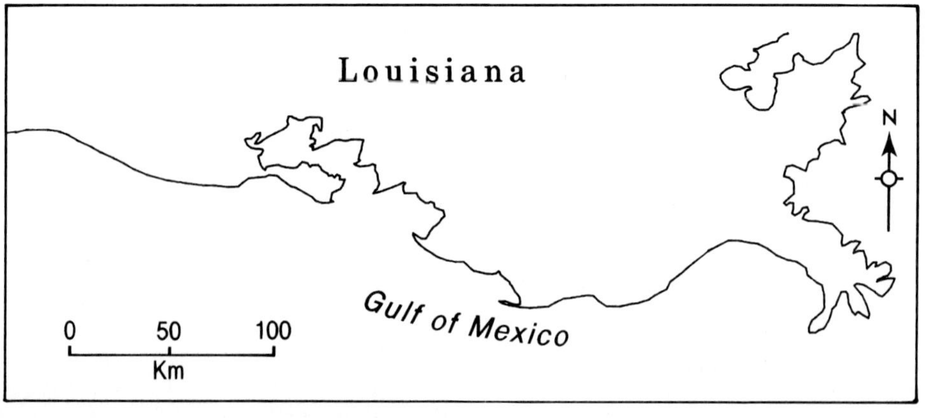

Figure 2.1 485 data points of the Louisiana coastline digitized from a map of scale 1:1,350,000.

For a reliable regression estimate, at least five chord step levels (five points on the regression plot) should be used.

TABLE 2.1 First 10 pairs from the List of 485 x, y Data Points for the Louisiana Coastline.

Longitude	Latitude
-93.87337	29.69477
-93.81975	29.72216
-93.72647	29.74524
-93.65149	29.7533
-93.55103	29.76092
-93.48736	29.76588
-93.41863	29.75689
-93.37266	29.76143
-93.31857	29.77211
-93.20638	29.77068

We demonstrate the technique on points digitized from a 1:1,350,000-scale map of the Louisiana coastline (Figure 2.1). The first 10 pairs from the list of 485 x,y data points are shown in Table 2.1. In addition to measuring a coastline, we also include examples of measuring D for a circle and a Koch snowflake curve for comparison. The results are presented in Figures 2.2a–c. The circle and the Koch curve were generated by computer routines. Table 2.2 summarizes the results from the application of this technique to several curves.

TABLE 2.2. Fractal Dimensions of Three Curves Measured by the Walking-divider Method

Object	Theoretical D	Measured D	R^2
Circle	1	1.0084	0.489
Louisiana coastline	$1 \leq D \leq 2$	1.2007	0.976
Koch snowflake	1.2619	1.2033	0.982

It is clear from measuring the circle and the Koch curve that the program returns fractal dimension values close to what we expect; minor discrepancies between the theoretical and the measured D values are due to the resolution used to represent these curves as a chain of discrete points. The Louisiana coastline as a whole yields a D that is comparable to that of the Koch curve, indicating a similar degree of irregularity for both curves.

```
$ run fract_line

    The Walking Divider Method
    for fractal curve measurement
    by Lam (1991), Qiu (1987), Shelberg (1982)

    ENTER DATA FILE NAME ===> circle.dat

          NO.OF POINTS =       360         MEAN DISTANCE =        1.0472
          TOTAL LENGTH =       375.9378    SINUOSITY =       358.0187

    ENTER THE STARTING CHORD LENGTH ==> 0.53
    ENTER NUMBER OF WALKS ==> 6

        WALKS    CL     LOG(CL)    STEPS    LENGTH   LOG(LENGTH)
        -----   ----    -------    -----    ------   ------------
          1   0.5300   -0.2757     708  375.9330     2.5751
          2   1.0600    0.0253     354  375.9288     2.5751
          3   2.1200    0.3263     177  375.9123     2.5751
          4   4.2400    0.6274      88  375.8534     2.5750
          5   8.4800    0.9284      44  375.6202     2.5747
          6  16.9600    1.2294      22  374.6760     2.5737
         ----------------------------------------------
         FRACTAL DIMENSION D = 1.0008
         R-SQ = 0.6495

          |-----------------------------------------------------------|
          |                                                           |
  2.5751  +    *         *         *                                |
          |                                  *                        |
          |                                                           |
          |                                            *              |
          |                                                           |
          +                                                           |
          |                                                           |
LOG       |                                                           |
          |                                                           |
(LENGTH)  |                                                           |
          +                                                           |
          |                                                           |
          |                                                           |
          |                                                           |
  2.5737  +                                                      *    |
          |                                                           |
          |----+---------+---------+---------+---------+---------+----|
            -0.2757                                            1.2294
                               LOG(CHORD LENGTH)

FORTRAN STOP
```

Figure 2.2 (a) Output from the walking-divider method applied to a circle.

Obviously, this analysis does not mean that the Louisiana coastline is a Koch snowflake—they don't look at all alike, nor are they formed by the same processes: one is a geometric construction, the other a set of line segments digitized from a hand-drawn map representing the coast at a given scale and resolution. The key to this analogy is that both

```
$ run fract_line

    The Walking Divider Method
    for fractal curve measurement
    by Lam (1991), Qiu (1987), Shelberg (1982)

    ENTER DATA FILE NAME ===> koch.dat

          NO.OF POINTS =      256          MEAN DISTANCE =        3.7257
          TOTAL LENGTH =      950.0417     SINUOSITY =        3.2062

    ENTER THE STARTING CHORD LENGTH ==> 1.9
    ENTER NUMBER OF WALKS ==> 6

        WALKS    CL     LOG(CL)    STEPS    LENGTH   LOG(LENGTH)
        -----   ----    -------    -----    ------   -----------
          1    1.9000    0.2788     424   807.2222     2.9070
          2    3.8000    0.5798     169   645.4702     2.8099
          3    7.6000    0.8808      79   606.1048     2.7825
          4   15.2000    1.1818      34   519.3861     2.7155
          5   30.4000    1.4829      16   486.8795     2.6874
          6   60.8000    1.7839       6   368.9306     2.5669
        ----------------------------------------------
        FRACTAL DIMENSION D = 1.2026
        R-SQ = 0.9608


             |-----------------------------------------------------------|
             |                                                           |
    2.9070 +    *                                                        |
             |                                                           |
             |                                                           |
             |                                                           |
             |             *                                             |
             +                                                           |
             |                       *                                   |
   LOG       |                                                           |
             |                                                           |
 (LENGTH)    |                                 *                         |
             |                                                           |
             +                                           *               |
             |                                                           |
             |                                                           |
             |                                                           |
    2.5669 +                                                       *     |
             |                                                           |
             |----+---------+---------+---------+---------+---------+----|
                 0.2788                                          1.7839
                                   LOG(CHORD LENGTH)

 FORTRAN STOP
```

Figure 2.2 (b) Output from the walking-divider method applied to the Koch snowflake.

objects represent geometric or geomorphologic processes that operate at a wide range of scales: infinitely wide in the case of the snowflake, considerably narrower, yet spanning a self-similar range from 1.5 to 200 kilometers, in the case of the coastline.

```
$ run fract_line

    The Walking Divider Method
    for fractal curve measurement
    by Lam (1991), Qiu (1987), Shelberg (1982)

    ENTER DATA FILE NAME ===> map1.bna

          NO.OF POINTS =        485          MEAN DISTANCE =        0.0271
          TOTAL LENGTH =         13.1395     SINUOSITY =         2.9949

    ENTER THE STARTING CHORD LENGTH ==> 0.014
    ENTER NUMBER OF WALKS ==> 8

        WALKS    CL      LOG(CL)    STEPS    LENGTH   LOG(LENGTH)
        -----   ----     -------    -----    ------   ------------
          1   0.0140   -1.8539     897    12.5779      1.0996
          2   0.0280   -1.5528     429    12.0259      1.0801
          3   0.0560   -1.2518     193    10.8186      1.0342
          4   0.1120   -0.9508      85     9.5584      0.9804
          5   0.2240   -0.6498      37     8.4421      0.9264
          6   0.4480   -0.3487      14     6.6245      0.8212
          7   0.8960   -0.0477       6     5.7750      0.7616
          8   1.7920    0.2533       2     5.0243      0.7011
        ---------------------------------------------
        FRACTAL DIMENSION D = 1.2007
        R-SQ = 0.9756

           |-----------------------------------------------------------|
           |                                                           |
           |                                                           |
  1.0996 +     *                                                       |
           |          *                                                |
           |                                                           |
           |               *                                           |
           |                                                           |
           |                    *                                      |
           +                                                           |
           |                                                           |
  LOG      |                          *                                |
           |                                                           |
(LENGTH)   |                                                           |
           |                                                           |
           +                                                           |
           |                               *                               |
           |                                    *                      |
           |                                                           |
           |                                                           |
  0.7011 +                                              *              |
           |                                                           |
           |                                                           |
           |----+----------+---------+---------+---------+---------+---|
               -1.8539                                          -0.3487
                               LOG(CHORD LENGTH)

FORTRAN STOP
```

Figure 2.2 (c) Output from the walking-divider method applied to the Louisiana coastline.

The R^2 computed from the regression indicates how well the resultant D characterizes the data. If the R^2 is low, D may not be an accurate measure of irregularity for the phenomenon. As with any regression analysis, it is essential to examine both R^2 and the regression plot before making conclusions. The regression plot may suggest that the curve (and presumably underlying process) may have a varying fractal dimension. We can easily observe from the Louisiana coastline that the degree of irregularity varies in different parts of the shore. Lam and Qiu (1992) have studied the relationships between the morphologic differences of the coast, their geomorphic processes, and the computed fractal dimensions. They divided the coastline into five subsections according to stages of landform development and measured their fractal dimensions. Their results indicate that shorelines dominated by deltaic (depositional) processes have higher fractal dimensions while shorelines dominated by marine processes have lower fractal dimensions. Their study therefore links morphometric measures such as fractal dimension with processes.

The dimension value derived from this method is influenced by at least three factors. First, the resultant fractal dimension of the coastline may vary according to the map scale from which the curve was digitized. Higher D values are expected from curves of higher resolution. Second, the initial STEP and the subsequent geometric sequence will affect the final D value. Because curves representing real *multifractal* phenomena (see Chapter 14) may reflect varying fractal dimensions with scale and location, an examination of the regression plot will show where distinct breaks of slope occur. Third, the result may also be affected by the location of the starting point.

Other Curve-Measurement Methods

The walking-divider method is exact, direct, and sensitive to the irregularity of the curve, but the method is time consuming because of the number of trigonometric functions involved. Several other techniques for measuring the dimensionality of curves have been used; these methods differ in terms of the sampling strategies, exactness, computational speed, and whether the original data are in raster or vector form. The equipaced polygon method, for example, minimizes computational time (Clark, 1986; Batty and Longley, 1987) and does not require interpolation of intermediate points. Instead, line lengths are computed by adding the distances first between two adjacent points, then the distances between two points that are spaced one point apart, then two points apart, and so on. For each line length, an average chord length is computed for use in the regression. This method is only an approximation of the walking-divider concept, but the results are more stable and the method is less sensitive to minor changes in the curve.

A method that combines the advantages of both the walking-divider and equipaced polygon methods has been proposed by Clark (1986), and others such as the cell-counting method (Goodchild, 1980) and the area–perimeter method (Kent and Wong, 1982; Woronow, 1981) are also used. A point of caution is that variations in D derived from different methods are not uncommon, so further studies and comparison are needed for valid and reliable interpretation of the resultant D values and associated scale ranges.

MEASUREMENT METHODS FOR SURFACES

Measuring the fractal dimension of a surface is a greater challenge than analyzing curves because adding a dimension makes the task of measurement much more difficult and ambiguous. Similar to curve measurement, various methods and algorithms have been developed, but may give different D values depending on the method.

Isarithm Method

Isarithms, such as contours, represent the intersection of a surface with a set of equally spaced horizontal planes. Using the argument that $D_{SURFACE} = D_{ISARITHM} + 1$, all we need do is to modify the divider technique to measure $D_{ISARITHM}$, which can be digitized from a map or derived from the corresponding digital surface, such as digital elevation model (DEM) data or digital remotely sensed image data. The following algorithm applies to regularly spaced digital surface data and was modified from Goodchild (1980) and Shelberg et al. (1983). A full listing of the FORTRAN program is included as Appendix 2.2, outlining the details of how the algorithm works.

In brief, the algorithm requires an input of a matrix of Z values, numbers of columns and rows defining the matrix (the number of rows does not have to equal the number of columns), an isarithmic interval for determining the number of isarithms to be examined, and the number of walks. The program then generates the different walk sizes following a geometric series, and the walk sizes are defined in units of grids (that is, 1, 2, 4, 8 grids, and so on). The program will automatically check if the maximum walk size exceeds the size of the grid matrix. If so, a smaller value for the number of walks must be used. The user also specifies the direction, row or column, from which the operation proceeds. The program computes the D for each isarithm, as well as the average of those D's that have R^2 greater than 0.9.

For each isarithmic value and each walk size (that is, chord STEP), the algorithm first classifies each cell below the isarithmic value as white and each above this value as black. It then compares each neighboring cell along the rows or columns and examines if the pairs are both black or both white. If they are of different colors, then there is an isarithm lying between the two neighboring cells and a boundary counter is incremented. The length of each isarithmic line is approximated by the total number of boundary pairs counted. The process continues until the number of walks has been completed. It is possible for a given walk size that no boundary cells are encountered. In this case, the isarithmic line is eliminated to avoid regression using fewer points than the given number of walks. This feature has been pointed out to be especially useful for fractal measurement involving Landsat images, because of the likely random occurrence of unusual spectral reflectance recorded at certain pixels. In effect, this feature ensures that the random noise in the images will not be taken into account in the computation process (Lam, 1990).

After counting the number of boundary cells per walk size for each isarithmic line, the algorithm performs a linear regression using their logarithms (Equation 2.2) and computes the corresponding D value using $D = 2 - b$. The surface's final fractal dimension is the average of the D values for those isarithms for which $R^2 \geq 0.9$.

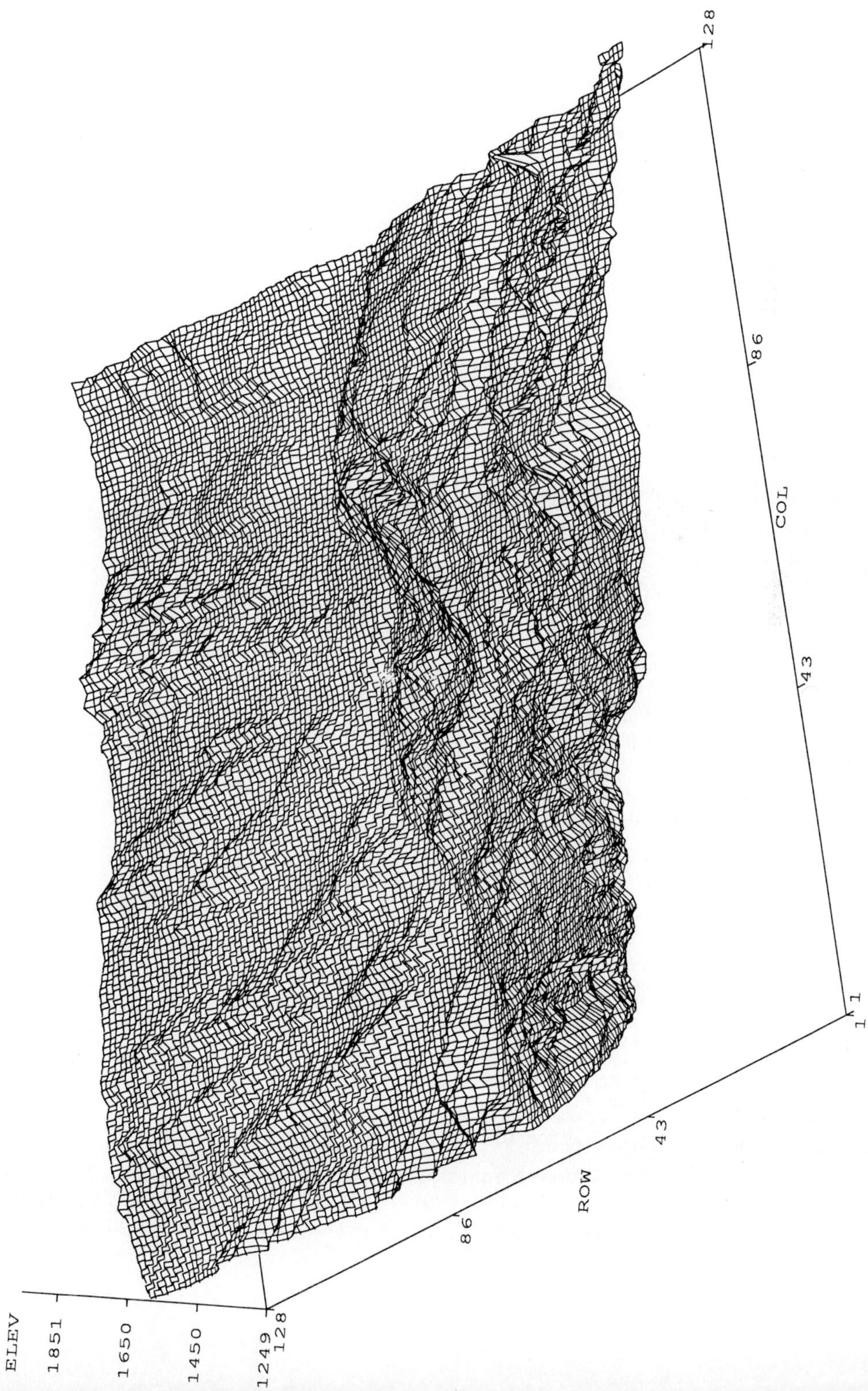

Figure 2.3 A digital elevation model (DEM) surface extracted from San Juan Mountains, Colorado.

Figure 2.3 shows an example of a digital surface extracted from the San Juan Mountains region of Colorado. The file consists of 128 rows by 128 columns, the pixel size is 1 - km^2, and the range of elevations is from 1249 to 1851 meters. Figure 2.4a and b is the output from applying the isarithm program to the sample data. The method results in an

```
$ run fract_dem

    The Isarithm Method
    for fractal surface measurement
    by Nina Lam (1991)

    ENTER THE INPUT FILENAME ===> colorado.dat
    NUMBER OF COLUMNS ===> 128
    NUMBER OF ROWS ===> 128
    MAXIMUM Z =     1851.000
    MINIMUM Z =     1249.000

    USE ROWS OR COLUMNS (R or C)? ===> r
    ENTER ISARITHM INTERVAL ===> 10
    INPUT NUMBER OF WALKS   ===> 6
    MAXIMUM WALK SIZE =    32

         CONTOUR  ELEVATION       B        D       RSQ

            1      1259.00     0.0000   0.0000   0.0000
            2      1269.00     0.0000   0.0000   0.0000
            3      1279.00     0.0000   0.0000   0.0000
            4      1289.00     0.0000   0.0000   0.0000
            5      1299.00    -0.7898   2.7898   0.5672
            6      1309.00    -0.5907   2.5907   0.6152
            7      1319.00    -0.5328   2.5328   0.7360
            8      1329.00    -0.5117   2.5117   0.7648
            9      1339.00    -0.4614   2.4614   0.7470
           10      1349.00    -0.4941   2.4941   0.8237
           11      1359.00    -0.4524   2.4524   0.8594
           12      1369.00    -0.4270   2.4270   0.8077
           13      1379.00    -0.4154   2.4154   0.8521
           14      1389.00    -0.3754   2.3754   0.8844
           15      1399.00    -0.3304   2.3304   0.9373
           16      1409.00    -0.3040   2.3040   0.9482
           17      1419.00    -0.2850   2.2850   0.9808
            .        ...         ...      ...      ...
            .        ...         ...      ...      ...
            .        ...         ...      ...      ...
           56      1809.00    -0.0153   2.0153   0.4286
           57      1819.00     0.0000   2.0000   0.0000
           58      1829.00    -0.0885   2.0885   0.6857
           59      1839.00     0.0000   2.0000   0.0000

    NO. OF ISARITHM LINES INCLUDED =   28
    NO. OF ISARITHM LINES NOT INCLUDED =   31
    DIMENSION =   2.2593
    R-SQ = 0.962421
    ISARITHM INT. =   10.00
    NUMBER OF WALKS USED:   6
    ROWS =     1,128
    COLUMNS =     1,128
    CALCULATION BY = r

    NEED SUMMARY FOR ISARITHM LINES(?)==> y
```

Figure 2.4(a) Simplified output listing from applying the isarithm method to the San Juan DEM data.

overall D value of 2.26 for the entire data set ($R^2 = 0.962$) and a value of $D = 2.32$ ($R^2 = 0.932$) for the 1449-meter (the mean) contour. The inclusion of the output (Figure 2.4)

```
  ENTER ISARITHM LINE NUMBER == > 4
                                                TOTAL
              ISARITHM LINE      NO. OF BOUNDARY CELLS      WALK SIZE
              -------------      ---------------------      ---------
                 1289.00                   52                   1
                 1289.00                   48                   2
                 1289.00                   46                   4
                 1289.00                   38                   8
                 1289.00                   32                  16
                 1289.00                    0                  32
  BY ROWS
 D= 0.0000
 R-SQ= 0.000000
  SUMMARY    4 OF   59

  INPUT NUMBER FOR NEXT SUMMARY ===> 20
                                                TOTAL
              ISARITHM LINE      NO. OF BOUNDARY CELLS      WALK SIZE
              -------------      ---------------------      ---------
                 1449.00                  383                   1
                 1449.00                  361                   2
                 1449.00                  322                   4
                 1449.00                  254                   8
                 1449.00                  166                  16
                 1449.00                  136                  32
  BY ROWS
 D= 2.3192
 R-SQ= 0.932156
  SUMMARY   20 OF   59

  INPUT NUMBER FOR NEXT SUMMARY ===> 57
                                                TOTAL
              ISARITHM LINE      NO. OF BOUNDARY CELLS      WALK SIZE
              -------------      ---------------------      ---------
                 1819.00                   21                   1
                 1819.00                   21                   2
                 1819.00                   21                   4
                 1819.00                   21                   8
                 1819.00                   21                  16
                 1819.00                   21                  32
  BY ROWS
 D= 2.0000
 R-SQ= 0.000000
  SUMMARY   57 OF   59

  INPUT NUMBER FOR NEXT SUMMARY ===> 0
FORTRAN STOP

                    COLORADO.DAT

                    1832
                    1841
                    1846
                    1851
                    1836
                    1837
                    1833
                    1826
                    1812
                     ..
                     ..
                     ..
```

Figure 2.4(b) Continuation of (a) and a listing of the first 10 elevation values of the San Juan DEM data.

together with the program listing (Appendix 2.2) illustrates how the algorithm actually works and what the results look like. Figure 2.4b also shows that the 1289 contour has been eliminated because zero boundary cells were encountered at walk size 32.

A number of factors will affect the D values resulting from the isarithm method. Real data are generally anisotropic (that is, they exhibit different characteristics when measured in different directions), so the orientation of the data with respect to such features as highways and fault lines will change the surface values. This means that D will vary depending on whether it is measured along rows, along columns, or in a non-cardinal direction. The maximum walk size used may also cause results at the large end of the regression to be unreliable. Furthermore, representing the surface D by the average D of those isarithms that have R^2's ≥ 0.9 is somewhat arbitrary. Some researchers choose to use the D of the mean isarithm instead, which would be valid if the surface were self-similar (Goodchild, 1980). Because a natural surface is not entirely self-similar, selecting the mean isarithm may produce a nonrepresentative fractal dimension. This issue of multifractality (and thus of heterodimensionality) is an important subject that is explored in some of the following research chapters.

In short, variations in D may occur using differing parameter inputs in the method. The detailed measurement of each isarithm in this method is considered advantageous since more information about the surface is inferred. Whether a surface is self-similar, multifractal, or anisotropic can be identified from the manner in which D varies with the isarithmic interval, the number of walks, and the row or column direction.

Variogram Method

In the Variogram method, fractal dimension is estimated from another important measurement of spatial complexity, the variogram. Variograms are functions describing how variance in surface height varies with distance, and they have been used extensively in *kriging*, a common spatial interpolation technique (Lam, 1983). For any continuous surface of dimension $2 \leq D \leq 3$, the expected value of the squared height difference between two points is given by

$$E[(Z_p - Z_q)^2] = k(d_{pq})^{2H} \tag{2.3}$$

where Z_p and Z_q are the elevations of the surface at points p and q, and d_{pq} is the distance between the two points. The parameter H describes the complexity of the surface and has values between 0 and 1; for large values of H, a surface shows *persistence* or a strong tendency to return to neighboring values. If the mean of the squared height difference (that is, variance) is computed for different distances, then D can be estimated from the slope b of the regression between the logarithms of variance and distance so that $D = 3 - H = 3 - b/2$ (Mark and Aronson, 1984).

An understanding of the relationship between H, D, and surface ruggedness is useful to fractal surface measurement and generation using the variogram. This correspondence is summarized in Figure 2.5. For the Brownian process $H = 0.5$, so $D = 2.5$; the expected variance between two points increases linearly with distance, resulting in a 45° line in the graph. For $H > 0.5$, the variance between nearby points is less than that in the Brownian process, but it increases exponentially and exceeds those in the Brownian process when distance is greater than 1 unit. On the other hand, if $H < 0.5$, variance increases exponentially for short distances and remains nearly constant for distant points. Consequently, an $H > 0.5$ (that is, $D < 2.5$) results in a surface of greater smoothness and persistence, showing a certain degree of trend, while the surface becomes rugged and antipersistent for $H < 0.5$ ($D > 2.5$). As H approaches 0, or D near 3.0, the surface becomes white noise with no persistence. For a range of surfaces with various H (hence various D), see Figure 3.7.

A FORTRAN program implementing the variogram method is included as Appendix 2.3. The program requires an input of x, y, z values either in the form of regular grids or irregular data points. It provides an option of sampling every nth point, rather than all the points for the computation of the variogram. In addition, the user specifies the number of distance groups used in the variogram. The program divides the distances between the points into a number of groups as specified, and the mean squared height difference is computed for each distance group. The program then computes the slope of the regression using a range of points as specified by the user. The output of the program includes the regression estimates, the D value, and a listing of distance values, number of pairs in each distance group, and their variances. These values can be input to other programs for plotting the regression. The same DEM surface from Colorado measured by the isarithm method above was also measured by the variogram algorithm. Figure 2.6 is the output of the program and the regression plot is shown as Figure 2.7. The D value computed from this method (D = 2.39) is higher than the D computed from the isarithm method (D = 2.26).

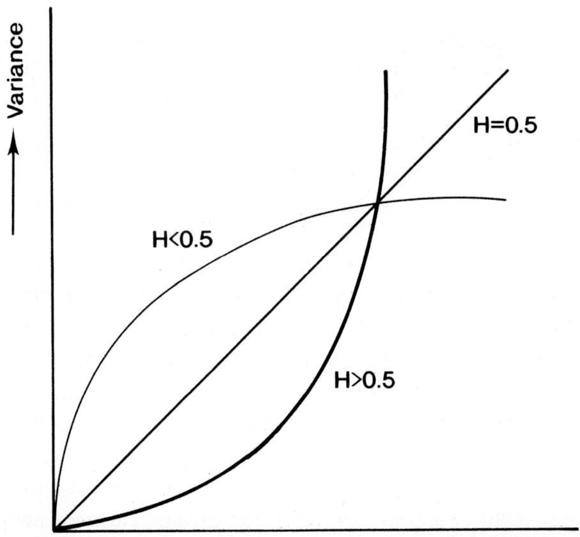

Figure 2.5 Relationship between variance and distance for different *H*. (Source: Lam, 1980.)

```
$ run fract_var

    The Variogram Method
    for fractal surface measurement
    by Lam (1991) and Qiu (1990)

ENTER INPUT FILENAME ==> colorado.dat
NUMBER OF COLUMNS ==> 128
NUMBER OF ROWS ==> 128
SELECT EVERY NTH POINT ==> 50
ENTER OUTPUT FILENAME ==> col3.out
    TOTAL NUMBER OF POINTS =    16384
    NUMBER OF POINTS SELECTED =       327
    NUMBER OF POINT PAIRS =     53301
    MINIMUM DISTANCE =       6.3246
    MAXIMUM DISTANCE =     170.4846
ENTER NUMBER OF DISTANCE GROUPS ==> 50
    POINT_PAIRS CLASSIFIED =    53301
    WRITING OUTPUT FILE .....
GROUP #   # OF CASES    LOG(DIST)    LOG(MEAN VARIANCE)
    1         906        0.9012          3.2028
    2         577        1.0511          3.3388
    3         835        1.1623          3.6120
    4        1346        1.2508          3.6013
    5         778        1.3243          3.6980
    .          ..         ...              ...
    .          ..         ...              ...
    .          ..         ...              ...
   47          17        2.2014          4.7914
   48           7        2.2103          4.9057
   49           6        2.2190          4.8341
   50           4        2.2275          4.9393
ENTER THE BREAKPOINT NUMBER ==> 50
    # POINTS INCLUDED IN REGRESSION =    50
    REGRESSION SLOPE =     1.2266
    FRACTAL DIMENSION =     2.3867
    R-SQUARE =    0.9802
    END OF TASK !
FORTRAN STOP
```

Figure 2.6 Simplified output listing from applying the variogram method to the San Juan DEM data.

A number of parameter inputs within a method may be specified, and the results from using different parameter values may vary considerably. This leads to the problem of determining whether the difference in the resultant dimension values for different surfaces is a result of true differences in the phenomena or a result of arbitrary decisions made during the measurement process. In brief, variations in resulting values of D may reflect differences in (1) data sampling, (2) parameter specification, and (3) methods. To assure comparability among the surfaces being examined, comparison of surfaces should be based on the results computed from one method and not from various methods.

Several other caveats should also be made. Studies have shown that variograms often do not behave linearly in the log–log plot at all scales. Most surfaces therefore are not strictly self-similar and the relationship between variance and distance becomes more and more unstable with increasing distance (Mark and Aronson, 1984, and Chapter 12). In most cases, the points at the end of the log–log plot will slope downward, leading to a $D > 3$, which is not strictly possible if we are to allow a surface to do no more than fill space. It is therefore useful to evaluate the log–log plot so that different D values may be computed for different scale ranges. Finally, Chapter 8 notes that the variogram method is not appropriate for detecting scale ranges.

There are, however, three important points in favor of the variogram technique. First, the method can be applied to both regular and irregular data. For regular grid data, a sample of points is often used to reduce the computational time. Second, for irregular polygonal data, centroids representing the polygons can be used to determine the variogram function. Third, when compared with the isarithm method, the variogram method is generally more reliable, involves less ambiguity, and is more direct and adaptable to irregularly spaced data.

Other Surface-measurement Methods

Clarke and Schweizer (1991) propose an elaborate technique called the *robust fractal estimator.* Their method uses the walking-diver method to compute for each cell the fractal dimension of each profile (vertical intersection) in both the north–south and east–west directions and places the average of the two in a new array. A fractal dimension of the

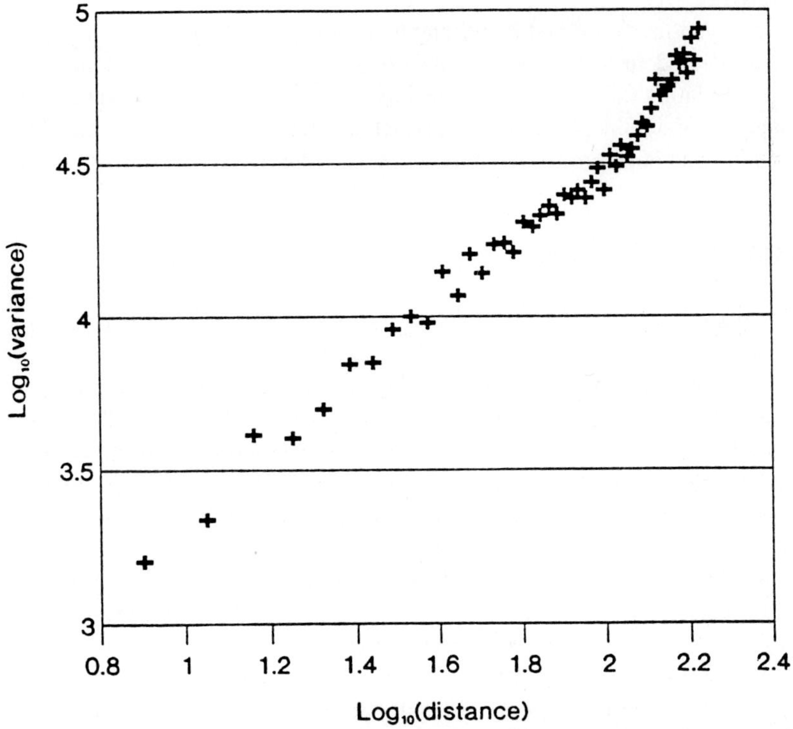

Figure 2.7 Regression plot of log distance and log variance from the variogram method, using the San Juan DEM data.

entire surface is obtained by combining the fractal dimensions of each cell using a weighted average and adding 1 to account for the additional topological dimension. This method produces results that are different from other methods and seems to consistently yield a lower fractal dimension. For example, in their study the same data set measured by the *robust fractal estimator* method yields a $D = 2.189$, compared with a $D = 2.976$ resulting from the variogram method. This leads to fundamental questions of whether fractal analysis of a vertical cut (profile) of a surface or of a horizontal cut (contour) are the same and whether a profile through a surface would have a fractal dimension exactly equal to that of the surface minus 1. Thinking speculatively, we should not be surprised that gravity–constrained vertical relief has a lower fractal dimension than horizontal contours!

The Fourier method uses the power spectrum derived from the surface (Burrough, 1981; Pentland, 1984; Clarke, 1988). The fractal dimension is determined from the slope of the regression line of the log–log plot of the power spectrum versus wavelength. The use of the Fourier power spectrum involves complicated computations and requires much computer time, and although the fast Fourier transform could be used to speed up the computation, it requires very large arrays with sizes that are powers of 2. Yet the Fourier method has an important property of invertibility, that is, an ability to transform back and forth between the power spectrum and the data without loss of information. With improved hardware and increasing availability of software, researchers ought to explore this method further. Other methods of surface measurement include such techniques as box counting (a variation of the technique used to characterize water pixels in Chapter 1), and the triangular prism method, which is a three-dimensional analog of the divider technique used in line measurement presented earlier in this chapter.

CONCLUSION

A major issue raised by this chapter is the question of what is being measured by these techniques. We naturally hope that D reflects underlying processes, and that variations in values over time and space mirror changes in the real world. The key problem is that the same process may lead to different patterns and that similar patterns are not necessarily caused by the same process. Nevertheless, morphometric measurement is a critical first step toward understanding spatial processes. Chapters 7 and 11 in this book elaborate on this central issue. The differences among these algorithmic approaches are often so significant, that thoughtless comparisons among, say, an isarithm D for one data set and a variogram D for another are not really valid (Roy et al., 1987; Clarke and Schweizer, 1991). Moreover, the techniques used to measure the underlying phenomenon (sampling frequency, map scale, accuracy, precision, and others) will usually influence results, as with any measurement technique.

In sum, fractal analysis is becoming more useful as its various methods yield consistent and reliable results, but technical problems of measuring fractals for scientific analysis remain. First, the fact that self-similarity exists only over certain ranges of scale makes it difficult to identify the break points. This affects the final D value that is used to characterize the curve or surface.

Second, surface measurement is more challenging than curve measurement because of the additional topological dimension involved. A more systematic and comprehensive appraisal of the major methods discussed above is needed. Until a thorough evaluation of the various methods is achieved, researchers should compare surfaces using results from one algorithm, as well as compare different algorithms using one data set.

Finally, an accurate measurement method is needed to reliably compare the various curve and surface generation methods. Ideally, the surface or curve generated with a certain D should lead to the same D when being measured by different methods; if the generated value does not agree with the measured value and if the reliability of the measurement method is uncertain, it is difficult to prove whether the generation method is suitable or not. Invertibility of fractal generation methods would be of significant value for simulation, but is difficult to confirm without a reliable measurement method. These observations take us to the subject of the next chapter.

REFERENCES

Batty, M., and Longley, P. A. 1987. Fractal-based description of urban form. *Environment and Planning* 14:123–134.

Burrough, P. A. 1981. Fractal dimensions of landscapes and other environmental data. *Nature* 294(19):240–242.

Clark, N. N. 1986. Three techniques for implementing digital fractal analysis of particle shape. *Powder Technology* 46:45–52.

Clarke, K. C. 1988. Scale-based simulation of topographic relief. *The American Cartographer* 15(2):173–181.

———, and Schweizer, D. M. 1991. Measuring the fractal dimension of natural surfaces using a robust fractal estimator. *Cartography and Geographic Information Systems* 18(1):37–47.

Goodchild, M. F. 1980. Fractals and the accuracy of geographical measures. *Mathematical Geology* 12:85–98.

Kent C., and Wong, J. 1982. An index of littoral zone complexity and its measurement. *Canadian Journal of Fisheries and Aquatic Sciences* 39:847–853.

Lam, N. S.-N. 1980. *Methods and Problems of Areal Interpolation.* Ph.D. dissertation. London, Ontario: Department of Geography, University of Western Ontario.

———. 1983. Spatial interpolation methods: A review. *The American Cartographer* 10:129–149.

———. 1990. Description and measurement of Landsat TM images using fractals. *Photogrammetric Engineering and Remote Sensing* 56(2):187–195.

———, and Qiu, H. L. 1992. The fractal nature of the Louisiana coastline. In *Geographical Snapshots of North America*, D. G. Janelle, ed. New York: Guilford Press.

Mark, D. M., and Aronson, P. B. 1984. Scale-dependent fractal dimensions of topographic surfaces: An empirical investigation, with applications in geomorphology and computer mapping. *Mathematical Geology* 16:671–683.

Pentland, A. P. 1984. Fractal-based description of natural scenes. *IEEE Transactions on Computer Vision and Pattern Recognition PAMI* 6(6):661– 674.

Roy, A., Gravel, G., and Gauthier, C. 1987. Measuring the dimension of surfaces: A review and appraisal of different methods. *Proceedings, Eighth International Symposium on Computer-Assisted Cartography (Auto-Carto 8)*, Baltimore, pp. 68–77.

Shelberg, M.C. 1982. *The Development of A Curve and Surface Algorithm to Measure Fractal Dimensions and Presentation of Results*. M. S. Thesis. Columbus Ohio: Department of Geography, The Ohio State University.

———, Lam, N. S.-N., and Moellering, H. 1983. Measuring the fractal dimension of surfaces. *Proceedings, Sixth International Symposium on Computer-assisted Cartography (Auto-Carto 6)*, Vol. 2, Ottawa, Canada, pp. 319–328.

———, Moellering, H., and Lam, N. S.-N. 1982. Measuring the fractal dimensions of empirical cartographic curves. *Proceedings, Fifth International Symposium on Computer-Assisted Cartography (Auto-Carto 5)*, Washington, D.C., pp. 481–490.

Woronow, A. 1981. Morphometric consistency with the Hausdorff–Besicovich dimension. *Mathematical Geology* 13:201–216.

APPENDIX 2.1

Listing of FORTRAN Program FRACT_LINE.FOR

THE WALKING-DIVIDER METHOD FOR FRACTAL CURVE MEASUREMENT

```
C***********************************************************************
C                                                                      *
C     The walking divider method for fractal curve measurement         *
C     This program was based in part on Mark Shelberg (1982) and has been  *
C     extensively modified by Hong-Lie Qiu (1987) and Nina Lam (1991)  *
C                                                                      *
C***********************************************************************
      CHARACTER*1 PLOT
      DIMENSION PLOT(71,25),PX(16),PY(16)
      REAL X(25000),Y(25000),HOL(16),VER(16),FACTOR,IX,IY
      BYTE INFILE(30)
      PARAMETER (IFN=11)
      CHARACTER*10 NAME1,NAME2,NAME3,NAME4,NAME5
      REAL*16 A,B,C,DIST,DISTA,DISTB,DISTC,F,X1,X2,X3,Y1,Y2,Y3,CL

      WRITE(*,'(/5X,A)') 'The Walking Divider Method'
      WRITE(*,'(5X,A)') 'for fractal curve measurement'
      WRITE(*,'(5X,A/)') 'by Lam (1991), Qiu (1987), Shelberg (1982)'
C
C  GET AND OPEN INPUT DATA FILE
C
      WRITE(*,'(/5X,A,$)') 'ENTER DATA FILE NAME ===> '
      READ(*,'(Q,30A1)') K,INFILE
         IF (K .EQ. 0) THEN
              STOP   '     INPUT FILE NAME ERROR !'
         ELSE
              INFILE(K+1) = '00'X
         ENDIF
         CALL ASSIGN(IFN,INFILE)
C
```

```
C    COUNT POINTS, CALCULATE TOTAL LENGTH AND MEAN DISTANCE BETWEEN TWO
C    ADJACENT POINTS, AND PRINT RESULTS.
C
          SUMCL = 0.0
          I = 0
          READ(IFN,*,END=999) IX,IY
          IX=ABS(IX)
          IY=ABS(IY)
          I=I+1
          X(1)=IX
          Y(1)=IY
       DO WHILE (.TRUE.)
          READ(IFN,*,END=999) IX,IY
          IX=ABS(IX)
          IY=ABS(IY)
          XC=IX
          YC=IY
          IF (XC .NE. X(I).OR.YC .NE. Y(I)) THEN
              I = I + 1
              X(I) = XC
              Y(I) = YC
              X1 = X(I)-X(I-1)
              Y1 = Y(I)-Y(I-1)
          DIST=SQRT(X1*X1 + Y1*Y1)
          SUMCL = SUMCL + DIST
          ENDIF
       ENDDO
  999        AVD = SUMCL/FLOAT((I - 1))
             STDIST =SQRT((X(1)-X(I))**2 + (Y(1)-Y(I))**2)
             SINUO = SUMCL/STDIST
       WRITE(6,101) I,AVD
  101  FORMAT(/11X,'NO.OF POINTS =',I8,8X,'MEAN DISTANCE =',F12.4)
       WRITE(*,103) SUMCL,SINUO
  103  FORMAT(11X,'TOTAL LENGTH = ',F12.4,3X,'SINUOSITY = ',F12.4)
       WRITE(*,'(/5X,A,$)') 'ENTER THE STARTING CHORD LENGTH ==> '
       READ(*,'(F)') FACTOR
       WRITE(*,'(5X,A,$)') 'ENTER NUMBER OF WALKS ==> '
       READ(*,'(I)') NW
C
C   MEASURE THE TOTAL LENGTHS OF THE LINE WITH DIFFERENT
C   DIVIDER OPENINGS
C
       WRITE(6,15)
       WRITE(6,16)
   15  FORMAT(/9X,'WALKS   CL    LOG(CL)   STEPS   LENGTH  LOG(LENGTH)')
   16  FORMAT(9X,'-----  ----  -------   -----   ------  -----------')
          CL =  FACTOR * 0.5
       DO 104 J = 1,NW
          CL = CL * FLOAT(2)
          K = 2
          IPNUM = 0
          X1 = X(K-1)
          Y1 = Y(K-1)
```

```
 20     X2 = X(K)
        Y2 = Y(K)
 21 IF (K .GE. I) GO TO 500
        DISTA=SQRT((X2-X1)*(X2-X1)+(Y2-Y1)*(Y2-Y1))
    IF (DISTA .LT. CL) GO TO 19
        XNEW = X1 + (CL/DISTA) * (X2 - X1)
        YNEW = Y1 + (CL/DISTA) * (Y2 - Y1)
        IPNUM = IPNUM + 1
        X1 = XNEW
        Y1 = YNEW
        GO TO 21
 19     X3 = X(K + 1)
        Y3 = Y(K + 1)
        X2 = X(K)
        Y2 = Y(K)
        DISTC=SQRT((X3-X1)*(X3-X1)+(Y3-Y1)*(Y3-Y1))
    IF (DISTC .LT. CL) GO TO 77
        DISTB=SQRT((X3-X2)*(X3-X2)+(Y3-Y2)*(Y3-Y2))
        DISTA=SQRT((X2-X1)*(X2-X1)+(Y2-Y1)*(Y2-Y1))
        IF(DISTB .EQ. 0.000000 .OR. DISTA .EQ. 0.000000) GO TO 777
        F = (DISTA*DISTA+DISTB*DISTB-DISTC*DISTC)/(2.0*DISTA*DISTB)
    IF (F .GT. 0.9999 .OR. F .LT. -0.9999) THEN
777     DIST = CL - DISTA
    ELSE
        C = ACOS(F)
        A = ASIN(DISTA*SIN(C)/CL)
        B = 3.141592654 - A - C
        DIST = DISTA * SIN(B) /SIN(A)
    ENDIF
        XNEW = X2 + (DIST/DISTB) * (X3 - X2)
        YNEW = Y2 + (DIST/DISTB) * (Y3 - Y2)
        IPNUM = IPNUM + 1
        X1 = XNEW
        Y1 = YNEW
    IF (K .LT. I) THEN
        K = K + 1
        GO TO 20
    ELSE
        GO TO 500
    ENDIF
 77     INDICA = I - 1
    IF (K .LT. INDICA) THEN
        K = K + 1
        GO TO 19
    ENDIF
500     FLENGTH = CL * IPNUM + SQRT((X(I)-X1)**2+(Y(I)-Y1)**2)
        PX(J) = LOG10(CL)
        PY(J) = LOG10(FLENGTH)
        WRITE(6,24) J,CL,PX(J),IPNUM,FLENGTH,PY(J)
 24 FORMAT(9X,I3,2F9.4,I7,2F10.4)
104 CONTINUE

        XD = 0.0
```

```
         YD = 0.0
         XY = 0.0
         XA = 0.0
         YA = 0.0
      DO 300 M = 1,NW
         XA = XA + PX(M)**2
         YA = YA + PY(M)**2
         XD = XD + PX(M)
         YD = YD + PY(M)
         XY = XY + PX(M)*PY(M)
  300 CONTINUE
         B = (XY-XD*YD/NW)/(XA-XD**2/NW)
         D = 1.0 - B
         ABOVE = (XY - XD*YD/NW) ** 2
         BELOW = (XA - XD*XD/NW) * (YA - YD*YD/NW)
         RSQ = ABOVE/BELOW
      WRITE(*,'(10X,A)')'-------------------------------------------------'
      WRITE(6,301) D
      WRITE(6,302) RSQ
  301 FORMAT(10X,'FRACTAL DIMENSION D = ',F6.4)
  302 FORMAT(10X,'R-SQ = ', F6.4//)
C
C   PRINT A SCATTERPLOT ON THE SCREEN
C
      DO 410 NY = 1,25
      DO 409 NX = 1,71
      IF (NX .NE. 11) GO TO 411
         PLOT(NX,NY) = '|'
         GO TO 409
  411 IF (NX .NE. 71) GO TO 412
         PLOT(NX,NY) = '|'
         GO TO 409
  412 IF (NX .LE. 10) GO TO 414
      IF (NY .NE. 1) GO TO 413
         PLOT(NX,NY) = '-'
         GO TO 409
  413 IF (NY .NE. 25) GO TO 414
         PLOT(NX,NY) = '-'
         GO TO 409
  414    PLOT(NX,NY) = ' '
  409 CONTINUE
  410 CONTINUE
         NAME1 = 'LOG'
         NAME2 = '(LENGTH)'
         NY = 12
      DO M = 1,8
         PLOT(M,NY) = NAME2(M:M)
      ENDDO
         NY = 14
         K = 0
      DO NX = 3,5
         K = K + 1
         PLOT(NX,NY)=NAME1(K:K)
```

```
      ENDDO
         NY = 1
      DO M = 1,6
         NX = 16 + 10 * (M - 1)
         PLOT(NX,NY) = '+'
      ENDDO
         NX = 11
      DO M = 1,4
         NY = 6 * M - 2
         PLOT(NX,NY) = '+'
      ENDDO
      DO M = 1,NW
         NX = NINT((PX(M)-PX(1))/((PX(NW)-PX(1))/50.0)) + 16
         NY = NINT((PY(M)-PY(NW))/((PY(1)-PY(NW))/18.0)) + 4
         PLOT(NX,NY) = '*'
      ENDDO
         WRITE(UNIT=NAME4,FMT='(F6.4)') PY(1)
         WRITE(UNIT=NAME5,FMT='(F6.4)') PY(NW)
         NY = 4
      DO K = 1,6
         NX = K + 3
         PLOT(NX,NY) = NAME5(K:K)
      ENDDO
         NY = 22
      DO K = 1,6
         NX = K + 3
         PLOT(NX,NY) = NAME4(K:K)
      ENDDO
      DO NY = 25,1,-1
         WRITE(6,422) (PLOT(NX,NY),NX=1,71)
      ENDDO
         WRITE(6,423) PX(1),PX(6)
         WRITE(6,424)
  422 FORMAT(1X,71A1)
  423 FORMAT(16X,F7.4,42X,F7.4)
  424 FORMAT(35X,'LOG(CHORD LENGTH)'///)
         CLOSE(11)
         STOP
         END
```

APPENDIX 2.2

Listing of FORTRAN Program FRACT_DEM.FOR

THE ISARITHM METHOD FOR FRACTAL SURFACE MEASUREMENT

```
C********************************************************************
C                                                                  *
C      The isarithm method for fractal surface measurement         *
C               by Nina Lam, 1991                                  *
C      This program was derived partly from Mark Shelberg (1982)   *
C      and has been modified extensively both in theory and        *
C      in program structure by Nina Lam (1991).                    *
C                                                                  *
C********************************************************************
C
      BYTE TITLE(30), NAME(30)
      PARAMETER (IFN=11)
      DIMENSION Z(256,256),IZ(256,256),B(100),R(100),NO(100),ICELL(20)
      DIMENSION YBAR(100),XBAR(100),YY(100,20),XX(100,20),DIMEN(100)
      DIMENSION ZC(100),TZ(256,256),A(100,20),IFLG(100)
      DOUBLE PRECISION Y,SUMY,SUMXY,SQY,SUMX,SQX,D1,D2,AM
      DATA ICC/'C   '/,JCC/'c   '/,IR/'R   '/,JR/'r   '/
      CHARACTER*1 AN

      WRITE(*,'(/5X,A)') 'The Isarithm Method'
      WRITE(*,'(5X,A)') 'for fractal surface measurement'
      WRITE(*,'(5X,A//)') 'by Nina Lam (1991)'
C****************************************
C
C     GET AND OPEN THE INPUT FILE
C
 1111 WRITE(*,'(/5X,A,$)') 'ENTER THE INPUT FILENAME ===> '
      READ(*,'(Q,30A1)') I, NAME
      IF (I.EQ. 0) THEN
      WRITE(*,'(///5X,A)') '******** ERROR IN FILENAME ********'
      GO TO 1111
      ENDIF
      OPEN(UNIT=IFN,FILE=NAME,STATUS='OLD')
C**********************************
C
C  READ NUMBER OF COLUMNS AND ROWS
C
      WRITE(*,'(5X,A,$)') 'NUMBER OF COLUMNS ===> '
      READ(*,'(I5)') ICOL
      WRITE(*,'(5X,A,$)') 'NUMBER OF ROWS ===> '
      READ(*,'(I5)') IROW
      DO IY=1,IROW
      DO IX = 1, ICOL
      READ(IFN,1112,END=999) Z(IY,IX)
 1112 FORMAT(F4.0)
      ENDDO
      ENDDO
```

```
C*********************************************
C
C  CALCULATE THE MINIMUM AND MAXIMUM Z-VALUES OF THE SURFACE
C
C
  999 IMINR = 1
      IMINC = 1
      IMAXR = IROW
      IMAXC = ICOL
      TZMIN=Z(IMINR,IMINC)
      TZMAX=Z(IMINR,IMINC)
      DO 1113 I= IMINR,IMAXR
      DO 1113 J= IMINC,IMAXC
      IF(TZMIN .GT. Z(I,J))TZMIN=Z(I,J)
      IF(TZMAX .LT. Z(I,J))TZMAX=Z(I,J)
 1113 CONTINUE
      ZMIN=TZMIN
      ZMAX=TZMAX
      WRITE(*,'(5X,A,F10.3)') 'MAXIMUM Z = ',ZMAX
      WRITE(*,'(5X,A,F10.3)') 'MINIMUM Z = ',ZMIN

C*********************************************
C
C  THE USER MAY CALCULATE D ALONG COLUMNS OR ROWS
C
    1 WRITE(*,'(/5X,A,$)') 'USE ROWS OR COLUMNS (R or C)? ===> '
      READ(5,2)IANS
    2 FORMAT(A4)
      IF((IANS .EQ. ICC) .OR. (IANS .EQ. JCC))GO TO 3
      IF((IANS .EQ. IR) .OR. (IANS .EQ. JR))GO TO 5
      WRITE(*,'(/5X,A)') '~~~~~PLEASE TRY AGAIN~~~~~~'
      GO TO 1
C
C  USE COLUMNS
C
    3 KFLAG=0
      ILR=IMINR
      IGR=IMAXR
      ILC=IMINC
      IGC=IMAXC
      DO 4 I=IMINR,IMAXR
      DO 4 J=IMINC,IMAXC
      TZ(I,J)=Z(I,J)
    4 CONTINUE
      GO TO 7
C
C  USE ROWS
C
    5 KFLAG=1
      ILR=IMINC
      IGR=IMAXC
      ILC=IMINR
      IGC=IMAXR
```

```
C
C  USE THE TRANSPOSE OF A MATRIX
C
      DO 6 I=IMINR,IMAXR
      DO 6 J=IMINC,IMAXC
      TZ(J,I)=Z(I,J)
    6 CONTINUE
C*********************************************
C
    7 WRITE(*,'(5X,A,$)') 'ENTER ISARITHM INTERVAL ===> '
      READ(5,*)ZINT
C
C  CALCULATE NUMBER OF ISARITHM LINES
C
       NCI=((ZMAX-ZMIN)/ZINT)+.5
       NINT=NCI-1.
C
C  THE NUMBER OF ISARITHM LINES CANNOT BE GREATER THAN 100
C  BECAUSE OF THE ALLOWABLE DIMENSION SPACE
C
      IF((NINT .GT. 0) .AND. (NINT .LE. 100))GO TO 8
      WRITE(*,'(5X,A)') '~~~~~ PLEASE TRY AGAIN ~~~~~'
      GO TO 7
C
C  THE NUMBER OF WALKS SHOULD BE AT LEAST 5 AND SHOULD NOT
C  EXCEED 1/2 OF THE SIZE OF THE DATA MATRIX (E.G. 2,4,8,16,32...)
C  THE MAXIMUM NUMBER OF WALKS IS SET TO BE 20
C
    8 WRITE(*,'(5X,A,$)') 'INPUT NUMBER OF WALKS  ===> '
      READ(5,*) IWALK
      ISIZE=2**(IWALK-1)
      WRITE(*,165) ISIZE
  165 FORMAT(5X,'MAXIMUM WALK SIZE = ',I4)
      ISIZE=2*ISIZE
      IF((ISIZE.GT.ICOL).OR.(ISIZE.GT.IROW)) GO TO 22
      IF(IWALK .LE. 20)GO TO 9
   22 WRITE(*,'(5X,A)') '~~~~~ PLEASE TRY AGAIN ~~~~~'
      GO TO 8
C
C  COMPUTE THE NUMBER OF ROWS
C
    9 NN=IGR-ILR+1
      DO 110 II=1,NINT
C
C  INITIALIZE VARIABLES FOR REGRESSION ANALYSIS
C
      SUMX=0.
      SQX=0.
      SUMY=0.
      SUMXY=0.
      SQY=0.
C
C  CALCULATE VALUE OF ISARITHM LINE
```

```
C
      ZC(II)=ZMIN+II*ZINT
      DO 30 I=ILR,IGR
      DO 20 J=ILC,IGC
C
C  COMPARE CELLS
C
      IF(TZ(I,J) .LE. ZC(II))GO TO 10
C
C  CELL IS ABOVE - BLACK
C
      IZ(I,J)=2
      GO TO 20
C
C  CELL IS BELOW - WHITE
C
   10 IZ(I,J)=1
   20 CONTINUE
   30 CONTINUE
C
C  DO UNTIL IRC EQUALS NUMBER OF WALKS
C
      IW=1
      DO 60 IRC= 1,IWALK
      S=0.
      IF(IRC.EQ.1) GO TO 42
      IW=2**(IRC-1)
   42 ICELL(IRC)=IW
      DO 50 I=ILR,IGR
      DO 40 J=ILC,IGC,IW
C
C  INCREMENT CELL SIZE
C
      JJ=J+IW
C
C  GO TO THE END OF EACH COLUMN
C
      IF(JJ .GT. IGC)GO TO 50
C
C  CALCULATE THE NUMBER OF BLACK AGGREGATES CONTAINING AT LEAST ONE
C  WHITE CELL
C
      IF(IZ(I,J) .EQ. IZ(I,JJ))GO TO 40
      S=S+1.
   40 CONTINUE
   50 CONTINUE
      IS=S
C
C  DO NOT INCLUDE ANY ISARITHM LINES WHICH HAVE 0 BOUNDARY
C  CELLS FOR ANY WALK
C
      IF(IS .EQ. 0)GO TO 105
      A(II,IRC)=S
```

```
C
C  TAKE THE MEAN NUMBER OF BOUNDARY CELLS
C
      AVGBC=S/FLOAT(NN)
C
C  LOG(NUMBER OF BOUNDARY CELLS)
C
      Y=ALOG10(AVGBC)
      YY(II,IRC)=Y
      SUMY=SUMY+Y
      SUMXY=SUMXY+Y*ALOG10(FLOAT(IW))
      SQY=SQY+(Y*Y)
   60 CONTINUE
      AM=FLOAT(IWALK)
      DO 65 KK= 1,IWALK
C
C  LOG(AGGREGATE WALK SIZE)
C
      IW=ICELL(KK)
      SUMX=SUMX+ALOG10(FLOAT(IW))
      XX(II,KK)=ALOG10(FLOAT(IW))
      SQX=SQX+(ALOG10(FLOAT(IW)))**2
   65 CONTINUE
      ICNT=0
C
C  IF EACH WALK CONTAINS THE SAME NUMBER OF BOUNDARY CELLS
C  THEN SET THE SLOPE=0 AND THE CORRELATION COEFFICIENT=0
C  THIS IS TO AVOID THE POSSIBILITY OF DIVIDING BY 0 OR TAKING
C  THE SQUARE ROOT OF A NEGATIVE NUMBER
C
      DO 70 LL=1,IWALK
      IF(A(II,1) .eq. A(II,LL))ICNT=ICNT+1
   70 CONTINUE
      IF(IWALK .NE. ICNT)GO TO 101
      B(II)=0.
      R(II)=0.
      GO TO 103
C
C  CALCULATE THE NUMERATOR AND DENOMINATOR FOR THE LINEAR
C  REGRESSION
C
  101 D1=SUMXY-(SUMX*SUMY/AM)
      D2=SQX-((SUMX*SUMX)/AM)
C
C  CALCULATE THE LINEAR REGRESSION (SLOPE) FOR EACH ISARITHM
C  LINE
C
      B(II)=D1/D2
C
C  CALCULATE THE CORRELATION COEFFICIENT FOR EACH ISARITHM LINE
C
      R1=D1/DSQRT(D2*(SQY-(SUMY**2/AM)))
      R(II)=R1*R1
```

```
C
C  CALCULATE THE DIMENSION FOR EACH ISARITHM LINE
C
 103  DIMEN(II)=2-B(II)
      IFLG(II)=1
      GO TO 110
C
C  ISARITHM INTERVAL IS TOO SMALL - INSUFFICIENT NUMBER OF
C  BOUNDARY CELLS ENCOUNTERED
C
  105 IFLG(II)=0
  110 CONTINUE
      BSUM=0.
      RSUM=0.
      NN=NINT
      WRITE(*,'(/10X,A/)') 'CONTOUR  ELEVATION     B       D      RSQ'
C
C  DO NOT INCLUDE THE ISARITHM LINES WHICH CONTAINED 0 NUMBER
C  OF BOUNDARY CELLS AND
C  AVERAGE ONLY THE ISARITHMS WITH RSQ >= 0.9
C
      DO 120 I=1,NINT
      WRITE(*,'(10X,I4,3X,F8.2,4X,3F8.4)') I,ZC(I),B(I),DIMEN(I),R(I)
      IF((IFLG(I).NE.0).AND.(R(I).GE.0.9))GO TO 112
      NN=NN-1
      GO TO 120
  112 BSUM=BSUM+B(I)
      RSUM=RSUM+R(I)
  119 CONTINUE
  120 CONTINUE
C
C  NUMBER OF ISARITHM LINES NOT INCLUDED
C
      NRCI=NINT-NN
C
C  NUMBER OF INCLUDED ISARITHM LINES
C
      NINT=NN
C
C  CALCULATE THE AVERAGE SLOPE
C
      AVGB=BSUM/FLOAT(NINT)
C
C  CALCULATE THE AVEAGE R-SQUARES
C
      TR=RSUM/FLOAT(NINT)
C
C  AVERAGE FRACTAL DIMENSION FOR THE SURFACE
C
      TDIMEN=2.-AVGB
      WRITE(*,'(//5X,A,I3)')'NO. OF ISARITHM LINES INCLUDED = ',NINT
      WRITE(*,124) NRCI
  124 FORMAT(5X,'NO. OF ISARITHM LINES NOT INCLUDED = ',I3)
```

```
C
C  PRINT OUT THE RESULTS
C
      WRITE(6,125)TDIMEN
  125 FORMAT(5X,'DIMENSION = ',F7.4)
      WRITE(6,127)TR
  127 FORMAT(5X,'R-SQ = ',F8.6)
      WRITE(6,152)ZINT
  152 FORMAT(5X,'ISARITHM INT. = ',F6.2)
      WRITE(6,153) IWALK
  153 FORMAT(5X,'NUMBER OF WALKS USED: ',I2)
      WRITE(6,160)IMINR,IMAXR
  160 FORMAT(5X,'ROWS = ',I3,',',I3)
      WRITE(6,170)IMINC,IMAXC
  170 FORMAT(5X,'COLUMNS = ',I3,',',I3)
      WRITE(6,180) IANS
  180 FORMAT(5X,'CALCULATION BY = ',A4)

C
C  PRINT OUT THE TABLE
C
      WRITE(*,'(//5X,A,$)') 'NEED SUMMARY FOR ISARITHM LINES(?)==> '
      READ(*,'(A)') AN
      IF (AN .EQ. 'N' .OR. AN .EQ. 'n') GO TO 888
      WRITE(*,'(//5X,A,$)') 'ENTER ISARITHM LINE NUMBER == > '
      READ(*,'(I)') IC
  415 WRITE(6,280)
  280 FORMAT(//1X,T39,'TOTAL',/,13X,'ISARITHM LINE',4X,
     *'NO. OF BOUNDARY CELLS',4X,'WALK SIZE')
      WRITE(6,290)
  290 FORMAT(1X,T14,13('-'),T31,21('-'),T56,9('-'))
      DO 320 II=1,IWALK
      IA=A(IC,II)
      WRITE(6,310)ZC(IC),IA,ICELL(II)
  310 FORMAT(15X,F8.2,12X,I6,16X,I2)
  320 CONTINUE
      IF(KFLAG .EQ. 1)GO TO 325
      WRITE(*,'(/5X,A)') 'BY COLUMNS'
      GO TO 327
  325 WRITE(*,'(/5X,A)') 'BY ROWS'
C
  327 continue
      write (6,328) dimen(ic)
  328 format (3x,'D= ',f6.4)
      write (6,329) r(ic)
  329 format (3x,'R= ',f8.6)
    ntotal=nint+nrci
      WRITE(6,330)IC,ntotal
  330 FORMAT(5X,'SUMMARY ',I3,' OF ',I3)
  334 WRITE(*,'(//5X,A,$)') 'INPUT NUMBER FOR NEXT SUMMARY ===> '
      READ(5,*) IC
      IF (IC .EQ. 0) GO TO 888
      IF (IC .LE. ntotal) GO TO 415
```

```
      WRITE(*,'(/5X,A)') '~~~~~ PLEASE TRY AGAIN ~~~~~'
      GO TO 334
  888 STOP
      END
```

APPENDIX 2.3

Listing of FORTRAN Program FRACT_LINE.FOR

THE VARIOGRAM METHOD FOR FRACTAL SURFACE MEASUREMENT

```
C**********************************************************************
C                                                                     *
C    The variogram method for fractal surface measurement             *
C    This program was originally developed by Hong-Lie Qiu (1990)     *
C    for measuring polygonal data.  The program was modified          *
C    extensively by Nina Lam to improve the program structure         *
C    and to include sampling, regression, and the measurement of      *
C    regular grid data.                                               *
C                                                                     *
C**********************************************************************
C
C  Variable declaration
C
      DIMENSION X(1000),Y(1000),Z(1000),DIST(500000),HVAR(500000)
      DIMENSION SUMZD(102),NCASE(102),GB(102),VL(102),DL(102)
      INTEGER*2 IFN
      PARAMETER (IFN=11)
      PARAMETER (NFO=12)
      BYTE NAME(30)

      WRITE(*,'(/5X,A)') 'The Variogram Method'
      WRITE(*,'(5X,A)') 'for fractal surface measurement'
      WRITE(*,'(5X,A//)') 'by Lam (1991) and Qiu (1990)'
C
C  Read the names of the input and output data files
C
      WRITE(*,'(A,$)') ' ENTER INPUT FILENAME ==> '
      READ(*,'(Q,30A1)') I, NAME
      IF (I .EQ. 0) THEN
         STOP ' ERROR IN FILENAME '
      ELSE
         NAME(I+1) = '00'X
      ENDIF
      CALL ASSIGN(IFN,NAME)

      WRITE(*,'(A,$)') ' NUMBER OF COLUMNS ==> '
      READ(*,'(I5)') ICOL
      WRITE(*,'(A,$)') ' NUMBER OF ROWS ==> '
      READ(*,'(I5)') IROW
      WRITE(*,'(A,$)') ' SELECT EVERY NTH POINT ==> '
      READ(*,'(I5)') IS
```

```
      WRITE(*,'(A,$)') ' ENTER OUTPUT FILENAME ==> '
      READ(*,'(Q,30A1)') I, NAME
      IF (I .EQ. 0) THEN
         STOP ' ERROR IN FILENAME '
      ELSE
         NAME(I+1) = '00'X
      ENDIF
      CALL ASSIGN(NFO,NAME)
C
C  Read X,Y, and Z values of all points; Select only a
C  sample of points for computing the variogram using
C  a regular sampling method
C
      UNIT=1
      NUM=0
      K=0
      DO 10 IY=1,IROW
      DO 10 IX=1,ICOL
      READ(IFN,100,END=150) ZZ
  100 FORMAT(F4.0)
      K=K+1
      IF(MOD(K,IS).NE.0) GO TO 10
      NUM=NUM+1
      Z(NUM)=ZZ
      X(NUM)=FLOAT(IY)*UNIT-(UNIT*0.5)
      Y(NUM)=FLOAT(IX)*UNIT-(UNIT*0.5)
   10 CONTINUE
C
C  Calculate distances and squared Z_value differences between
C  any two of the selected points
C
  150 WRITE(*,'(5X,A,I8)') 'TOTAL NUMBER OF POINTS =',K
      WRITE(*,'(5X,A,I8)') 'NUMBER OF POINTS SELECTED =',NUM
      NP = 0
      MAXICR = NUM -2
      ICR = 1

      DMAX=-1.0
      DMIN=999999.0
  200 CONTINUE
      DO LP = 1, NUM-ICR
         NP = NP + 1
         SIDEA = (X(LP)-X(LP+ICR)) ** 2
         SIDEB = (Y(LP)-Y(LP+ICR)) ** 2
         DIST(NP) = SQRT(SIDEA + SIDEB)
         HVAR(NP) = (Z(LP)-Z(LP+ICR)) ** 2

C  Find the maximum and minimum

      IF (DIST(NP) .GT. DMAX) THEN
          DMAX = DIST(NP)
          MAXNP = NP
```

```
      ELSE
          IF (DIST(NP) .LT. DMIN) THEN
              DMIN = DIST(NP)
          ENDIF
      ENDIF
      ENDDO
C
C  Check if all combinations are included
C
      IF (ICR .LE. MAXICR) THEN
         ICR = ICR + 1
         GO TO 200
      ENDIF
      WRITE(*,'(5X,A,I10)') 'NUMBER OF POINT PAIRS =', NP
      WRITE(*,'(5X,A,F12.4)') 'MINIMUM DISTANCE =', DMIN
      WRITE(*,'(5X,A,F12.4)') 'MAXIMUM DISTANCE =', DMAX
C
C  Determine the number of distance groups and their boundaries
C
      WRITE(*,'(A,$)') ' ENTER NUMBER OF DISTANCE GROUPS ==> '
      READ(*,'(I)') NG
         RANG = (DMAX - DMIN) / FLOAT(NG)
      DO J = 1, NG+1
         SUMZD(J) = 0.0
         NCASE(J) = 0
         GB(J) = DMIN + RANG*(J-1)
      ENDDO
      GB(NG+1)=GB(NG+1)+1.0
C
C  Group the distance array, find out number of cases falling within
C  each group, sum up the squared z-value differences.
C
      DO L = 1, NP
         DO M = 1, NG
            IF (DIST(L) .GE. GB(M) .AND. DIST(L) .LT. GB(M + 1)) THEN
               NCASE(M) = NCASE(M) + 1
               SUMZD(M) = SUMZD(M) + HVAR(L)
            ENDIF
         ENDDO
      ENDDO

      NPC = 0
      DO M = 1, NG
         NPC = NPC + NCASE(M)
      ENDDO
      WRITE(*,'(5X,A,I8)') 'POINT_PAIRS CLASSIFIED =',NPC
C
C  Compute the mean variance and mid-point distances for each group.
C  Write the result to an output file
C
      WRITE(*,'(5X,A)') 'WRITING OUTPUT FILE .....'
      WRITE(NFO,360)
      WRITE(*,360)
  360 FORMAT(' GROUP #   # OF CASES   LOG(DIST)   LOG(MEAN VARIANCE)')
```

```
      DO M = 1, NG
        DMP=GB(M)+RANG*0.5
        DL(M)=ALOG10(DMP)
        IF (NCASE(M) .NE. 0) THEN
           AVEZD = SUMZD(M) / FLOAT(NCASE(M))
           VL(M) = ALOG10(AVEZD)
        ELSE
           VL(M)= 0.0
        ENDIF
      WRITE(NFO,370) M,NCASE(M),DL(M),VL(M)
      WRITE(*,370) M,NCASE(M),DL(M),VL(M)
      ENDDO
  370 FORMAT(3X,I3,5X,I6,5X,F10.4,5X,F10.4)
C
C  Input the breakpoint number and compute the regression
C  up to and including that point
C
      WRITE(*,'(A,$)') ' ENTER THE BREAKPOINT NUMBER ==> '
      READ(*,'(I5)') KK
      SX=0.0
      SY=0.0
      SXY=0.0
      SQX=0.0
      SQY=0.0
      K=0
      DO 400 M=1,KK
         IF(VL(M).EQ.0.0) GO TO 400
           SX=SX+DL(M)
           SY=SY+VL(M)
           SXY=SXY+DL(M)*VL(M)
           SQX=SQX+DL(M)**2
           SQY=SQY+VL(M)**2
           K=K+1
  400 CONTINUE
      D1=(FLOAT(K)*SXY)-(SX*SY)
      D2=(FLOAT(K)*SQX)-SX**2
      B=D1/D2
      FD=3.0-(B/2.0)
      D3=(FLOAT(K)*SQY)-SY**2
      R=D1/(D2*D3)**0.5
      RSQ=R**2

      WRITE(*,'(5X,A,I5)') '# POINTS INCLUDED IN REGRESSION =',K
      WRITE(*,'(5X,A,F10.4)') 'REGRESSION SLOPE =',B
      WRITE(*,'(5X,A,F10.4)') 'FRACTAL DIMENSION =',FD
      WRITE(*,'(5X,A,F10.4)') 'R-SQUARE =',RSQ
C
C  Close the input and output files
C
      CLOSE(IFN)
      CLOSE(NFO)
      WRITE(*,'(5X,A)') 'END OF TASK !'
      STOP
      END
```

CHAPTER 3

FRACTAL SIMULATION AND INTERPOLATION

Nina Siu-Ngan Lam
Department of Geography and Anthropology, Louisiana State University, Baton Rouge

Lee De Cola
U. S. Geological Survey, Reston, VA

Simulation is useful in understanding spatial processes and portraying the forms they give rise to. Interpolation allows the analyst to portray limited data in a realistic fashion. This chapter introduces the use of the fractal model in simulating such spatial objects as curves and surfaces. We argue that randomness, representing uncertainty and complex spatial relationships, is an essential component in the simulation process. The midpoint subdivision method for curve simulation and the shear displacement method for surfaces were selected for detailed discussion with examples. FORTRAN code for each of the methods is provided in Appendixes 3.1 and 3.2.

INTRODUCTION

The sheer technical virtuosity of Mandelbrot's mathematical presentation launched a revolution in computer graphics; in fact, Mandelbrot has been heard to claim that one of his most significant contributions to IBM's profitability (he is an IBM fellow) is the use of the company's computers to render the Mandelbrot set! Fractal simulation has become part of our computer culture. Filmmakers use fractals to generate extraterrestrial landscapes, computer programmers produce videotapes and disks containing simulated fractal sets, composers write fractal music, and fractal T-shirts, postcards, and cartoons are everywhere. Articles in scholarly and popular journals appear regularly across every major discipline.

Fractals have been used as models to simulate terrestrial, extraterrestrial, and other imaginary objects, including coastlines (Carpenter, 1981; Dutton, 1981; Batty, 1989), terrain (Goodchild, 1980), trees (Batty, 1985), cities (Batty and Longley, 1986), clouds, ocean floors, particle growth, planets, dragons, and many other computer graphic applications (Mandelbrot, 1983; Peitgen and Richter, 1986; Peitgen and Saupe, 1988).

But what is all the excitement about? Why do we need fractal graphics, what are they good for, and how do they satisfy our abiding quest for the understanding of spatial patterns and spatial processes? In particular, how can geographers and geoscientists use the simulation possibilities of fractals? This chapter focuses on how fractal curves and surfaces are simulated and how these simulated curves and surfaces can be used, with a special emphasis on geographical applications.

In Chapter 2 we argued that measurement is the first, inductive step toward scientific understanding and that simulation often follows. The ability to simulate an object accurately depends on our knowledge about the object and how well our model represents it. This knowledge is derived largely from the results of measurement. At the same time, a simulation model can further our understanding of forms and underlying spatial processes if it successfully mimics reality. Figure 3.1 is a simple diagrammatic sketch of the relationship between the two stages of scientific inquiry: measurement and simulation.

Simulation is a deductive process that is useful in affirming our understanding of the objects and images we have measured. If we understand the process leading to the form of a real object and if we are able to simulate this form closely, then it is possible to utilize simulation to interpolate and predict values for the whole based on a small sample of data, thus reducing data and saving computer storage and processing time. In effect, simulation provides us with a laboratory for spatial experimentation. This chapter highlights some major applications of the fractal simulation capability, again with a focus on geographical applications.

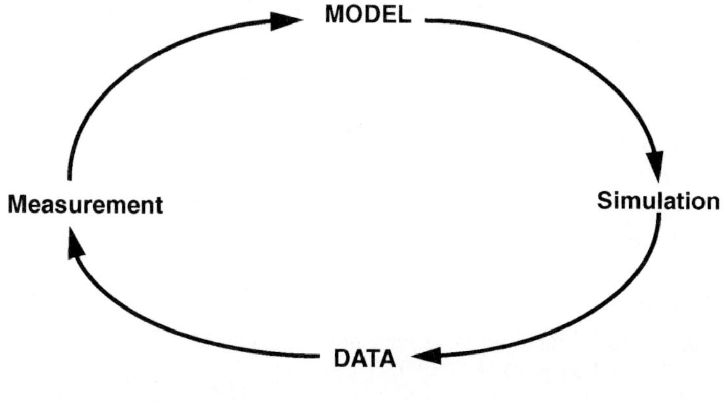

Figure 3.1 The recursive relationship between measurement and simulation.

Analytical uses of simulated graphical data are important for three reasons. First, simulated data may be used as hypothetical data sets for investigating complex real-world geographical problems. Simulated fractal surfaces have been used as test data sets to examine the performance of various spatial interpolation methods (Lam, 1982; 1983) and to assess the efficiency of a quadtree data structure (Mark and Lauzon, 1984). Second, fractal curve simulation has been used as an inverse of curve generalization by adding details to the generalized curve (Carpenter, 1981; Dutton, 1981; Jiang, 1984). Third, fractal surfaces have been suggested as null-hypothetical landscapes to which various geomorphic, biologic, or anthropogenic processes can be compared (Goodchild and Mark, 1987; Loehle, 1983).

From a practical point of view, reliable and efficient simulation methods for curves, surfaces, and other objects provide tremendous savings in computer storage and processing time, because we only need store algorithms and parameters in order to re-create the original data as needed. Fractal techniques can be used quite efficiently for this purpose, particularly if the images or objects have some degree of self-similarity. Considerable success in applying fractals to image compression has been reported in the field of computer graphics (Barnsley and Sloan, 1988).

METHODS FOR CURVE SIMULATION

The simplest way to simulate fractals is through construction, as was illustrated in Chapter 1. The various Koch curves can be generated by a set of parameters that can be used as a model for simulating such real-world curves as boundaries, rivers, contours, or coastlines. Yet a real coastline that looked like a Koch curve would be extraordinary because ideal fractals are strictly self-similar, perfectly repetitive, and completely deterministic (compare Figures 1.6 and 2.1). There is a large degree of uncertainty in nature, however, because of our incomplete knowledge of spatial phenomena and their underlying processes. To produce boundaries and surfaces that look real, therefore, an aspect of randomness should be included in the simulation process: although fractal constructions provide infinite detail, monotonous self-identity rather than stochastic self-similarity tends to destroy the illusion of reality.

But how much randomness should we include in the simulation process? The issue of randomness relates to the issue of invertibility mentioned in Chapter 2; we want a natural curve or surface generated by a specific fractal technique using a given D value to yield an empirical D close to the given D value. Because randomness tends to create forms that deviate somewhat from the given value, the issues are how much deviation should we expect and whether this deviation can be detected by the fractal measurement methods. These issues are not new: randomness has long been used to lend realism to simulation, and the standard random number algorithms provided in computer libraries are deterministic, sensitive to initial conditions, and not analytic—all necessary criteria of chaotic systems (see Figure 1.4).

In the case of regular curve simulation, for example, we have already shown in Chapter 2 that the walking-divider measurement method illustrated does yield D values similar, but not exactly equal, to their expected values. For the Koch curve and circle used in Chapter 2, the computed D values are 1.2033 and 1.0084 which are comparable to their expected values of 1.26 and 1, respectively. Now we turn to the problem of simulating more realistic data.

Carpenter Midpoint Subdivision Method

The first interpolation technique illustrated is Carpenter's (1981) method of midpoint subdivision. The approach is considered exact in that a set of existing control points is retained throughout the process. The technique then adds randomly placed midpoints to the line segments connecting the points from a previous iteration. The incorporation of an aspect of randomness in Carpenter's method is considered desirable for simulating natural boundaries.

The subdivision method is most commonly used because of its simplicity and flexibility. The method adds details by recursively subdividing a curve and moving subcurves into different locations. The location of the midpointis determined by a function related to the dimensionality or ruggedness of the simulated curve. Figure 3.2 illustrates how an original line AD is further subdivided at recursion levels 1 and 2. Statistical self-similarity is maintained by ensuring that the deviation between the interpolated and the actual midpoint be proportional to the distance between the end points. The points produced by this method are not true fractal sets but only approximates. The dimension of the curve generated by this subdivision method may be determined by applying the walking-divider method presented in Chapter 2.

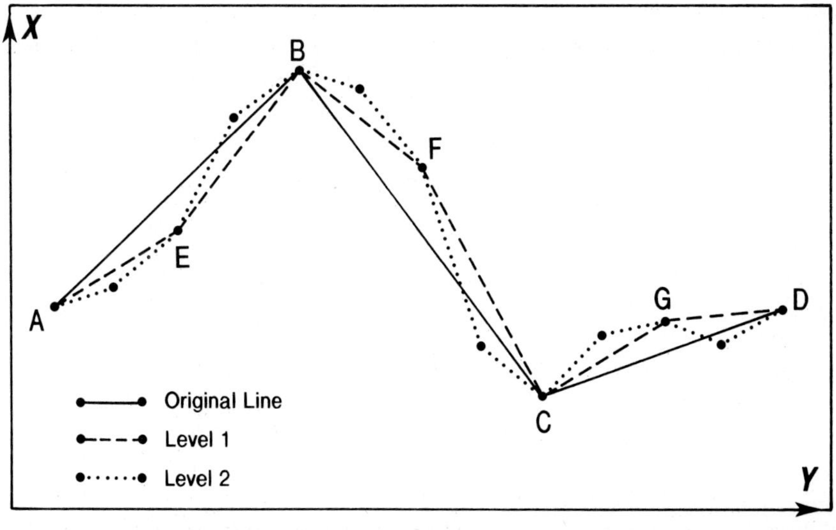

Figure 3.2 The Carpenter midpoint subdivision method. Point E has $X_E = 1/2 * (X_A+X_B)$, $Y_E = 1/2 * (Y_A+Y_B) + RR$, where RR is a stochastic component controlled by the length of segment AB, the scale factor, the roughness factor, and a Gaussian random number (see text and Appendix 3.1).

Appendix 3.1 is a FORTRAN program that uses Carpenter's approach. The program requires an input of the x,y coordinates of the control points and the following parameters: a roughness factor with values ranging from 0 to 1, a scale factor specific to the scale and resolution of the sample data, the number of iterations (that is, final recursion level), and a seed value for the random generator to generate uniform variates between -1 and +1. In addition to generating midpoints between each line segment for every recursion level, the program also computes the increase in line length for each level and estimates the fractal dimension of the final curve based on the average increase in line length, that is, the similarity ratio, where:

$$D = \ln(2) / \ln(2/\text{ratio}) \tag{3.1}$$

For example, if the length of the generated line increases by an average ratio of 1.3 each time it is subdivided, then its similarity ratio is 1.3; applying Equation 3.1, D becomes 1.61 (Carpenter, 1981). A simple manipulation of Equation 3.1 indicates that, in order for a generated line to have a dimension value not exceeding 2.0, the similarity ratio must not exceed 1.41. Alternatively, the fractal dimension of the generated line can also be deter-

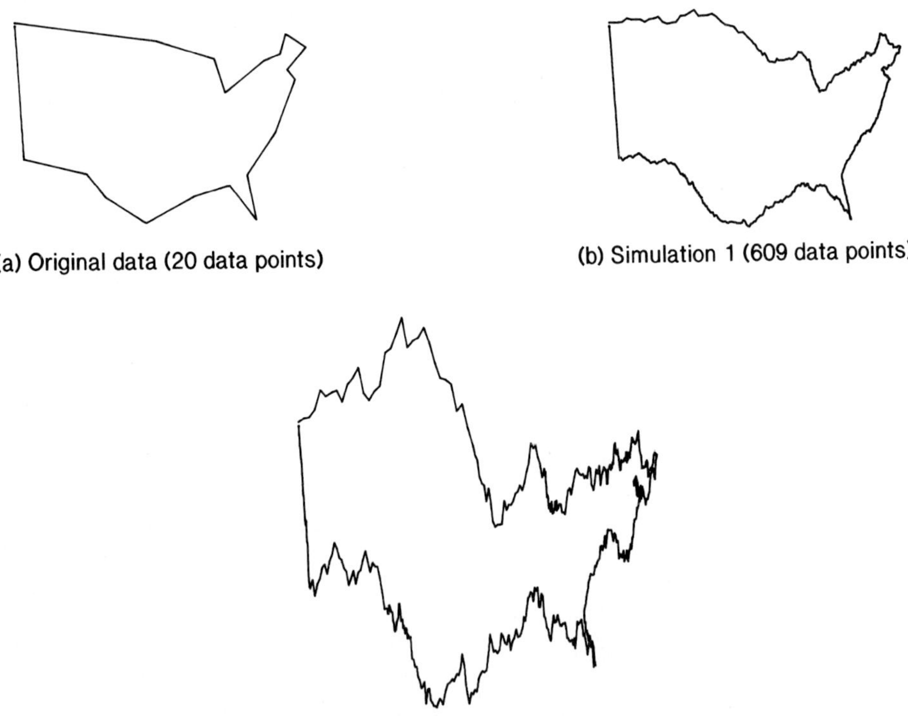

(a) Original data (20 data points)

(b) Simulation 1 (609 data points)

(c) Simulation 2 (609 data points)

Figure 3.3 (a) Original U.S. outline defined by 20 data points, (b) as simulated by midpoint subdivision, with roughness factor = 0.2, estimated $D = 1.07$, and (c) roughness factor = 0.8, estimated $D = 1.48$.

mined by the walking-divider method as discussed in Chapter 2. Finally, the program writes the final series of x,y coordinates in ASCII form so that the data can be displayed by a graphics package.

To demonstrate the line-generation algorithm, a set of 20 sample points was selected from the United States outline, shown in Figure 3.3. This curve is used to illustrate the effect of the method on a readily recognizable shape and not to suggest a way of reproducing the complexity of the actual U.S. boundary! Two different roughness factor values were applied, with all other parameters held constant to facilitate comparison. Figure 3.4 contains the output listings. The plots of the original 20 sample points and the two simulated U.S. country outlines are also shown in Figure 3.3. For a recursion level of 6 and a scale factor of 2.0, the fractal dimensions for the two simulated boundaries were estimated as 1.07 and 1.48 for roughness factors of 0.2 and 0.8, respectively. As expected, the simulated boundaries deviate from the real boundary since only a very small number of control points were used; more control points would obviously improve the fidelity of this interpolation.

The central assumption of the algorithm is that the displacement of the midpoint in the y dimension is proportional to the square root of the distance between the two end points in the x dimension, adjusted by the roughness factor. The higher the roughness factor is, the more rugged the curve and the higher the fractal dimension. The ultimate dimensionality of the generated curve depends on the specific scale and resolution of the data. Therefore, it is necessary to include a scale factor, and some experimentation with optimal scale factors is required. For the present example, a scale factor of 2.0 seems reasonable.

This method (or any other interpolation technique) can also be used to simulate completely artificial curve data merely by feeding the program a set of random locations. The fractal dimension can be determined from the methods presented above (Equation 2.2 or 3.1), but, unlike the surface simulation method discussed below, this technique does not provide a mechanism whereby a direct D value can be specified at the beginning of the simulation process.

Dutton Midpoint Displacement Method

Another midpoint subdivision method, proposed by Dutton (1981), uses three data points at a time to simulate a curve (Figure 3.5). It proceeds by first computing the midpoints P_2' and P_4' of the two line segments defined by the three data points P_1, P_2, and P_3, the lengths of these two line segments, and, in order to maintain self-similarity, the ratio of these two lengths. The interior point (P_2) is then displaced to P_3' according to the length fraction and the angle between the two segments. The smaller the angle, the farther away the interior point is relocated and the more irregular the line will become.

Four parameters control this curve-simulation process. The first two, *sinuosity dimension* (*SD*) and *uniformity coefficient* (*UC*), determine the location of the interior point. The sinuosity dimension corresponds to the angle specified above and therefore specifies the amount of waviness of the output curve. Dutton claims that *SD* is equivalent to Mandelbrot's fractal dimension (D), with values ranging from 1 to 2. In the actual algo-

```
$ run line_gen

    The random midpoint displacement method
    for fractal curve generation
    by Nina Lam (1992)
ENTER INPUT FILENAME ==> usline.dat
ENTER OUTPUT FILENAME ==> usline.out
INPUT THE ROUGHNESS FACTOR (0-1) ==> 0.2
INPUT A SCALE FACTOR (e.g. 5.0) ==> 2.0
INPUT RECURSION LEVEL (2-8) ==> 6
INPUT A SEED VALUE FOR RANDOM GENERATOR
       (A LARGE INTEGER 1-2147483647) ==> 12345678
    NUMBER OF CONTROL POINTS READ =    20
    NUMBER OF RANDOM NUMBERS GENERATED =  589
    THE FIRST 20 RANDOM NUMBERS ARE:
     0.9092      -0.8299      -0.6143       0.3099       0.7585
    -0.1908      -0.6160      -1.1507      -0.1419      -0.6816
     0.9877       0.0754       0.5312       0.8258       0.3618
    -1.1017       0.5672       0.2138      -0.2104      -0.2333
    RECURSIVE LEVEL =     6
    THE ROUGHNESS FACTOR IS =  0.20
    TOTAL NUMBER OF POINTS RESULTED =  609
    THE LINE LENGTHS OF EACH LEVEL ARE :
           70.14
           71.54
           73.29
           75.98
           80.05
           88.62
    ESTIMATED SIMILARITY RATIO =  1.0483
    ESTIMATED FRACTAL DIMENSION =  1.0731

$ run line_gen

    The random midpoint displacement method
    for fractal curve generation
    by Nina Lam (1992)
ENTER INPUT FILENAME ==> usline.dat
ENTER OUTPUT FILENAME ==> usline1.out
INPUT THE ROUGHNESS FACTOR (0-1) ==> 0.8
INPUT A SCALE FACTOR (e.g. 5.0) ==> 2.0
INPUT RECURSION LEVEL (2-8) ==> 6
INPUT A SEED VALUE FOR RANDOM GENERATOR
       (A LARGE INTEGER 1-2147483647) ==> 12345678
    NUMBER OF CONTROL POINTS READ =    20
    NUMBER OF RANDOM NUMBERS GENERATED =  589
    THE FIRST 20 RANDOM NUMBERS ARE:
     0.9092      -0.8299      -0.6143       0.3099       0.7585
    -0.1908      -0.6160      -1.1507      -0.1419      -0.6816
     0.9877       0.0754       0.5312       0.8258       0.3618
    -1.1017       0.5672       0.2138      -0.2104      -0.2333
    RECURSIVE LEVEL =     6
    THE ROUGHNESS FACTOR IS =  0.80
    TOTAL NUMBER OF POINTS RESULTED =  609
    THE LINE LENGTHS OF EACH LEVEL ARE :
           70.14
           90.88
          108.08
          130.42
          159.81
          214.92
    ESTIMATED SIMILARITY RATIO =  1.2524
    ESTIMATED FRACTAL DIMENSION =  1.4807
```

Figure 3.4 Output from the midpoint subdivision method, using two different roughness factor values, 0.2 and 0.8. The outlines are shown in Figure 3.3.

rithm however, the angle between the two segments is specified rather than a numeric value of *SD*, and the exact relationship between *SD* and the angle is not defined.

The uniformity coefficient has values ranging from -1 to +1. It determines how much sinuosity is applied to the generated curve. If *UC* = +1, then the point will be displaced to the location that *SD* implies. If *UC* = 0.5, then the point is moved half of the distance. If *UC* = 0.0, then *SD* has no effect and the point will not be displaced. If a negative value of *UC* is specified, the new point is displaced in an opposite direction.

The two other parameters, *straightness tolerance* (*ST*) and *smoothness tolerance* (*SM*), determine, respectively, the maximum and the minimum lengths of a segment allowed to be subdivided. These two parameters are particularly useful in simulating political boundaries where long linear segments are quite common. After each iteration, the output curve will have about twice the number of points as the input curve. The process repeats until the desired detail is reached.

Dutton's algorithm has been applied to natural and political boundaries, such as U.S. state boundaries, with acceptable results (Dutton, 1981). He has shown through his fractal curve simulation algorithm that U.S. state boundaries can be regenerated fairly well based on a filtered (that is, generalized) digital file that has 1055 points. This is a 90% savings in computer storage and time when compared with the 11,541 coordinate pairs of the original digital file. This method, however, is considered approximate because it does not retain the location of the original control points except for the first and the last coordinates. The method is also considered deterministic, but the exact relationships among fractal dimension and the various parameters of the method have not been clearly defined.

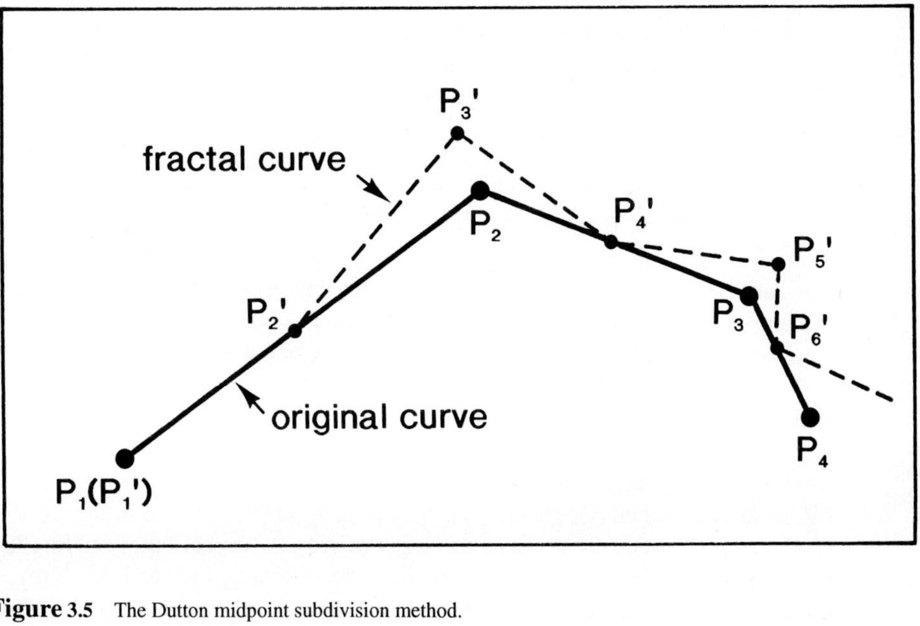

Figure 3.5 The Dutton midpoint subdivision method.

Another advantage of this method is that it offers parameters that can be used to control the appearance of the output curve. Further criteria, such as the straightness tolerance and smoothness tolerance, could be added to the Carpenter method to help simulate political boundaries, as in the example used here. A modified algorithm combining the advantages of both methods would be useful.

Other Curve-simulation Methods

Three other approaches to simulating fractal curves deserve mention. The shear displacement method (Klinkenberg and Clarke, 1992) cuts a line segment at a randomly selected location and displaces the left and right parts of the line segment vertically in opposite directions. The amount of displacement is determined by a Gaussian distribution, and the location of the cutting point is determined by a uniform distribution. The process is repeated until sufficient detail is achieved. This method can be easily extended to two dimensions (discussed below) where the shear lines are determined by the points and their angles of intersections in surface simulation, but this method is computationally expensive.

In the modified Markov method, the next point is generated by using a weighted sum of the previous values plus a random variable (Carpenter, 1981). This method tends to create lines in one direction and cannot be made to pass through two or more specified points. Moreover, it cannot generate the same set of points if the control points are given in a reverse order. The method is mathematically complex, and extension of this method to surfaces remains an unsolved problem.

Fractal curves can also be created by using an inverse Fourier transform of a spectrum whose amplitude decreases with the reciprocal square root of frequency, using white noise for the phase component (Carpenter, 1981). This method is mathematically complex and computationally expensive, and although the technique of fast Fourier transform (*FFT*) can be utilized for efficient computation, the method is limited to curves of lengths that are powers of 2.

METHODS FOR SURFACE SIMULATION

The same midpoint subdivision method for simulating curves can be extended to a higher dimension for simulating surfaces, but, as we might expect, the addition of another dimension to the problem requires more steps than in the one-dimensional case.

Shear-displacement Method

A method for surface simulation that is widely used and easy to implement is the shear-displacement method (Goodchild, 1980). The method starts with a surface of zero altitude (flat plain) represented by a matrix of square grids. A succession of random lines across the plain is generated, and for each line the surface is faulted vertically to form a cliff. The process is repeated until several cliffs are created between adjacent sample points. The set of random fault lines generated has the property that their points of intersection follow a Poisson distribution, while the angles of intersection are distributed uniformly between 0 and $2\pi\pi$. The heights of the cliffs are controlled by the parameter *H* in such a way that the variance between two points is proportional to their distance scaled by the variogram parameter *H* (Equation 2.3).

As in the midpoint subdivision method, *H* describes the persistence of the surface and has values between 0 and 1, and the fractal dimension of the simulated surface can be determined by $D = 3 - H$. The value $H = 0.5$ (or $D = 2.5$ for surfaces) results in a Brownian surface. The relationship between variance and distance for different *H* was discussed in Chapter 2.

A FORTRAN program based on the shear-displacement method, modified from Goodchild's (1980) original algorithm, is included in Appendix 3.2. The program requires the following input: the number of rows and columns defining the output matrix, the number of random cuts (that is, cliffs), an *H* value, and a seed value for the random-number generator. Figure 3.6 is an example output listing, and Figure 3.7 illustrates the four surfaces simulated by the program using four different *H* values. All the surfaces shown have 50 rows and 50 columns, and use 300 random cuts. The seed value for the random-number generator was fixed intentionally to demonstrate more clearly the effect of *H*, and hence *D*, on the complexity of the simulated surfaces. It is clear from Figure 3.7 that as *D* increases surface complexity increases.

In addition, we measured the fractal dimensions of the four generated surfaces using the isarithm method described in Chapter 2 to compare the differences between the two sets of dimensions, the ones used in the simulation process (*H* values) and the ones measured. The results are listed in Table 3.1. Discrepancies between the simulated *D* and the

```
$ run surf_gen

    The shear displacement method
    for fractal surface generation
    modified by Nina Lam (1991)

ENTER THE MATRIX SIZE (N BY N) ==> 50
ENTER THE H VALUE (0-1) ==> 0.2
ENTER THE NUMBER OF CUTS (e.g.NS=300) ==> 300
ENTER A SEED VALUE FOR RANDOM GENERATOR
     (A LARGE INTEGER 1-2147483647) == > 34782911
ENTER OUTPUT FILENAME ==> d8.dat

    INPUT PARAMETERS N,NS,H,ISEED ARE:
       50         300    0.2      34782911
    THE FIRST 20 RANDOM NUMBERS ARE:
     0.3572        0.4320        0.7952        0.2440        0.9977
     0.7732        0.0474        0.0623        0.3185        0.7514
     0.5147        0.8328        0.3639        0.2738        0.9152
     0.9657        0.4212        0.8133        0.0135        0.9772

    GENERATING AND WRITING X,Y,Z
    FILE WRITTEN AND TASKS COMPLETED
FORTRAN STOP
```

Figure 3.6 Output from the shear-displacement method.

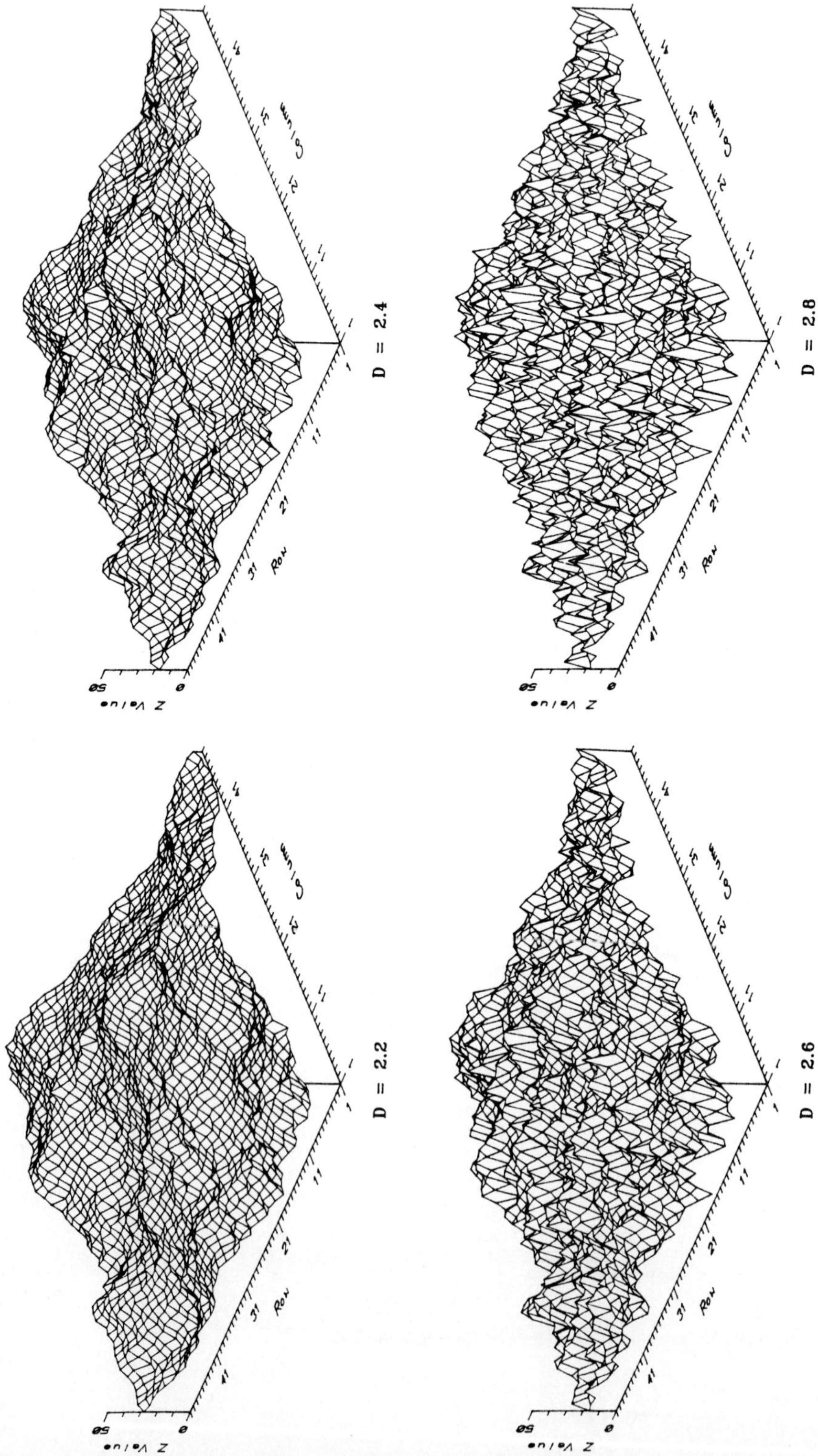

Figure 3.7 Examples of shear-displacement surfaces.

measured D were found; the isarithm measurement consistently overestimates the *D* used by the shear displacement method.

Table 3.1 Comparison between Given *D* (or *H*) and Measured *D* for Four Generated Surfaces

Surface	*H*	Measured *D*	R^2	*Z* min	*Z* max
D=2.8	0.2	2.84	0.98	-5.48	7.05
D=2.6	0.4	2.72	0.98	-8.11	12.01
D=2.4	0.6	2.51	0.96	-11.25	26.64
D=2.2	0.8	2.34	0.97	5.97	59.17

Other Surface-simulation Methods

An obvious generalization of the midpoint subdivision method for surface generation is to start with a grid that is subdivided into half the grid size in two steps. The heights of the midpoints of all squares are computed by interpolation from their four neighbors and adding a random offset. The offset in each stage is determined by a scaling factor controlled by a Gaussian random variate and the scaling factor *H*. For squares, the scaling factor is $2^{-0.5H}$ (Peitgen et al., 1992). As we discussed in Chapter 2, for surfaces $D = 3 - H$, and for lines $D = 2 - H$ (Goodchild, 1980).

Two other methods for fractal surface generation are the successive random addition method introduced by Voss (1985) and the Fourier filtering method discussed in Peitgen and Saupe (1988). The former method involves successive interpolation of the vertical *z* values of the center points from their four neighbors. This method addresses the nonstationarity problem of the midpoint displacement technique by displacing all of the points (instead of only the midpoints) so the surfaces generated by this process are self-affine (Feder, 1988). The Fourier filtering method assumes the surface is composed of a series of sine and cosine waves of increasing frequencies and decreasing amplitude; the method becomes one of choosing the appropriate Fourier coefficients and then performing a two-dimensional inverse transform (Clarke, 1988).

Additional control parameters can be incorporated within each of the surface-generation methods. The addition of visualization techniques (for example, using colors to suggest relief or water bodies) results in fractal surfaces that can be rendered to look like natural landscapes, such as those displayed in Peitgen and Saupe (1988).

In summary, an important aspect of fractal landscape simulation is that the characteristics of these surfaces are relatively easy to specify, so they can serve as simple geometrical tools for the description of more complicated surfaces.

CONCLUSION

This chapter has reviewed and illustrated some of the considerable research on the simulation of realistic and useful spatial data that has been done since Mandelbrot's work in the 1970s. Scientists are now able to make use of a wide range of these techniques for creating esthetically pleasing imaginary landscapes, interpolating sparse data, and exploring spatial processes. Nevertheless, a number of questions remain. For example, how do we know whether the curve or surface being generated is of the dimension that we specified at the beginning of the simulation process? This query returns us to the previous question of how these curves and surfaces should be measured, a subject discussed in Chapter 2. Because fractal measurement is basically a sampling process, perfect agreement between the dimension used to simulate an object and its measured dimension is difficult to achieve. As always, visual inspection, is a useful way to validate the simulated and interpolated results from fractals. Although difficult to implement, the property of invertibility for both fractal simulation and fractal measurement methods should be of value both conceptually and practically, and such properties require further study.

REFERENCES

Barnsley, M. F., and Sloan, A. D. 1988. A better way to compress images. *Byte* 13:215–223.

Batty, M. 1985. Fractals—geometry between dimensions. *New Scientist* 15(1450):31–35.

———. 1989. Geography and the new geometry. *Geography Review* 2(4):7–10.

———, and Longley, P. 1986. The fractal simulation of urban structure. *Environment and Planning A* 18:1143–1179.

Carpenter, L. C. 1981. Computer rendering of fractal curves and surfaces. *Proceedings of ACM (Association for Computing Machinery) SIGGRAPH Conference*, Seattle, Washington, July 1980.

Clarke, K. C. 1988. Scale-based simulation of topographic relief. *The American Cartographer* 15(2):173–181.

Dutton, G. H. 1981. Fractal enhancement of cartographic line detail. *The American Cartographer* 8:23–40.

Feder, J. 1988. *Fractals*. New York: Plenum Press.

Goodchild, M. F. 1980. Fractals and the accuracy of geographical measures. *Mathematical Geology* 12:85–98.

———, and Mark, D. M. 1987. The fractal nature of geographic phenomena. *Annals of the Association of American Geographers* 77(2):265–278.

Jiang, R.–S. 1984. *An Evaluation of Three Curve Generation Methods*. M.S. Thesis. Columbus, Ohio: Department of Geography, The Ohio State University.

Klinkenberg, B., and Clarke, K. C. 1992. Exploring the fractal mountains. In *Automated Pattern Analysis in Petroleum Exploration*, I. Palaz and S. K. Sengupta, eds. New York: Springer-Verlag.

Lam, N. S.–N. 1982. An evaluation of areal interpolation methods. *Proceedings, Fifth International Symposium on Computer-assisted Cartography (Auto-Carto 5)* 2:471–479.

———. 1983. Spatial interpolation methods: A review. *The American Cartographer* 10:129–49.

Loehle, C. 1983. The fractal dimension and ecology. *Speculations in Science and Technology* 6(2):131–142.

Mandelbrot, B. B. 1983. *The Fractal Geometry of Nature*. San Francisco: W.H. Freeman & Co.

Mark, D. M., and Lauzon, J. P. 1984. The space efficiency of quadtrees: An empirical examination including the effects of two-dimensional run-encoding. *GeoProcessing* 2:367–83.

Peitgen H.-O., and Richter, P. H. 1986. *The Beauty of Fractals*. New York: Springer-Verlag.

———, and Saupe, D. eds. 1988. *The science of fractal images*. New York: Springer-Verlag.

———, Jürgens, H., and Saupe, D. 1992. *Fractals for the Classroom. Part I. Introduction to Fractals and Chaos*. New York: Springer-Verlag.

Voss, R. F. 1985. Random fractal forgeries. In *Fundamental Algorithms for Computer Graphics*, R. A. Earnshaw, ed., pp. 805–835. New York: Springer-Verlag.

Appendix 3.1

Listing of FORTRAN Program LINE_GEN.FOR

RANDOM DISPLACEMENT METHOD FOR CURVE GENERATION

```
C***********************************************************************
C                                                                      *
C The random midpoint displacement method for curve generation        *
C          by Nina S.-N. Lam, 1991.                                    *
C The program is based on the algorithm described in Carpenter (1992) *
C and Peitgen, et al. (1992).                                          *
C                                                                      *
C***********************************************************************

      DIMENSION X1(2560),Y1(2560),X2(2560),Y2(2560),RR(2540),TD(8)
      INTEGER*4 ISEED
      INTEGER*2 IFN
      PARAMETER (IFN=11)
      PARAMETER (NFO=12)
      BYTE NAME(30)

      WRITE(*,'(/5X,A)') 'The random midpoint displacement method'
      WRITE(*,'(5X,A)') 'for fractal curve generation'
      WRITE(*,'(5X,A//)') 'by Nina Lam (1992)'

C
C  Read the names of the input and output files
C
      WRITE(*,'(A,$)') ' ENTER INPUT FILENAME ==> '
      READ(*,'(Q,30A1)') I,NAME
      IF(I.EQ.0) THEN
        STOP ' ERROR IN FILENAME '
      ELSE
        NAME(I+1)='00'X
```

```
      ENDIF
      CALL ASSIGN(IFN,NAME)

      WRITE(*,'(A,$)') ' ENTER OUTPUT FILENAME ==> '
      READ(*,'(Q,30A1)') I,NAME
      IF(I.EQ.0) THEN
        STOP ' ERROR IN FILENAME '
      ELSE
        NAME(I+1)='00'X
      ENDIF
      CALL ASSIGN(NFO,NAME)
C
C  Set parameters for curve generation
C
      WRITE(*,'(A,$)') ' INPUT THE ROUGHNESS FACTOR (0-1) ==> '
      READ(*,'(F)') RF
      WRITE(*,'(A,$)') ' INPUT A SCALE FACTOR (e.g. 5.0) ==> '
      READ(*,'(F)') S
      WRITE(*,'(A,$)') ' INPUT RECURSION LEVEL (2-8) ==> '
      READ(*,'(I)') LEVEL
      WRITE(*,'(A,$)') ' INPUT A SEED VALUE FOR RANDOM GENERATOR '
      WRITE(*,'(A,$)') '        (A LARGE INTEGER 1-2147483647) ==> '
      READ(*,'(I)') ISEED
C
C  Read x,y control points, max.=20 (with 8 levels)
C
      N=0
      DO WHILE(.TRUE.)
        READ(IFN,*,END=99) X,Y
        N=N+1
        X1(N)=X
        Y1(N)=Y
      ENDDO
   99 CONTINUE
      WRITE(6,500) N
  500 FORMAT(/5X,'NUMBER OF CONTROL POINTS READ =',I5)
C
C  Generate Gaussian random numbers with mean=0
C
      NR=(N-1)*(2**(LEVEL-1)-1)
      DO I=1,NR
        RR(I)=RAN(ISEED)+RAN(ISEED)+RAN(ISEED)-1.5
      ENDDO
      WRITE(6,540) NR
  540 FORMAT(5X,'NUMBER OF RANDOM NUMBERS GENERATED =',I5)
      WRITE(6,541)
  541 FORMAT(5X,'THE FIRST 20 RANDOM NUMBERS ARE:')
      WRITE(6,'(5F12.4)') (RR(I),I=1,20)
C
C  Start recursive subdivision and compute line length
C
      LL=1
      DO 30 L=1,LEVEL
```

```
      IF(L.EQ.1) GO TO 40
      K=1
      DO I=1,N-1
        XMID=0.5*X1(I)+0.5*X1(I+1)
        DD=ABS(X1(I+1)-X1(I))
        D=RF*S*SQRT(DD)*RR(LL)
        YMID=0.5*Y1(I)+0.5*Y1(I+1)+D
        X2(K)=X1(I)
        Y2(K)=Y1(I)
        K=K+1
        X2(K)=XMID
        Y2(K)=YMID
        K=K+1
        LL=LL+1
      ENDDO
      X2(K)=X1(N)
      Y2(K)=Y1(N)
      N=K
      DO I=1,N
        X1(I)=X2(I)
        Y1(I)=Y2(I)
      ENDDO
 40   DSUM=0.0
      DO I=1,N-1
        DX=X1(I)-X1(I+1)
        DY=Y1(I)-Y1(I+1)
        D=SQRT(DX*DX+DY*DY)
        DSUM=DSUM+D
      ENDDO
      TD(L)=DSUM
 30 CONTINUE

    WRITE(6,550) LEVEL
550 FORMAT(5X,'RECURSIVE LEVEL =',I5)
    WRITE(6,580) RF
580 FORMAT(5X,'THE ROUGHNESS FACTOR IS =',F6.2)
    WRITE(6,560) N
560 FORMAT(5X,'TOTAL NUMBER OF POINTS RESULTED =',I5)
C
C  Print line lengths; compute the average increase ratio and
C  estimate the fractal dimension
C
    WRITE(6,570)
570 FORMAT(5X,'THE LINE LENGTHS OF EACH LEVEL ARE :')
    WRITE(6,510) (TD(I),I=1,LEVEL)
510 FORMAT(5X,F12.2)
    RATIO=0.0
    DO I=1,LEVEL-1
      D=TD(I+1)/TD(I)
      RATIO=RATIO+D
    ENDDO
    RATIO=RATIO/FLOAT(LEVEL-1)
    DIMEN=LOG(2.0)/LOG(2.0/RATIO)
```

```
      WRITE(6,520) RATIO
  520 FORMAT(5X,'ESTIMATED SIMILARITY RATIO =',F8.4)
      WRITE(6,590) DIMEN
  590 FORMAT(5X,'ESTIMATED FRACTAL DIMENSION =',F8.4)
      DO I=1,N
        WRITE(NFO,592) X1(I),Y1(I)
      ENDDO
  592 FORMAT(2F12.2)
      END
```

Appendix 3.2

Listing of FORTRAN Program SURF_GEN.FOR

SHEAR DISPLACEMENT METHOD FOR SURFACE GENERATION

```
C************************************************************************
C                                                                       *
C The Shear Displacement method for surface generation                  *
C            modified by Nina Lam, 1991.                                *
C This program was modified from the algorithm by Goodchild (1980)      *
C                                                                       *
C************************************************************************

      DIMENSION TTH(2000),RSTH(2000),ADD(2000),XM(2000),STH(2000)
      DIMENSION RR(10000)
      INTEGER*4 ISEED
      REAL*8 CUM
      COMMON H,H2(1000),NH
      PARAMETER (NFO=12)
      BYTE NAME(30)

      WRITE(*,'(/5X,A)') 'The shear displacement method'
      WRITE(*,'(5X,A)') 'for fractal surface generation'
      WRITE(*,'(5X,A//)') 'modified by Nina Lam (1991)'
C
C  Read input parameters and output filename
C
      WRITE(*,100) ' ENTER THE MATRIX SIZE (N BY N) ==> '
  100 FORMAT(A,$)
      READ(*,'(I)') N
      WRITE(*,100) ' ENTER THE H VALUE (0-1) ==> '
      READ(*,'(F)') H
      WRITE(*,100) ' ENTER THE NUMBER OF CUTS (e.g.NS=300) ==> '
      READ(*,'(I)') NS
      WRITE(*,100) ' ENTER A SEED VALUE FOR RANDOM GENERATOR'
      WRITE(*,100) '        (A LARGE INTEGER 1-2147483647) == > '
      READ(*,'(I)') ISEED
      WRITE(*,100) ' ENTER OUTPUT FILENAME ==> '
      READ(*,'(Q,30A1)') I,NAME
      IF(I.EQ.0) THEN
```

```
          STOP ' ERROR IN FILENAME '
        ELSE
          NAME(I+1)='00'X
        ENDIF
        CALL ASSIGN(NFO,NAME)
        WRITE(6,'(//5X,A,$)') 'INPUT PARAMETERS N,NS,H,ISEED ARE:'
        WRITE(6,'(2I10,F6.1,I12)') N,NS,H,ISEED
C
C  Generate random numbers with U(-1,1)
C
        NR=5000
        DO I=1,NR
          RR(I)=RAN(ISEED)
        ENDDO
        WRITE(6,187)
  187   FORMAT(5X,'THE FIRST 20 RANDOM NUMBERS ARE:')
        WRITE(6,'(5F12.4)') (RR(I),I=1,20)
C
C  Start generation
C
        KK=0
        H=2.0*H
        NH=N*1.4142*2+2
        DO I=1,NH
          S=I/2.0-0.5
          H2(I)=S**H
        ENDDO

        DO 3 K=1,NS
   33     CONTINUE
          KK=KK+1
          TH=2.0*3.14159*RR(KK)
          KK=KK+1
          R=RR(KK)*N*1.4142
          KK=KK+1
          I=1+INT(TH*2.0/3.14159)
          GO TO (31,32,33,34),I
   31     CR=1.1412*N*COS(3.14159/4.0-TH)
          IF(R.GT.CR) GOTO 33
          GOTO 35
   32     CR=N*COS(TH-3.14159/2.0)
          IF(R.GT.CR) GOTO 33
          GOTO 35
   34     CR=N*COS(TH)
          IF(R.GT.CR) GOTO 33
   35     TTH(K)=TAN(TH)
          STH(K)=SIN(TH)
          RSTH(K)=R/SIN(TH)
          ADD(K)=RR(KK)*2.0-1.0
    3   CONTINUE

        WRITE(6,101)
  101   FORMAT(/5X,'GENERATING AND WRITING X,Y,Z')
```

```
    DO 4 I=1,N
      X=X+1.0
      Y=0.0
      YM=I-0.5
      DO K=1,NS
        XM(K)=RSTH(K)-YM/TTH(K)
      ENDDO
    DO 4 J=1,N
      XCOL=J-0.5
      CUM=0.0
      DO K=1,NS
        S=(XM(K)-XCOL)*STH(K)
        Q=TWOH(S)
        CUM=CUM+Q*ADD(K)
      ENDDO
      Y=Y+1.0
  4 WRITE(NFO,5) X,Y,CUM
  5 FORMAT(1X,2F10.0,5X,G12.5)
    WRITE(6,102)
102 FORMAT(5X,'FILE WRITTEN AND TASKS COMPLETED')
    STOP
    END

    FUNCTION TWOH(S)
      COMMON H,H2(1000),NH
      S1=ABS(S+0.5)
      S2=ABS(0.5-S)
      IH1=INT(S1*2.0)+1
      IH2=IH1+1
      H1=H2(IH1)+2.0*(S1-INT(S1))*(H2(IH2)-H2(IH1))
      IH1=INT(S2*2.0)+1
      IH2=IH1+1
      HH1=H2(IH1)+2.0*(S2-INT(S2))*(H2(IH2)-H2(IH1))
      TWOH=H1-HH1
    RETURN
    END
```

CHAPTER 4

A FRACTAL PARADIGM FOR GEOGRAPHY?

Lee De Cola
U. S. Geological Survey, Reston, VA

Nina Siu-Ngan Lam
Department of Geography and Anthropology, Louisiana State University, Baton Rouge

Mandelbrot has said that until he discovered chaos his life was chaotic (Albers and Alexanderson 1985). Can the fractal paradigm of diversity, irregularity, and complexity bring focus, regularity, and simplicity to geography? This is neither possible nor necessarily desirable, but the chapters of this book testify to the usefulness of the fractal idea to geographic thought, research, and technique. This chapter concludes the introductory part of the book with a consideration of the role of geometry in geography, followed by a brief review of the geography contained within Mandelbrot's magnum opus, *The Fractal Geometry of Nature* (1983). We then cite a few of the problems that must be resolved if the fractal approach is to be fully incorporated into geographic thought, and we conclude with a few thoughts about the prospects for a fractal paradigm in geography.

GEOGRAPHY, GEOMETRY, AND CARTOGRAPHY

We began Chapter 1 with a rough taxonomy of spatial objects, arguing that the concept of stochastic self-similarity provides geographers with a new way of measuring, analyzing, and ultimately understanding the complexity of the world (Pickover, 1990). The chapter also demonstrated how the fractal approach allows us to characterize complexity in some kinds of geographical data. The next two chapters then presented techniques whereby fractal ideas allow scientists to create and analyze complex objects. Although the usefulness of the methods should speak for themselves, it remains to assess the present and future role of fractal concepts in geographic science.

We have intended to demonstrate that fractal geometry, as indicated by the dictionary definition at the beginning of Chapter 1, provides geographers with a rigorous approach to form and dynamics principally through the manipulation of scale and resolution. The definition, while incomplete and colloquial, helps us understand what happens when geography does what its etymological roots require by "carving up" the world and then looking at the results at different scales. This new geometry provides geographers with several new approaches to the problems of measuring, analyzing, and representing the earth. Fractal geometry does not require a strict Cartesian space, so concepts of space, time, and distance can be extended to comprehend the rich complexity of the real world. Moreover, fractal geometry applies the extremely flexible concepts of set theory to analyze the representational topology and analysis of measurements. Furthermore, it has abstracted the notion of measure to be the key fundamental attribute of objects. Finally, it requires the analyst to be explicit and rigorous as never before about the concepts of scale, resolution, and size.

Table 4.1 is an attempt at an evolutionary portrait of geography. As in Chapter 1, we begin with objects that exist in multidimensional space/time that can be represented at scales that preserve some aspects of the objects' attributes. The second column of the table gives a few examples of the wide range of objects that are the subject matter of geography. Any geographic discussion about a real object is always founded on some typical physical scale range, in that the object of discourse (or at least its bounding set) is regarded as having a real location and some maximum extent. Yet, even though we may be able to locate an object and set a maximum for its size, specifying size and counting the numbers of its constituent parts, critical elements of any truly rigorous discussion, are often quite difficult. And questions of form are even more difficult to settle.

The geometric approach to the world is outlined in the third column of the table. Classical geometry, while having its origins in the measurement of things in the real world ("geo"), typically works within a Euclidean space of N dimensions, examining the nature and relationships of abstract figures (for example, points, lines, and polygons) whose fractal dimension is some integer. Unlike geography, however, geometry lacks scale because its objects are regarded as being idealizations that have neither characteristic length nor absolute size. Furthermore, although classical geometry can assign various attributes to its objects, it cannot characterize form except by establishing correspondences (for example, equality and similarity) between a given object and some simple ideal object (for example, a line or ellipse).

Cartography has linked geography with geometry by projecting four dimensional real world objects upon two dimensional maps. Classical geometry provided cartography with an analytical framework beyond mere description by transforming real things into abstract figures. That these objects may be digitized into geographic information systems (GIS) models and rendered on computer screens does not alter the fact that mapped objects remain basically elementary topological entities.

It is true that cartographic scale forces an explicit comparison between the complex and irregular geographic world of things and the ideal realm of entities and that generalization enables geographers to form a link between ideal tractable objects and the form and behavior of real things. Yet cartography, and geographic analysis as a whole, faces a tension between the need to faithfully represent things in the real world and the confines of a highly formal geometric language that requires that all representations of things share the characteristics of simple figures. In short, the geometric tool often acts as a filter.

TABLE 4.1 Key Concepts in Geography, Geometry, Cartography, and Fractals

Concept	Geography	Classical Geometry	Cartography	Fractal Geometry
Space	Real	Euclidean	Projection	Metric
Dimensions	Space–time = 4	Space = N	Map = 2	N-space
Objects	Things Person River City Mountain	Figures Points Line Polygon Pyramid	Entities Position Chain of lines Polygon Contour	Sets Dusts Curves, networks Clusters, islands Surfaces
Scale	Typical length	Similarity	Scale ratio	Similarity ratio Scaling Resolution
Attributes	Size Number Location Form	Length, area, etc. Number Position Similarity	Length, area, etc. Number Position Shape	Measure Fractal dimension

The ideas of this table can be seen, for example, in the different views it gives of the notion of a river. Geography has long studied the phenomenon of rivers in their various types, places, forms, and roles, but has only infrequently attempted a rigorous characterization of what precisely a river might be. Geometry allows rivers to be formalized as curves and cartography (as well as GIS) uses this formalization to create rivers as chains of lines. Fractal geometry returns some of the original phenomenological richness to the notion of rivers by allowing them to be represented simply as curves, as networks, and even as sets of observations (water versus nonwater) that can be visualized at multiple scales and analyzed in terms of multiple representations. In a sense, this new geometry returns us to an older discourse of complexity.

A GEOGRAPHIC TOUR OF THE FRACTAL GEOMETRY OF NATURE

Beyond reading this book, there is no better way to understand the geographic relevance of the fractal paradigm than by surveying the geographic content of *The Fractal Geometry of Nature* (Mandelbrot, 1983). Anyone with more than a passing interest in the subject of fractals simply must read this book. The discussion that follows makes reference to 19 of Mandelbrot's 42 chapters as having some relevance to geographic theory. While a systematic view of the fractal nature of the world at geographic scales (say 10^0 to 10^6 meters) has yet to be written, this guide may help that future author.

In spite of the book's title, Mandelbrot set out not to describe nature, but to present an "essay" (his word) on how fractals shed light on natural phenomena. As he says, "my work is concerned primarily with shapes in the real space one can see" (p. 193). And although the book was written to give people a new outlook on the world, Mandelbrot is a mathematician, not a geographer. As natural history, the book is itself fractal: irregular, complicated, and self-similar. It can therefore be organized in many ways, including geometrically (which is what Mandelbrot in part attempted to do), systematically (in terms of the major ideas of systems theory, such as hierarchy, dynamics, flows, and others), or epistemologically (in terms of various disciplines).

We have tried here to organize ideas from the perspective of what could loosely be called earth science, and within this science from large-scale astrophysics to small-scale biology, concluding with a discussion of cartography and a brief miscellany. It is hoped that this brief geography of the magnum opus will assist first readers in appreciating the book's value to spatial scientists and that readers familiar with the work will be stimulated to return to it for more valuable insights.

Astronomy

Galaxies and the universal density of matter are discussed in Chapter 9 of *The Fractal Geometry of Nature* (*FGN* 9), a bold discussion of the fractal nature of the universe. The idea is that matter is neither uniformly nor randomly scattered in the universe, but appears in fractal sets of widely varying local density. This theme appears again in a discussion in *FGN* 22 of the hierarchical nature of clusters and galaxies. Craters and their size distributions are explored in *FGN* 33 which deals as well with galaxies. It is argued that craters have the COUNT-STEP distribution we introduced in our Chapter 1. It is quite simple to simulate a lunar landscape using randomly placed filled circles whose diameters are 1/RAND, where RAND is a uniformly distributed random variable.

Geology

Strata are referred to in *FGN* 23 in the sense of "layered" hierarchical phenomena. The link (pursued in this book in Chapter 9) between dusts and real strata is not developed, but no doubt would have been had Mandelbrot thought of it. Earthquakes are a rich source of fractal examples (a brief note appears in the "update to the second edition"): they occur along highly irregular faults and at irregular intervals directly related to magnitudes. Terrain can be modeled fractally as Brownian motion that is random but spatially autocorrelated (tied to history), as in *FGN* 28, which shows by now classic pictures of simulated landscapes. It is important to recognize that terrains are (mathematical) mappings $\mathbf{R}^2 \rightarrow \mathbf{R}$, where $\mathbf{R}$ is the set of real numbers.

Meteorology

Clouds are intermittent cascades that *FGN* 10 argues result from the turbulent interaction of temperature and water vapor. Although they are nearly everywhere—and the bane of land imaging—we cannot put our finger on them either physically or analytically. Length, area, and volume measurements of clouds and rain areas are demonstrated to reveal fractality in *FGN* 12. Rain areas are also discussed in *FGN* 12 and the biographical *FGN* 40 as having hyperbolic distributions (COUNT–SIZE). It should always be borne in mind that throughout Mandelbrot's discussions real phenomena are measured in terms of some outer size limit, some resolution, and some threshold. The term "rain area" is therefore less likely to be misunderstood than the word "cloud." Finally, the Gulf Stream is also discussed in *FGN* 10 as a manifestation of turbulence.

Hydrology

Rivers are a rich source of fractal examples. In *FGN* 9 we are asked to imagine the "monstrous" journey made while keeping a river system always to our left, impossible to do if the river system is space-filling, a problem touched on in Chapter 7 of this book. This problem is further analyzed in *FGN* 12 on length–area–volume relations, relating rivers and basins. Finally, *FGN* 17 discusses branching processes in trees and rivers (see our Chapters 6 and 7), and *FGN* 36 looks at random-walk simulations of rivers. Flows are the other hydrologic topic discussed, viewed in *FGN* 27 as a time series resulting from dynamic weather events and the geometric structure of river networks.

Biology

Trees and lungs are discussed in *FGN* 16 and 17 as networks like rivers but in three-dimensional space. The interesting problem in the function of lungs is to create a structure in three-dimensional space that can be regarded as a network of one-dimensional lines associated with a two-dimensional surface for the exchange of what are regarded as zero-dimensional particles. This basic problem is common to many fractal systems, including central places (see our Chapter 10).

Cartography

Coastlines, which can be related to the geomorphological issues of lakes and the hydrologic problems of rivers, are perhaps the most frequently cited example of the use of fractals in geography. This problem is introduced in *FGN* 5, which asks the classic question about coastline length that must be answered in terms of resolution (see our Chapter 2).

Coastlines and contours are also the focus of *FGN* 28, which presents the now famous fractal simulations of landscape relief. Streets, a mapping problem related to coastlines, are simulated in *FGN* 31. Road networks are often designed with a specific hierarchy, although their use is also a function of location, condition, and time (see our Chapter 11). Mandelbrot mentions the topology created by streets and their associated polygons. Regions are two-dimensional sets whose boundaries and areas are a function of resolution. Contours of terrains can become islands (*FGN* 13) (see our Chapter 14) and lakes (*FGN* 29), depending respectively on whether "land" is in the minority or the majority. Islands can also be viewed as connected sets of cells (*FGN* 23), as well as the interior polygons of self-intersecting random walks (*FGN* 25), and all may conform to a COUNT–SIZE rule.

Miscellany

Time series appear frequently in the book, not only because they can be treated as continuous yet not necessarily differentiable functions, but also because Mandelbrot's earliest work was in this area. Economic geography is touched on in the analysis of prices in *FGN* 37. Size distributions also appear in many places as a manifestation of scaling (Schroeder, 1991) in income (and city size) distributions (*FGN* 38). Central places can be regarded as a manifestation of intermittent (*FGN* 10) economic activity in which trade is distributed over smaller and smaller regions.

FUTURE OF THE FRACTAL PARADIGM

This review of *The Fractal Geometry of Nature*, as well as the research chapters of this book, should serve to reassure the skeptical of the importance of the fractal paradigm to geography as well as to science in general. The history of science may be viewed broadly as an evolutionary process from the simple through the complicated to the complex. Science began by the construction of simple models of processes that were presumed to underlie the "booming, buzzing confusion" of the real world. Euclidean geometry and Newtonian mechanics served this epistemology quite well. The rise of statistical thinking in the late eighteenth century provided scientists with tools for the investigation of complicated data presumed to conform to simple models but overlaid with noise that, it was assumed, could be reduced by the successive addition of explanatory factors. In the late twentieth century, scientists are beginning to confront complexity directly by examining complicated forms in space and richly detailed dynamic processes that are regarded as the result of multiscale interactions that cannot be resolved (note the term) into simple signals overlaid with complicated noise. In this view, to paraphrase Mies van der Rohe (1959), nature is "in the details" at all scales.

The future of science is a matter for speculation, but there seems to be three scenarios. Fractal ideas may be rejected and disappear from thinking and from the literature. This seems unlikely given their power and influence in a wide range of fields. Or fractal thinking may establish itself as the key paradigm of geographic thought, certainly broader than a hypothesis and even more fundamental than a theory. This, too, seems unlikely; geography is too broad a field to be unified by any one idea.

A third scenario is that the fractal ideas of scaling, self-similarity, complexity, and the rest will be incorporated into geographic thought and become part of our language, used and elaborated self-consciously at first, but eventually without a second thought. It is helpful to recall the fate of another and quite similar aspiring paradigm of a generation ago, General Systems Theory (von Bertalanffy 1953). GST was intended to be a kind of grand theory, encompassing ideas of process and form from the social and natural sciences, and its ideas of equifinality, allometry, systems analysis, and others have become either an explicit part of our language or embedded in our thinking.

Recent literature in the field of fractals testifies to a continuing process of intellectual growth. The last few years have brought several important books on the subjects of fractals in space and chaos in time. While many of the key ideas of these new sciences of complexity predate Mandelbrot's *Fractal Geometry of Nature*, his book was the first full and aggressive statement of their importance. We should take seriously Mandelbrot's somewhat disingenuous expression of modesty that he was standing on the shoulders of a number of original thinkers—from children to unsung scholars of the seventeenth century. The power of these ideas continues to reverberate in all the sciences.

Consider a few of these recent works. James Gleick (1987) popularized many of the key ideas of temporal chaos and spatial fractality, serving to install these terms in the language of informed people everywhere. Many geographers reading this book must have felt some excitement to note that new tools were evolving for the analysis of everyday complexity and that great attention is being bestowed on the developers of those tools. A physicist, Jens Feder became the first scientist to synthesize and gather most of the important ideas and research in fractal analysis (Feder, 1988). The book made accessible to numerate readers most of the important analytical tools scientists needed to make their own contributions to fractal research. In the same year, Barnsley (1988) presented an elaborate tutorial for the use of fractal ideas in the graphical representation of complexity, and then Falconer (1990) provided an extremely useful set-theoretic foundation for the analysis of fractal data. As geographers read these books and use their ideas in discussions, lectures, and writings, fractal concepts are becoming part of their thought processes. Certainly, there is no turning back.

Geographers will continue to elaborate upon these ideas and incorporate them in their research because geographers are preoccupied with the understanding and representation of complex phenomena, and fractal ideas are certainly the leading paradigm in this area. Just as the Gaussian process of multiple causes leading to a normal distribution was once the leading model for the description of data at one scale, so is the fractal process of self-organization at multiple scales likely to become a leading paradigm of scientific application. Indeed, this may explain why an increasingly large number—some reports claim over 50%—of the papers appearing in the physics literature contain the word "fractal" somewhere in their text.

THE RESEARCH AGENDA

But evolution of fractal ideas toward a secure place in geography is not likely to be smooth. What are the constraints in this process? First, fractal techniques are not yet sufficiently consistent in notation or approach to be a unifying theme in geographical analysis. For example, there are many ways to find the fractal dimension of a phenomenon, and the answers are not always the same. This problem is perhaps due as much to inconsistencies among researchers as to fundamental flaws in the theory (for example, we often ascribe the fractal dimension of the boundary to its enclosed region), but there are confusing ambiguities among such approaches as box and divider methods.

Second, as the previous chapters and the research chapters illustrate, fractal analysis needs to be better integrated with such venerable multiscale techniques as spatial autocorrelation, variograms, Fourier analysis, the modifiable area unit problem, and so forth (see Gaile, 1990, for an overview of the place of fractals in analytical technique). There is an understandable tendency among junior scientists to gather under the banner of new ideas in part out of frustration with the inadequacies of the old, but as well out of a need to overthrow established regimes. In the case of what some might call a fractal "revolution" existing technique needs to be integrated with and even incorporated into contemporary ideas.

Third, fractal theory is still largely descriptive rather than inferential. Although new models allow us to fit functions to data from a wide range of scales, they provide less understanding of the processes that give rise to these data. It is certainly exciting to be able to simulate landscapes, rivers, and clouds, but these models do not as yet provide us with sufficient explanation of process to be regarded as more than descriptive tools. The fact that our models are so much more powerful today gives cause for optimism that explanation will be soon in coming, but for the moment fractal models are too much a description of form rather than an inference of process. Nevertheless, we should always be open to the possibility that concern with process may be a search for simple models that may be inappropriate to the understanding of complexity.

Finally, a great deal more work needs to be done before the fractal model is fully established. Each of us has lists of frustrations: we seek more methodical treatments of such problems as intersections and projections, codimensions, multifractals, the integration of chaos theory, and a coherent theory of form. This work is being done, and it will lead to fruitful contributions in the fields of mathematics as well as the geosciences. But these gaps constrain progress in applied areas at the moment.

Having said all this, what is the research agenda for the near future? Obviously, more of the kind of work presented here is needed and is coming. The theme of this book is that rigorous concepts of space, dimensionality, scale, self-similarity, and complexity form the organizing principles of fractal research, and each of these concepts points to new and exciting avenues of investigation. Fractal models will continue to be used to represent empirical data, occasionally inappropriately, but usually in ways that reveal important regularities underlying the apparent irregularity of the world. And fractal processes will continue to be used for simulating time series, regions, surfaces, and so forth. This simultaneous induction and deduction (Figure 3.1) constitute the essential dialogue of science.

From the perspective of analysis, data exploration must focus on data at different levels of scale, resolution, hierarchy, aggregation, and so on. Data from one level may be useful for the limited description of the way a phenomenon is imprinted on a particular region, but data across a range of scale levels have become mandatory for a deeper understanding of the phenomenon. Systems for remote sensing, geographic information, spatial analysis, and cartography need to provide images at multiple resolutions, objects at multiple hierarchies, models for modifiable areas, and features at multiple scales—in time as well as in space. When this is accomplished no simple tasks will be possible to present multilevel statistics for the characterization of complexity.

Finally, the fractal paradigm will enable geographers to better address a central problem of our time, that of linking up the *local* realm of individuals and the groups they form within a *regional* realm of action and policy in a way that harmonizes with a *global* realm of natural process. Multilevel data clearly demonstrate changes at the global scale that may be felt as threats or opportunities by individuals but can only be addressed by action at the group (community, nation) level. Obviously, the fractal model is more than just a technique, but less than revelation, yet it is heartening that many scientists are finding in the study of complex processes and forms at many scales an opportunity for dialogue across disciplines. It is perhaps no accident that we are developing powerful techniques for multilevel understanding at a time when the world is perceived to be in crisis at many scales.

REFERENCES

Albers, D. J., and Alexanderson, G. L. eds. 1985. *Mathematical People*. Boston: Birkhauser.

Barnsley, M. F. 1988. *Fractals Everywhere*. New York: Academic Press.

Falconer, K. J. 1990. *Fractal Geometry: Mathematical Foundations and Applications*. New York John Wiley & Sons.

Feder, J. 1988. *Fractals*. New York: Springer-Verlag.

Gaile, G. L. 1990. Whither spatial statistics? *The Professional Geographer* 42(1):95–100

Gleick, J. 1987. *Chaos: Making a New Science*. New York: Viking Press.

Mandelbrot, B. B. 1983. *The Fractal Geometry of Nature*. San Francisco: W.H. Freeman & Co.

Mies van der Rohe, Ludwig 1959, N.Y. Herald Tribune, June 29.

Pickover, C. A. 1990. *Computers, Patterns, Chaos and Beauty*. New York: St. Martin's Press.

Schroeder, M. 1991. *Fractals, Chaos, Power Laws*. New York: W.H. Freeman & Co.

von Bertalanffy, L. 1953. *General Systems Theory*. New York: George Braziller.

Part II

RESEARCH AND APPLICATIONS IN PHYSICAL GEOGRAPHY

CHAPTER 5

FRACTALS AND GEOSTATISTICAL METHODS IN LANDSCAPE STUDIES

Peter A. Burrough
Institute of Geographical Services, University of Utrecht, Utrecht, The Netherlands

ABSTRACT

The conventional, qualitative view of landscapes used to be that they could be modeled completely by uniform regions delineated by crisp boundaries and smooth contour lines. Quantitative descriptions of the spatial and temporal aspects of the earth's surface have revealed that differences are not just limited to the landscape scale, but occur at all scales. Complex, multiscale phenomena can be described in two complementary ways. *Fractals* attempt to model multiscale variation in terms of self-similarity and scaling: variations seen at one level of resolution are similar in pattern to those seen at other levels of resolution. Fractal behavior can be experimentally examined in terms of length–step size, area–perimeter relations, power spectra, variograms, and the Korcak–Zipf law. True fractals yield linear graphs on double log plots of these relations; the slopes of the graphs give experimental estimates of the fractal dimension D, of the phenomenon being studied. Examination of estimated D values for natural phenomena ranging from impact craters to landforms, rock porosity, caves, coast and lake outlines, river patterns, geological faults, soil variation, mineral abundance, cloud forms, and windblown sands suggests that, when the driving process for the phenomenon can be scaled and it operates in a single medium, then the products of the process may exhibit strong fractal properties. This seems to be the case for clouds, some windblown sands, impact craters on Mars, river patterns in relation to drainage basin areas, and caves in limestone. For landforms in general, mineral abundances, and soil properties, the picture is less clear, and it seems that most variation is the result of several different processes that operate at different times to produce a multiscale pattern, but not a true fractal. To date, most fractal studies in landscapes have been descriptive and have not explored hypothesis testing or the use of fractal parameters such as D for problems such as interpolation from point samples. The methods of *geostatistics* (for example, kriging) were specifically developed for estimating mineral reserves and for interpolation, and they are practical tools to be used in single-scale or multiscale situations.

INTRODUCTION

Most qualitative descriptions of the variations of the earth's surface, such as thematic maps of soil, geology, or landform, have treated the world as though it were made up of smooth, uniform regions that can be delineated by crisp boundaries or smooth contour surfaces. The gradual realization that this Platonic model of single-scale atomic objects was at best only an approximate model of the real world has come about as scientists have attempted to underpin the former, largely intuitive and qualitative descriptive models with quantitative observations (see, for example, Beckett and Webster, 1971). Far from supporting the status quo, these quantitative observations have revealed greater insights into the nature of the spatial variation of geographical phenomena and have shown that in many cases the smooth polygons and contour envelopes of standard thematic maps belie a complexity that exceeds the capabilities of classical statistics for their description and explanation. Quantitative descriptions of the spatial and temporal variations of many aspects of the earth's surface, ranging from coastlines to the surfaces of mineral grains, have revealed that their differences are not just limited to a single scale; variation occurs over all scales, from that of the whole earth to that of atoms and molecules. The studies have also revealed tantalizing evidence that sometimes, but by no means always, patterns of variation at one scale seem to be a scaled-down version of differences at another scale. This could mean that variations at different scales were somehow linked, which, if true, might reveal new insights into the underlying processes.

This chapter reviews two approaches to handling the complex, multiscale spatial variation of geographical phenomena in landscapes and examines the relations between them. The first approach uses fractals as a means of describing observed variations. The methods by which fractal dimension can be estimated are reviewed, and the value of the fractal concept is examined in the light of several case studies taken from the literature, ranging from the description of landforms and rock surfaces to the distribution of mineral reserves, soil properties, and a variety of porous media.

The second approach is that of geostatistics, the use of spatial covariance functions for estimating the amounts and spatial distribution of attributes of the earth's surface from point samples (David, 1977; Davis, 1986; Journel and Huijbregts, 1978; Matheron, 1971; Webster and Oliver, 1990). Geostatistical methods are based on extensions of classic statistical theories involving Gaussian distributions, multivariate regression, and spatial autocovariance. They are used widely for mapping and for estimating the amounts and spatial distribution of minerals, chemical properties of rocks, soil, and water from point samples taken from boreholes and soil samples. As we shall see, both geostatistics and fractals use spatial correlation structures and the two-dimensional or three-dimensional Gaussian random fields for handling the irregular but structured variation that occurs in nature. Whereas fractal studies have largely been used for qualitative description, geostatistical methods are concerned with providing practical tools for predicting the most likely pattern of variation of mineral reserves, groundwater pollution, or soil fertility. The two approaches are complementary, although this is not always apparent.

FRACTALS

Studies of fractals in the geographical sciences, here taken to include geology, soil science, physical geography, and cartography, have been mostly concerned with the shapes, sizes, and spatial distributions of ore bodies, lakes, islands, caves or pores, and soil (Burrough, 1981, 1984; Goodchild and Mark, 1987; Peitgen and Saupe, 1988; Turcotte, 1986, 1989; Unwin, 1989). This section presents a review of work using fractals in the following areas: (1) macroscopic and supermacroscopic surfaces (landforms) resulting from the chemical and physical processes of weathering and soil tillage; (2) three-dimensional holes and surfaces in rock pores and rock bodies (for example, caves and coral reefs); and (3) spatial distribution and size distribution of islands, ore bodies, and mineral and chemical properties of rocks and soil, including those caused by pollution.

Before proceeding further, it is well to pause and ask just why fractal concepts should be of value in geographical sciences. Fractals are a means of describing the complicated irregular features of variation at all spatial scales. This complexity, expressed in the form of "roughness," can be described quantitatively by the fractal dimension D. Besides, fractals embody important ideas of self-similarity, in which the spatial behavior or appearance of a system is largely independent of scale. Many real phenomena in rocks, soil, and landscapes appear to share these attributes. Mandelbrot revived work of Richardson (1961) that showed that the length of a coastline or a boundary between two countries such as Spain and Portugal is a direct function of the scale at which the length is measured. Håkanson (1978) showed how the "normalized" length of closed geomorphic lines, such as lake outlines and coastlines, could be represented functionally independent of scale. Apart from scale, however, all representations of a given coastline drawn at different scales appeared statistically to be similar in wiggliness. Zonneveld (1973) pointed out how, when examining geomorphological features on aerial photographs, their real scale could not be deduced from form alone.

This scale independence of pattern and form is also apparently present in thematic maps of soil and geology. If the pattern of soil or geology of an area is mapped at a small scale, say 1:1,000,000, the surveyor will divide the landscape into mapping units that express the major sources of variation at that scale. If, subsequently, a part of the same area is mapped at a larger scale, say 1:100,000, new details will be resolved in areas that were previously accepted to be uniform. The process can continue at still larger and larger scales, even down to microscopic scales in which different mineral accumulations are identified in thin sections of rocks or soil. It is remarkable that the relative distribution of boundaries in the resulting maps often appears not to be directly dependent on scale, but remains more or less constant over several scales (Burrough, 1983a, 1986). Indeed, the cumulative distribution of boundaries on all maps (that is, the chance that boundary spacings occur at given distances) can often be described adequately by a simple exponential function:

$$P(x) = 1 - e^{-\alpha x} \tag{5.1}$$

in which $P(x)$, the probability varies between 0 and 1, x indicates position, and the average boundary intensity depends only on the single scaling parameter α (Figure 5.1). To be fair, this distribution does not describe the cumulative probability distribution of all boundaries on all maps, and Burgess and Webster (1984) have shown that in some cases either the gamma distribution or the Weibull distribution gives a better fit. Both the gamma distribution and the Weibull distribution can be reduced to an exponential distribution, however. Burgess and Webster suggest that an a priori knowledge of the cumulative probability distribution of boundaries could help to optimize natural resource mapping by providing an indication of where and how often investigations should be made to locate the boundaries between different kinds of soil or rock.

A second application concerns the size distribution of ore bodies, mineral reserves, or other important deposits. It has often been observed that, whereas many small occurrences of, say, an ore body, may be found in a given area, there will be correspondingly fewer large occurrences. A knowledge of the exact relationship between ore-body size and frequency of occurrence would be extremely useful when estimating mineral reserves or planning exploratory investigations, because the chance of success of finding a deposit of a given size or quality could be estimated beforehand. Placing too much reliance on a statistical model, however, could be dangerous (see the section on Mineral Reserves).

A third major potential use of fractals concerns the problems of surface roughness and inexactness that occur in applications as diverse as the transport of fluids in porous media and the measurement of soil pH. In these situations it is commonly observed that the degree of roughness (in a surface) or inexactness (in a measurement) is a function of scale and not of experimental technique. Very often, attempts to increase the precision of measurements by working at greater detail only result in the discovery of uncertainties at the new scale.

Therefore, fractal ideas could be of value by providing (1) a theoretical basis for observed features of self-similarity and scale independence, (2) a quantitative description of surface roughness or complexity that could aid mapping and interpretation of mineral resources or assist explanation of processes such as chemical weathering, and (3) a null hypothesis against which real-world data can be tested and thereby increase the number of ways of making objective tests between different sets of real-world data.

Mandelbrot (1977, 1982) introduced the term *fractal* specifically for temporal or spatial phenomena that are continuous but not differentiable and that exhibit partial correlations over many scales. The term fractal, strictly defined, refers to a one-dimensional series, a two-dimensional surface, or a three-dimensional volume in which the Hausdorff–Besicovitch dimension exceeds the topological dimension. A continuous mathematical

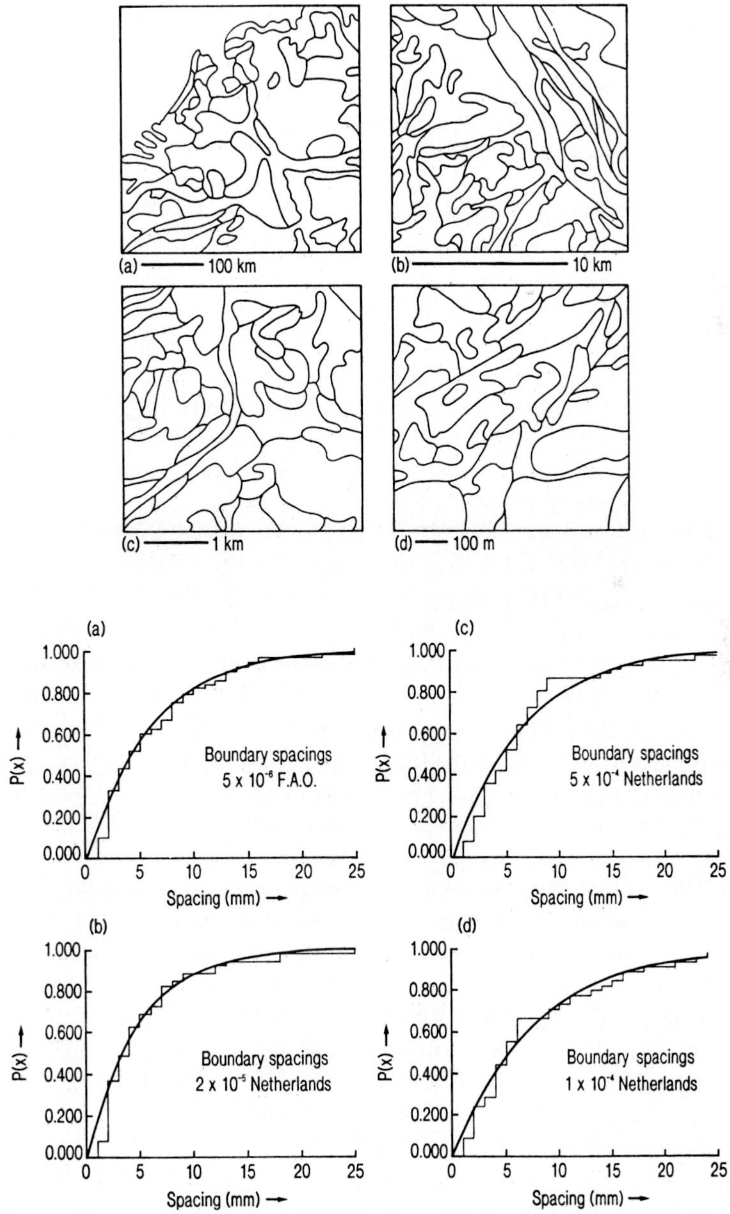

Figure 5.1 The upper four boxes show the outline boundary patterns of soil as mapped in northwest Europe. Map fragment (a) contains part of the North Sea. Each map fragment in the series (b) to (d) is contained in a part of the previous map fragment that has been remapped at a larger map scale. The apparent statistical self-similarity of the four map fragment line patterns is shown by the cumulative probability diagrams below. These probability diagrams suggest that for all four map fragments the chance of encountering a boundary after traversing a given distance on the printed map is much the same. In this respect the line patterns resemble fractals. Reproduced by permission of Oxford University Press from Burrough (1986), *Principles of Geographical Information Systems for Land Resources Assessment*.

function (such as a polynomial that might be used to model the concentration of a chemical element over space) is differentiable because it can be split up into an infinite number of straight lines or planar areas. A nondifferentiable series cannot be so resolved because every attempt to split it up into smaller parts results in the resolution of still more "wiggliness." For a linear fractal function, the Hausdorff–Besicovitch (or fractal) dimension D may vary between 1 (completely differentiable) and 2 (so irregular that it effectively takes up the whole of a two-dimensional space), while for surfaces the corresponding range for D is between 2 and 3.

The second important characteristic of fractals is *self-similarity*; when examined at increasingly higher resolution, increasing amounts of detail are resolved that are scaled versions of the variations seen at lower resolution. Mandelbrot (1977, 1982) and Peitgen and Richter (1986) have elegantly demonstrated many beautiful fractal patterns based on simple deterministic models and fractal sets based on complex numbers. Although these models reveal a wealth of complexity at all scales, the patterns they produce contain regularities that are unrealistic for the properties encountered in geographical sciences. These properties can be better modeled by a stochastic fractal in which the scaling and self-similarity properties of the fractal are embodied not in exact patterns but in statistical properties such as trail lengths, area–perimeter ratios, spatial autocovariances, and rank-order or frequency distributions.

Brownian Fractal Functions

Mandelbrot (1977, 1982) and Mandelbrot and Wallis (1969) suggested originally that a family of Brownian fractional noises (sic) (fBn) are suitable for modeling the stochastic properties of self-similarity encountered in the geographical sciences. Brownian linear motion (otherwise known as the Bachelier–Wiener–Levy process) is the simplest example of a random process having the properties of nonstationarity, a variance that is proportional to the length of the series, and statistical self-similarity. For a series of equally spaced spatial or time intervals having location x, the sequence of increments of differences $[Z(x+\varepsilon) - Z(x)]$ in the value of the property $Z(x)$, is a normally distributed Gaussian process with zero mean and variance σ^2. Brownian motion is said to be a statistically self-similar process in the sense that, if the intervals are divided by a positive but arbitrary ratio r and rescaled by the ratio r^H, where $H = 0.5$, the rescaled function has a probability distribution that is identical to that of the original function.

The semivariance (one-half the variance of increments) of the Brownian line function is defined by

$$\gamma(h) = 1/2E[Z(x+h) - Z(x)]^2 = h^{2H} \tag{5.2}$$

where $Z(x)$, $Z(x+h)$ are the values of the function at points x and $x+h$, respectively, separated by a Euclidean distance h, and H is a scaling power parameter, which for normal Brownian motion has the value 0:5. A graph of semivariance (or variance) against sample spacing is called a *variogram* (or semivariogram). By allowing H to vary between 0 and 1, Mandelbrot showed that a whole family of statistically self-similar variations could be produced that cover the range of spatial intricacy from smooth curves to plane-filling noise. These series can be simulated by replacing the normally distributed Gaussian error term in the equation for Brownian motion,

$$Z(x + \varepsilon) = Z(x) + \varepsilon \tag{5.3}$$

by a discrete fractional Gaussian noise (dfGn). These are stationary Gaussian random processes normalized to zero mean and unit variance defined as having the covariance

$$C(hH) = [|h + 1|^{2H} - 2|h|^{2H} + |h - 1|^{2H}] \tag{5.4}$$

where h is the sampling interval, and $0<H<1$ is the scaling parameter as before. When $H > 0.5$, the fractional noise has a positive correlation over the series, expressing the persistence of positive or negative values. When $H = 0.5$ or zero, the covariance of the dfGn, is zero and the fractional noise is identical to the usual Gaussian normally distributed, uncorrelated noise. When $0< H <0.5$, the correlation between "past" and "future" values of the noise is negative, indicating *antipersistence*, or, in other words, the alternation of positive and negative values of the dfGn.

When these noises are used to generate Brownian fractal lines, the positive or negative autocorrelation in the noise controls the form of the resulting functions. Large values of H (>0.5) result in smoothly varying functions in which long-range fluctuations dominate; small values of H (<0.5) result in fluctuations in which short-range variations dominate (Figure 5.2). Note how the slopes of the variograms of the various functions reflect the value of H, and thus of D because the semivariance $\gamma(h)$ varies as $(4 - 2D)$ (Figure 5.2 insets). The fractal dimension of the Brownian fractal process is given by $(4-m)/2$, where m is the slope of the semivariogram. Figure 5.3 shows how the slope of the variogram of Brownian fractal lines varies with D in direct and double logarithmic forms.

When extended to two dimensions, the same process can be used to generate Brownian fractal surfaces (Mandelbrot, 1977, 1982; Goodchild, 1980; Voss, 1985). Because these Brownian fractal surfaces are isotropic, the Brownian fractal series presented in Figure 5.2 can be considered as sections of a Brownian fractal surface.

If the value of H is chosen correctly, the Brownian fractal surfaces render striking images that suggest the outlines of mountains and lake and island shorelines, although not all geographers have been convinced of the veracity of the model (for example, Scheidegger, 1970). These properties have led to fractal methods being extensively used for simulating landscapes in science fiction films (Fournier et al., 1982) or for providing ideal models of landscape for investigating theoretical aspects of the mapping process (Goodchild and Dubuc, 1987). A Brownian fractal surface has the property that if a waterline is drawn across it (that is, it is flooded to an arbitrary depth), the distributions of the sizes of islands and lakes with respect to their rank order of size will be hyperbolic and not normal or lognormal.

The hyperbolic distribution seems to have a real place in nature. Certain geographical objects appear to exhibit a hyperbolic size distribution of the number of objects of given size $N(A)$ that are larger than size a. This relation is known as the Korcak empirical law: the probability of an area A exceeding a minimum area a is given by $\Pr(A>a) = ka^{-B}$ (Mandelbrot, 1975, 1977). For area measurements, $B = (D/2)$.

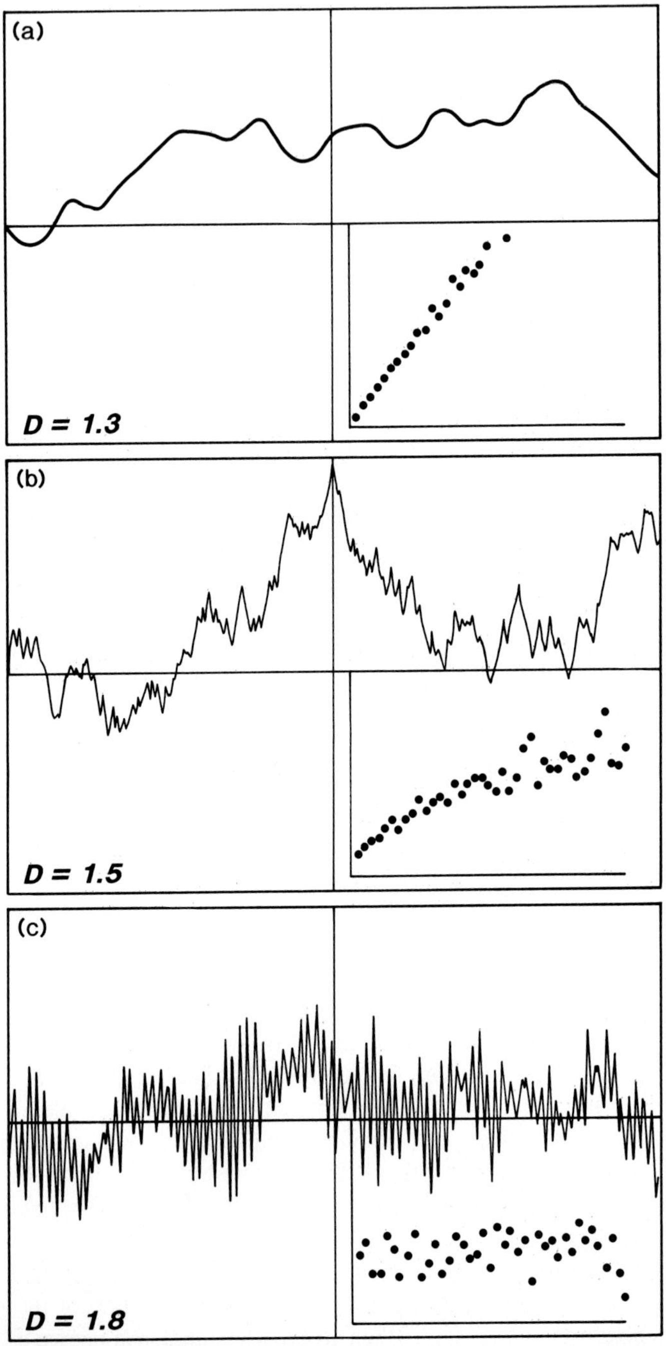

Figure 5.2 Three examples of Brownian fractal trails. The insets show the variograms of each trace. Note that the smoother the trace is, the smaller the fractal dimension.

The Pareto distribution is closely related to the Korcak empirical law. The Zipf law, which is a special case of the Pareto distribution in which "the biggest unit is twice as big as the second largest, three times as big as the third largest, and so on...," is given by the series x, $x/2$, $x/3$, ... x/n. The Zipf relation is expressed mathematically as $Sm/Sn = (n\ m)^k$, where Sm is the size of item of rank m, Sn is the size of item of rank n, and K is a proper fraction. For length measurements the number of units longer than L is given by

$$N(L) = N(L_0) \cdot \left(\frac{L}{L_0}\right)^{-D} \tag{5.5}$$

where $N(L_0)$ is the number of units longer than the reference length L_0. Brownian fractals fit the Zipf and Korcak relations. We should note that the log normal, gamma, and Weibull distributions do not describe truly self-similar fractals because the plots of size Sn against x are not linear on a log–log graph.

Other Models of Stochastic Fractals

While the Brownian fractal motion is the most widely used stochastic model, other models exist and should be mentioned here. Berry and Lewis (1980) have shown that the Weierstrass–Mandelbrot fractal function (WMF),

$$W(t) = \sum_{n=-\infty}^{\infty} \frac{[1 - exp\ (ib^n t)] \cdot exp\ (i\phi_n)}{b^{(2-D)^n}} \tag{5.6}$$

where $1<D<2$, $b>1$, and ϕ_n represents an arbitrary phase, also produces complex traces with statistical moments that have the same properties as the Brownian fractal motion, that is, a power spectrum $P(w)$ that varies approximately as $w^{(5-2D)}$, and a variogram that var-

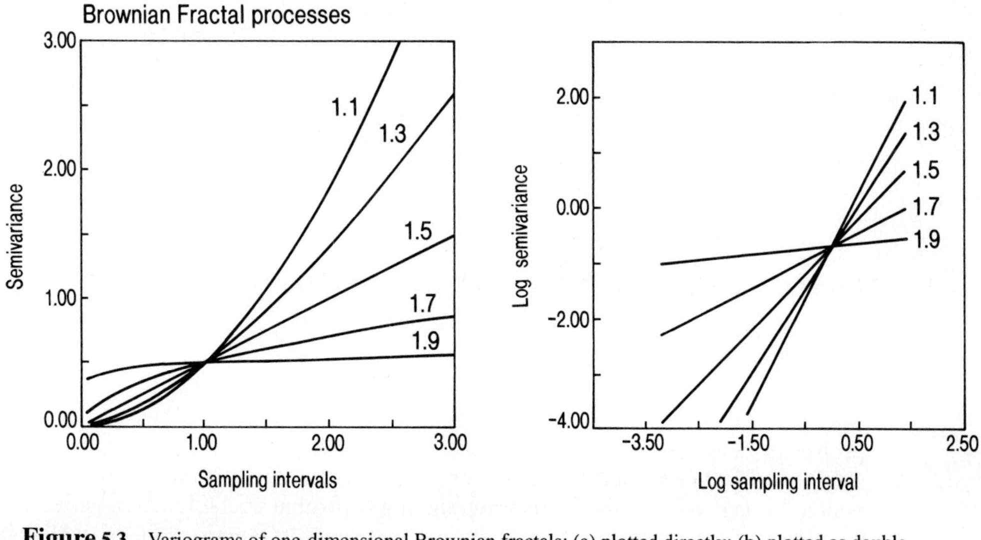

Figure 5.3 Variograms of one-dimensional Brownian fractals: (a) plotted directly; (b) plotted as double logarithms.

ies as t^{4-2D} at the origin. Ausloos and Berman (1985) have extended these ideas to allow flexible two-dimensional simulation of fractal surfaces (see also Jones et al., 1989).

Both the Brownian fractal function and the Weierstrass–Mandelbrot function assume complete self-similarity. Put another way, the models assume contributions to the spatial variation at all possible spatial scales. It is not unreasonable to suppose that spatial variation could be concentrated in a number of discrete but logarithmically ordered scales in which the balance of the weights of the variation at the different scales reflects variations in fractal dimension. Burrough (1983b) proposed such a nested model, which assumes that the value of a chemical or physical property Z at a point x_i is the sum of a number of independent, spatial random functions, denoted by RF_j, that operate over discrete spatial ranges r_j. Each random function could in principle be identified with a single physical or chemical process (such as mineral-grain weathering, surface-splash erosion, rock type, and geological formation). It is understood in the model that, once the value of a given realization of RF_j has been determined, that value remains constant for the whole of the spatial range r_j. It is also assumed that each RF_j is distributed with mean μ_j and variance σ^2; the contribution of a given RF_j to the value of property Z at point x_i is given by weight β_j so that

$$Z(x_i) = \sum_{j=0}^{q} \beta_j RF_{ji} \tag{5.7}$$

where RF_{ji} is the value of the random function RF_j at x_i. By convention, RF_0 expresses contributions of nonspatially correlated noise from possible measurement errors at each data point x_i.

The variogram of the one-dimensional nested model is the weighted sum of the variograms of each contributing scale and

$$\Upsilon(h) = \beta_0^2 + \frac{h}{r_2}\beta_1^2 + \frac{h}{r_1}\beta_2^2 + \ldots + \frac{h}{r_q}\beta_q^2 \tag{5.8}$$

in which the ranges r_j of the nested random functions RF_j are initially assumed to increase logarithmically ($r_j = 1, 2, 4, 8, \ldots$) and the β_j^2 give the weights of each nested random function. This model allows the value of the fractal D value to vary between 1.5 and 2.0 and for D to vary with scale (Figure 5.4).

METHODS FOR DETERMINING FRACTAL DIMENSIONS

Isotropic Fractals

To see whether a given complex distribution pattern, surface, or series has true fractal properties, it is necessary to have simple and reliable methods for estimating their apparent D values. One method was mentioned in the discussion of the semivariogram above, in which it was noted that the semivariogram of a Brownian fractal function varies as $h^{(4-2D)}$. Consequently, the value of D can be determined directly from the slope of the line by plotting the semivariance against sample spacing on double logarithm paper. Because the power spectrum of a Brownian fractal $P(w)$ is proportional to $(w^{-(5-2D)})$, D can be obtained similarly from the power spectrum (which is the Fourier transform of the autocovariance function, which is itself related to the semivariance or variance of increments).

The fractal dimension of a Brownian fractal trail, a Brownian fractal area, or a fractional Brownian volume can be determined by measuring the length, area, or volume using a measuring device that can be set to different sizes. For example, a complex one-dimensional trail can be measured with a set of dividers set to separations of 1, 2, 4, 8, 16 mm, and so on. The length of the trail $L(i)$ is proportional to $i^{(1-D)}$. For two or three dimensions, the area or volume can be estimated by the number $N(i)$ of disjoint squares (or cubes of appropriate dimension) of size i needed to completely cover the set of points contained in the area or volume. The number of squares or cubes $N(i)$ varies as i^{-D}, and the method is known as *functional box counting analysis* (Figure 5.5) (Lovejoy et al., 1987).

The fractal dimensions of islands and lake outlines and other two-dimensional fractal shapes can be obtained by estimating the area–perimeter ratio for given intervals i. According to Mandelbrot (1977, 1982), the estimated area A varies as $P^{2/D}$, where P is the estimated perimeter. Clarke (1986) presents a computer program for estimating the fractal dimension of topographic surfaces using a three-dimensional equivalent of the linear walking-dividers method. Kennedy and Lin (1986) provide a FORTRAN program for determining the fractal dimension of closed forms. Longley and Batty (1989) present four different algorithms for measuring the fractal dimensions of closed lines.

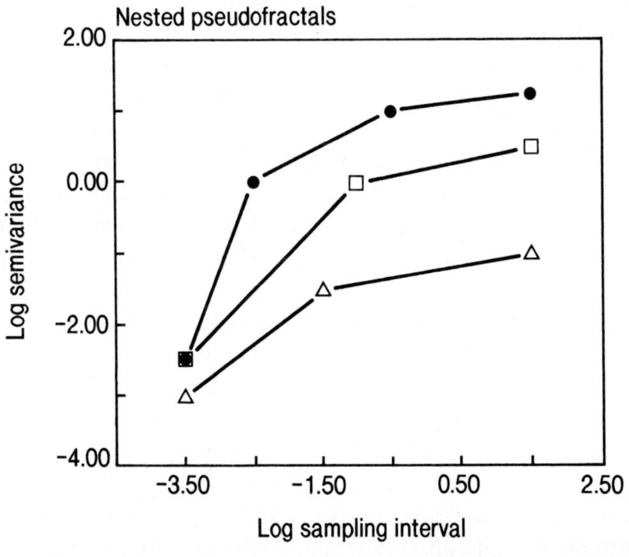

Figure 5.4 log–log variograms of pseudofractals generated by a one-dimensional nested process. The position of the breaks of slope give an indication of the spatial scales of the superimposed, discrete random functions.

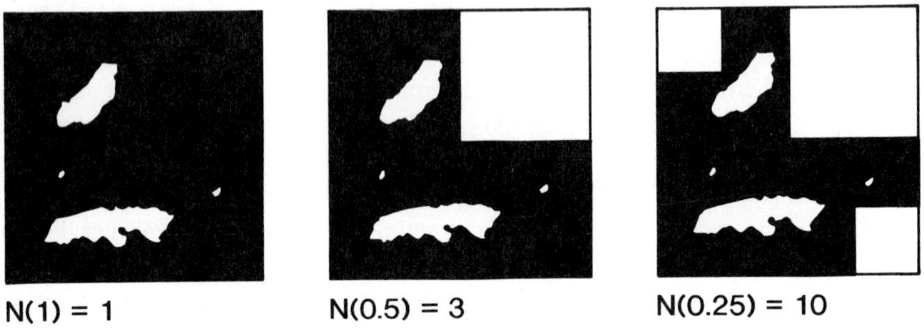

Figure 5.5 Functional box counting allows fractal dimensions to be estimated from the number of boxes of area L. At resolution 1, N(*1*) = 1 (left), N(*0.5*) = 3 (center), and N(*0.25*) = 10 (right).

Nonisotropic Fractal Variation

Because ideal fractal surfaces are isotropic, all the above methods take no account of layering or directional anisotropy. Semivariances can be estimated in different directions from several transects or from two-dimensional data (Journel and Huijbregts, 1978; Webster, 1985), and the slopes of the double log plots of the variograms may reveal information about the variation of D with direction. Such variation could reveal important information about the spatial patterns being studied.

Lovejoy et al. (1987) show how the functional box-counting technique can be modified to take account of stratification and layering (vertical anisotropy) in atmospheric fields such as clouds. They define an *elliptical dimension* that characterizes the flattening of clouds as the size of the cloud field increases. They show that such estimates of elliptical dimension provide quantitative measures of stratification for different kinds of atmospheric fields (for example, rain fields or wind fields), which should be accounted for in quantitative models of weather prediction. If these methods are useful for modeling clouds of water vapor, then they should also be useful in modeling the movement and dispersion of other gases, fluids, vapors, and colloids (including pollen). Table 5.1 summarizes the methods that can be used to estimate D values of real data.

Estimating Fractal Dimensions for Real Data

We can estimate the fractal dimensions of real data, assuming that the real data have a probability distribution that matches that of the Brownian fractal model. As already suggested, knowledge of the fractal dimension of a given geochemical phenomenon could be useful either for scientific description or when estimating spatial patterns or amounts of resources. Table 5.2 lists the apparent fractal dimensions of a range of geochemical and geographical data selected from the literature. These results are presented without comment as to their veracity and the appropriateness of the fractal model.

TABLE 5.1 Methods of Estimating D from Sampled Data

Method	Relationship	Estimate of D
1. Length of trail, dividers method	$L(\lambda)=1K^{1\text{-}D}$ λ is step size K is a constant	Plot log L against log λ; slope of line is 1 - D
2. Area–perimeter	$A = kP^{2/D}$ A is estimated area P is estimated perimeter k is constant	Plot log A against log P; slope is $2/D$.
3. Functional box counting	$N(L) \propto L^{-D}$ $N(L)$ is number of boxes of size L	Plot log $N(L)$ against log_L; slope is -D
4. Variogram	$2\,\gamma(h) = h^{4\text{-}2D}$ h is sampling interval	Plot log $\gamma(h)$ against log h; slope is (4-2D)
5. Power spectrum	$P(w)=w^{-(5\text{-}2D)}$ $P(w)$ is power w is frequency	Plot log $P(w)$ against log w; slope is -(5-2D)
6. Korcak–Zipf empirical law for islands	$Nr\,(A>a)=Fa^{(D/2)}$ $Nr\,(A>a)$ is the number of islands larger than size a, F is a constant	Plot log $Nr\,(A>a)$ against log a; slope is -(D/2)
7. Functional box counting	$N\,(L) \propto L^{-D}$ $N\,(L)$ is the number of disjoint squares (or cubes) of side L to cover the set	Plot log $N\,(L)$ against log L; slope is -D

APPLICATIONS OF FRACTALS IN GEOCHEMISTRY AND GEOGRAPHICAL SCIENCES

Landform, Rock, and Mineral Grain Surfaces

In testing Mandelbrot's assertion that Korcak's empirical law applied to island and lake outlines, Goodchild (1982) digitized with an accuracy of 0.25 mm (0.01 in.) all the lake shores and contours of Random Island, a small glaciated island off the coast of Newfoundland, using a 1:50,000 scale topographic map. All lakes greater than 0.0625 mm^2 (0.01 in^2) were included (153 lakes). Goodchild estimated D values for the coastline, contours, and the lake outlines using the length–step size method. He found that for the shoreline and both contour sets the data were well fitted by the linear log–log fractal model, but for the lake outlines the log–log plot was sigmoid in shape. In other words, the estimates of D for the lake outlines varied over the range of step sizes used. Grid-cell chain counts (a variant of the functional box-counting method) were also used to estimate D; the results are shown in Table 5.3. Goodchild also estimated D from area–perimeter relationships and showed that for the contour envelopes and lake outlines these estimates were inconsistent with the length–step size results. However, all estimates of D showed a constant increase with altitude.

TABLE 5.2 Estimated Fractal Dimensions for Geographical Data

Location	Property	Estimated D	Reference
Wales	Sodium content of soil	1.7–1.9	Burrough, 1981
Australia	Soil phosphorus	2.0	Burrough, 1981
	Soil pH	1.5	
	Soil potassium	1.6	
France	Iron ore deposits	1.4–1.9	Burrough, 1981
Wales	Soil strength	1.76–1.97	Armstrong, 1986
	Microtopography	1.55–1.91	
England	Soil pH	1.6–1.8	Burrough, 1983a
North America	Landforms	2.13–2.83	Mark and Aronson, 1984
Worldwide	Distribution of metero-logical stations	1.75	Lovejoy et al., 1986
North America	Rock surfaces	0.82–1.68	Brown and Scholz, 1985
Worldwide	Soil surfaces	2.43–2.99	Avnir et al., 1985
North America	Fault surfaces	1.001–1.0066	Aviles et al., 1987
Borneo	Limestone caves	1.074–1.546	Laverty, 1987
Norway	Coastal forelands	1.10–1.98	Elliot, 1989

Goodchild also compared the form of the relation between rank and area for the contour envelopes and the lake outlines. The results for Random Island showed that the plots for log (rank) versus log (area) [log $Nr(A > a)$ versus log (a)] were *not* straight, either for contour envelopes or for the lake outlines. The estimates of D varied over the part of the graph used and they were not consistent with the estimates made by the other methods. The distribution of lake areas fitted a log normal distribution much better than they fitted the hyperbolic one.

Goodchild concluded that, although various aspects of the topography of Random Island exhibited the same type of behavior as a Brownian fractal surface, the island as a whole differs significantly from that model. Some aspects, such as the increase of apparent D with altitude, could be explained by the increasing dominance of chemical and physical (frost shattering) weathering processes as one goes inland from the coast to higher altitudes; along the coast the physical processes of sediment deposition and tidal scour would serve to smooth the topography, producing lower D values. Although there were many points where the topography of Random Island and a Brownian fractal surface were similar, Goodchild concluded that there were also so many points where the model did not match that "no geographer would accept [the fractal Brownian model] as a new piece of the coast of Newfoundland."

TABLE 5.3 Variation of Estimates of D with Method for Random Island[a]

Method	250-Foot Shoreline	500-Foot Contour	Lake Contour	Outline
Trail length 1.11 (dividers)	1.11	1.19	1.31	1.3–1.8
Trail length (cell counting)	1.13	1.33	1.54	1.61
Area–perimeter	1.14	1.17	1.19	1.30

a. Source: Goodchild, 1982.

Goodchild's study is by no means the last word on fractals and landform. More recently, Mark and Aronson (1984) and Culling and Datko (1987) have attempted to characterize landforms in terms of estimated fractal dimension. Mark and Aronson, whose work is further reviewed in the section on Geostatistics and Mapping below, came to conclusions similar to Goodchild, that distinct scales are present in a landscape that can be associated with identifiable geomorphic processes. In contrast to Goodchild's findings that D was dependent on estimation method, Culling and Datko claim that two independent methods gave consistent results for 15 different, albeit small, localized areas. Culling and Datko analyzed smoothly varying landscapes in southern England, whereas Goodchild and Mark and Aronson were more concerned with large areas of rugged, mountainous terrain.

There have been several recent studies on the fractal nature of natural rock surfaces and on the roughness of natural geological faults. Brown and Scholz (1985) made a broad-bandwidth study of several natural rock surfaces, using digital surface profilers to get data as accurate as possible. They used fast Fourier transform methods to estimate D values from the power spectrum of surface roughness. Casual examination of their results suggests that self-similar behavior occurs over distances from 1 meter to micrometers (a range of 10^6), but more careful analysis of the results showed that the slope of the power spectra is not constant for a given surface and tends to decrease gradually as one moves from high to low frequency. In many spectra, clear but subjectively interpreted breaks of slope could be seen, which they interpret as transitions from surfaces at the scale of within smooth mineral grains to surfaces at the scale of different grains. A second change of slope was often associated with the size of the rock outcrop scale. They concluded that the rock surfaces were fractally nonhomogeneous, with no single fractal dimension able to describe the whole surface. Consequently, no simple scaling laws apply to the whole rock surface domain, but Brown and Scholz concluded that fractal descriptions of surface roughness within transition zones were useful measures for the numerical generalization of realistic model surfaces.

The relations between apparent fractal dimension and scale were also revealed by Aviles et al. (1987) and Power et al. (1987) in their studies of the roughness of the San Andreas fault zone in California. They attribute these differences in estimated fractal dimension D to the different kinds of physical processes that produce shear and jaggedness over large (several kilometers) and short (tens to hundreds of meters) distances. They note that it is possible that those faults having large D values may be too irregular to rupture in a single great earthquake.

At microscopic and molecular scales, Avnir et al. (1985) and Farin and Avnir (1987a and b) have shown that fractal ideas also apply to mineral surfaces, albeit again within certain definable and recognizable transition zones or domains. They note that the characteristic reaction dimension of a mineral or a catalyst surface may not necessarily be the same as the topologically determined fractal dimension because of steric effects, chemical selectivity, surface morphology change during reaction (for example, by solution), and trapping and reactions in pore volumes. Dicke and Burrough (1988) showed that the reaction of spider mites to different *semiochemicals* (chemicals that affect behavior) could be distinguished by the fractal dimension of the animals' trails over a uniform surface. Hayward et al. (1989) and Whalley and Orford (1989) discuss the use of fractals for analyzing two-dimensional outlines of mineral grains.

The general consensus emerging from these studies (and others such as Culling, 1989; Elliot, 1989; Unwin, 1989) is that the estimated fractal dimension is a useful parameter for *distinguishing between surfaces that have resulted from different processes*. Estimated fractal dimensions, however, may vary with both scale and measurement technique, so it is important to conduct all comparisons at similar scales. The fact that few surfaces seem to be truly self-similar, although possessing variation at all scales, is a clear indication that the Brownian fractal model is too simple to explain all facets of reality at once.

Estimating Mineral Reserves and Pollutant Levels

Mineral Reserves. Recently, Paliwal et al. (1986) reported an attempt to assess the potential reserves of lead and zinc in India by using Zipf's law to extrapolate from knowledge of existing deposits to the possible complete, but currently unknown, set of all Indian lead and zinc deposits. In doing so, they were following the example of several other exploratory geologists who had used the ideas behind Zipf's law without necessarily explicitly recognizing the association with fractals (Folinsbee, 1977; Rowland and Sampey, 1977; Rudenno, 1981; Tapp and Moniuszko, 1978). They used data from 24 known deposits and ranked them according to size. The total known reserves were estimated at 21.346 million tons of lead and zinc.

When the data were plotted on double log paper (Figure 5.6), the data were irregular and appeared to fit no known distribution. However, Paliwal et al. (1986) reasoned that, if there were still major deposits to be found, their approximate sizes could be estimated by shuffling the data to match the hyperbolic distribution. To do this successfully, they had to assume that there is still a major deposit of 18.7 million tons (twice the current largest deposit) waiting to be found. Also they had to assume that major deposits at ranks 3–8, 11–17, and 19–21 and more minor deposits at ranks 23–26, 28–29, 31–34, and 36–51 had yet to be found. When the data were adjusted to fit Zipf's law (Figure 5.6), the estimate of the total amount of undiscovered resource amounted to approximately 66 million tons of lead and zinc, a grand total of 87 million tons. They concluded that only 25% of the total resources for the country had been discovered, and they made major recommendations about how further exploration should be conducted. Their conclusions seem very optimistic.

Let us now consider what the prognoses of lead and zinc reserves are if the actual distribution is log normal rather than hyperbolic, or in other words, that the distribution is not a true fractal. As before, it is necessary to assume that some deposits have not yet been discovered. The gaps can be filled between the current rank 1 and 2 deposits by inserting six new orebodies of 6.5, 5.5, 4.5, 3.5, 3.0 and 2.5 million tons. The total lead + zinc then amounts to 46.8 million tons, or just over twice the current known amount. Figure 5.7 shows that the cumulative distribution of these "adjusted" data fits a log normal model well.

Howrath et al. (1980) have cautioned that Zipf's law "leads to grossly over-optimistic predictions influenced by the size of the largest known deposit..." Certainly, both the Zipf and the log normal fitting can be described as Procrustean—the data have been stretched to fit a convenient theoretical bed. Given the uncertainties associated with the original data (which surely cannot be better than ±10%), it seems unwise to accept or

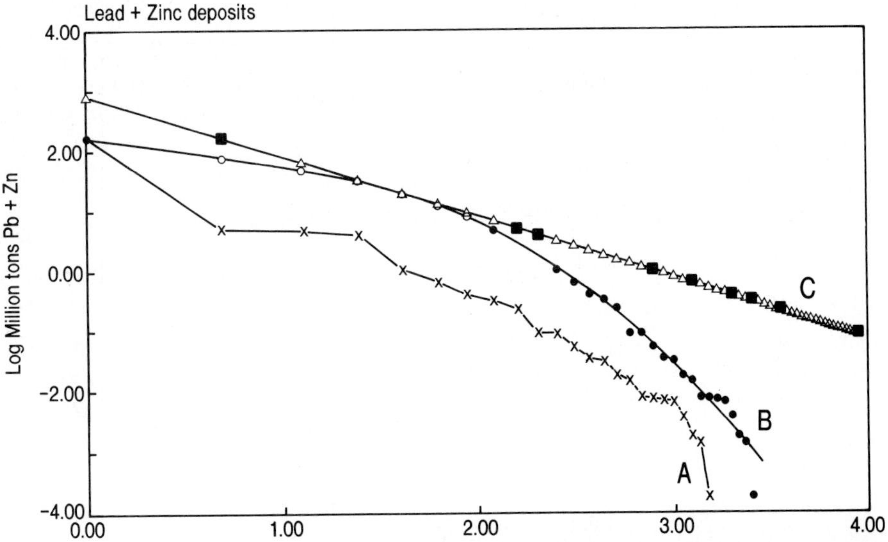

Figure 5.6 Lead+zinc abundance data stretched to fit theoretical distribution models. (A) Original data ×; (B) data stretched to fit a log normal distribution (• actual data, **o** missing data); (C) data stretched to fit Zipf's law (shaded square represents actual data; triangle represents missing data)

reject either model as totally unlikely; but considering the large number of new deposits that must be added to make a fit, a sober judge would conclude that here the hyperbolic model (the true fractal distribution) is the more unlikely and possibly overestimates the total lead and zinc reserves by almost 100%. As Henley (1987) has pointed out, we need sensible procedures for selecting the most rational or appropriate technique or model for mineral reserve estimation. The best possible estimates of reserves are vital for economic and strategic reasons; yet in the absence of large amounts of expensive data, there seems at present no sound a priori test for preferring one theoretical distribution model to another. We have yet to discover, for example, what the relationship might be between any given geological or soil-forming process and the resulting true patterns of distribution and their possible fractal dimension.

Soil Pollution. As with mineral reserve estimation, in soil pollution studies, we wish to estimate as accurately as possible the amounts of dangerous materials present in the ground. The reasons are not to determine whether an economically extractable resource is present, but to decide whether a given piece of land is suitable or unsuitable for a particular kind of activity. Soil sampling and laboratory analysis of heavy metals, chlorinated hydrocarbons, and other organic and inorganic chemicals are costly, and so investigators are always pressured to make authoritative statements about the extreme and average levels of pollutants from a small set of samples. Their job would be made easier, and their statements about the estimated levels of pollutants present would be more accurate, if it were known which kind of distribution—normal Gaussian, log normal, or Zipf fractal—

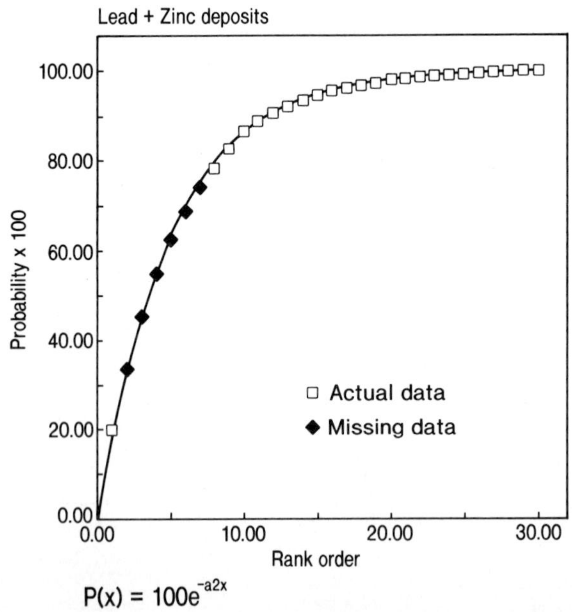

Figure 5.7 Cumulative probability diagram of lead+zinc data stretched to fit a log normal distribution.

fitted the data best. Many soil-pollutant levels are not normally distributed, but are positively skewed, and so investigators often assume log normality. If the distribution was of the Zipf fractal type, then, as with the mineral reserves, it might be possible to make sensible statements about the chances of still larger occurrences of pollutants being found.

Figure 5.8 shows Zipf plots for data from six soil-pollution studies carried out in the Netherlands and in the United States of America. All data, with the exception of the zinc data from Hackfort, were not normally distributed, but instead they were highly positively skewed. The Zipf plots show that the evidence for accepting the fractal distribution is not strong, with only the levels of arsenic at the Hackfort site and the levels of zinc at Geertruidenberg showing any kind of affinity for log–log linearity over all ranks. The reasons for this nonfractal behavior are not difficult to imagine. Soil pollution is often a result of human activity that operates at a given spatial scale and intensity. True fractal behavior will only arise if the process is self-similar over many scales, and this scaling is unlikely when the pollution process is dominated by activities at a single, human scale. This is best emphasized by the zinc data from Geertruidenberg, where the soil surrounding an electricity power pylon was sampled for zinc levels. Because of weathering and distribution of the zinc by rainwater, the highest levels of pollutants are found nearest the pylon and the lowest farther away. Any idea of scaled self-similarity in the form of smaller pylons near the base of the large one does not make physical sense. Again, it is wise to carefully consider the nature of the physical processes leading to pollution before jumping to conclusions about the kind of distribution that the data should follow.

Application of Fractals to Porous Media

Porous media are of tremendous importance in many aspects of geoscience. At the microscopic scale, chemical reactions in soil take place at the interface between water, air, mineral grains, and organic complexes; fluids pass through soils and rock pores. At the macroscopic scale, fluids are stored in reservoirs or move through rock fissures. Caves in limestone are the expression of solution reactions involving dissolved carbon dioxide that enlarge initial weaknesses along joints, fractures, or rock homogeneities. A major theoretical and practical difficulty with heterogeneous porous media is the problem of relating laboratory-scale and laboratory-determined dispersion coefficients to the field scale. There is a rapidly increasing literature reporting investigations into fractal approaches to porous media problems (Chelidze, 1982; Dawe et al., 1985; Geffen et al., 1983; Katz and Thompson, 1985; Koplik et al., 1983; Lenormand and Zarconne, 1985; Rivier et al., 1985; Robinson, 1983; Sen et al., 1982; Stanley, 1984; Wilke et al., 1985; Williams and Dawe, 1986).

In recent years, independent of the fractal movement, soil physicists have been applying scaling methods in an attempt to rationalize the heterogeneity and nonlinearity they encounter when studying the movement of water through soil (Miller, 1980). The methods have been applied with considerable success in situations where the movement of soil water is constrained only by the physical dimensions of the porous media. For example, scaling soil-water movement from a fine silt to a coarse sand allows demonstrations and laboratory experiments to be speeded up greatly. Scaling methods are less successful, however, when the adsorption and movement of water is governed by the chemical and

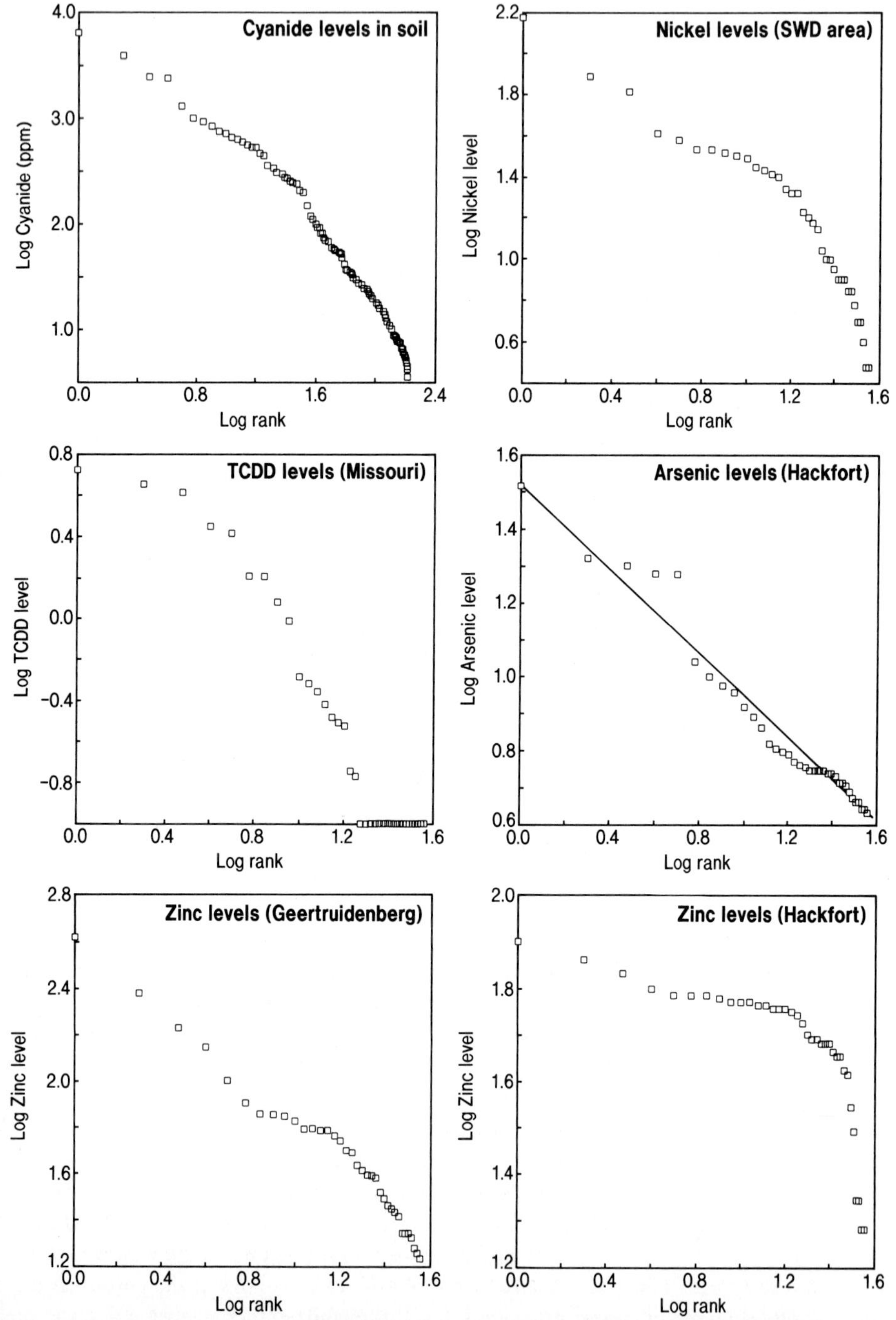

Figure 5.8 Double logarithmic plots of pollution levels versus rank for six sites in the Netherlands and the United States. None of the sites suggests a strong adherence to Zipf's law.

physical properties that result from the kinds of soil particles as well as their sizes. For example, the finest soil particles (<2 micrometers) are very often clay colloids whose physical behavior is strongly dependent on their crystalline structures. This behavior is often well understood, and it is not necessarily always helpful to adopt a fractal approach (see comments by Avnir, 1986).

Fractal Dimensions of Caves. Caves can be regarded as "superpores" created by the chemical action of carbon-dioxide-rich water on limestone rocks. Curl (1986) presents an interesting study of the fractal dimensions and geometries of caves. He shows that the lengths of all caves in many parts of the United States are distributed hyperbolically rather than log normally, as required by self-similar objects possessing fractal geometry. Curl develops the idea of linked modular elements (standard spherical cave explorers having a characteristic length) on the lines of the Lovejoy et al. (1987) functional boxes and ellipses in order to estimate cave lengths. He finds that the expected conditional distribution of modular element sizes in a cave, given length and modules, is also distributed hyperbolically. Estimated fractal dimensions of caves are about 2.8, which is similar to that of the Menger sponge (Mandelbrot, 1982). Curl presents an algorithm for estimating modular element sizes from cave surveys, which should enhance the current range of methods for analyzing information about caves.

Curl's work is supported by a recent paper by Laverty (1987) that shows that both plan lengths and total surveyed lengths of caves measured in Sarawak and in Spain conform to the self-similar fractal behavior, with estimated D values ranging between 1 and 1.5 over a range of measurement resolutions from 1 to 100 meters.

Fractal Dimensions of Coral Reefs. In many senses, coral reefs can be regarded as the antithesis of caves. Chemically, they are formed by deposition rather than by solution of limestone; coral rocks are the precursors of the limestone in which caves form. Coral reefs present a large complex surface to the outer world (although they also contain important perforations), whereas caves present their surfaces to a dark, inner world. Bradbury and Reichelt (1983) presented a study of the fractal dimension of the edge of a coral reef at 2-m depth over distances ranging from 10 cm to 100 m. They collected data by making field measurements using specially constructed dividers or, in other words, by a variant of the walking-dividers method. The method chosen is especially useful for estimating D in the field. They chose step lengths L along the coast and then determined how many similar but smaller steps of length $L/10$ fitted in L.

Unfortunately, they chose to analyze their data by computing the log of the variance of the number of substep lengths against substep length, which, as Mark (1984) pointed out, was inappropriate. Mark showed that the proper estimates of D are given by $\log_{10}(S)$, where S is the mean number of substeps. This analysis yielded an estimated D for the 2-m-deep coral contour of 1.13 to 1.16, which is much less than Bradbury and Reichelt's first estimate of 1.9 to 2.0, and is rather smooth as coastlines go. Intuitively, we would expect coral reefs to be much rougher, but also that the degree of roughness might well vary with scale. Individual corals are frequently intricate and well formed; the coral reef as a whole may be smoothed by wave action. As with soil particles, there is clearly a large distinction to be made between the chemical, physical, and biological processes that operate at different scales.

Fractals and Drainage Networks

Drainage networks have strong fractal characteristics, exhibiting scaling and self-similarity, at least in their general behavior (Mandelbrot, 1982). Gray (1961) showed that the relationship of the length of the principal stream L_p to the watershed area A was more or less constant for 47 watersheds over five orders of magnitude and could be modeled by

$$L_p = 140A^{0.568} \tag{5.9}$$

with L_p in miles and A in square miles. Also, L_p is related to the length of the drainage basin by

$$L_p \approx L^{D\mathrm{min}} \tag{5.10}$$

with the scaling exponent $D_H = D_{\mathrm{min}}/2$.

Recently, the advent of automated drainage basin analysis using digital elevation models (DEMs), such as the studies by Band (1986) and Jensen and Domingue (1988), has revived interest in the fractal study of drainage networks (Thornes, 1990; Stark, 1991). Stark (1991) has shown how a self-avoiding mode of invasion percolation for drainage network evolution results automatically in drainage networks that have a $D_H = 0.565$ (a fractal dimension of $D_{\mathrm{min}} = 1.13$), which is almost identical to Gray's empirical results. Stark's invasion model seems to be closer to reality than other stochastical models of drainage basin development. It is sensitive to local topography and to spatial variations in the strength of the materials in the landscape. Furthermore, the model can incorporate features such as regional tilting without affecting the fractal properties of the streams.

GEOSTATISTICS AND MAPPING

Interpolation and Geostatistics

Unlike properties such as landform, which in principle can be determined over the whole surface to any level of detail required, the spatial variation of chemical properties of soil and rocks can only be interpolated from samples taken at specific, discrete locations. The leap from known sites to predicting values at unvisited locations, and also to estimating total abundances within an area, must be done using methods of interpolation. The success of these methods depends greatly on the degree to which the kind of variation to be mapped can be modeled by the theory behind the particular interpolation method used.

Geostatistical techniques were originally developed for mineral reserve estimation, but their use in the geographical sciences and more recently in geographical information systems (see Burrough, 1986) has been largely for interpolating quantitative data to whole surfaces from point observations. Most interpolation techniques used in standard geographical information systems are based on the concept of underlying ideal, smoothly varying surfaces that can approximate the true variation as closely as possible. The methods used include exact and smoothing splines (Dubrule, 1983, 1984), global and local regression surfaces (Ripley, 1981; Davis, 1986) and inverse-distance local weighting interpolators (Shepard, 1968). Reviews of interpolation methods are given by Agterberg (1982), Burrough (1986), and Lam (1983), among others.

Interpolation methods that use exact surface-fitting methods ignore the important role of local uncertainties and irregularities in the variation of the property being sampled. These variations arise from laboratory or measurement errors and from spatial variations occurring between the sample points that cannot be observed directly. Dissatisfaction with the ability of exact techniques to cope with the stochastic aspects of variation in minerals led Matheron (1971) and his co-workers to develop geostatistical methods for interpolation. These methods, known collectively as *regionalized variable theory*, use information about the average statistical aspects of the variation of the property being studied in order to deduce "optimal" weights for interpolating (predicting) the value of the property at unsampled points. If the property is interpolated at enough points on a fine, regular grid, contour lines can then be drawn indicating how the levels of the property are thought to vary over the area in question.

Geostatistics has developed into an important branch of quantitative geoscience, with major applications in mineral exploration (Journel and Huijbregts, 1978; David, 1977), groundwater (Delhomme 1978), soil science (Nielsen and Bouma, 1985; Webster, 1984; Webster and Oliver, 1990; Verly et al., 1984), and other relevant fields.

Geostatistical methods are based on the recognition that the spatial variation of any geological, soil, or hydrological property, known as a *regionalized variable*, is too irregular to be modeled by a smooth mathematical function, but can be better described by a stochastic surface. The interpolation proceeds by first exploring and then modeling the stochastic aspects of the regionalized variable. The resulting information is then used to estimate the weights for interpolation.

The success of *kriging*, as optimal interpolation is often known, rests on the validity of certain important assumptions about the statistical nature of the variation. These assumptions, which are the basis for kriging, are contained in regionalized variable theory.

Regionalized variable theory assumes that the spatial variation of any variable can be expressed as the sum of three major components. These are (1) a structural component, associated with a constant mean value or a constant trend; (2) a random, spatially correlated component; and (3) a random noise or residual error term.

Let x be a position in one, two, or three dimensions. Then the value of a variable Z at x is given by

$$Z(x) = \mathrm{m}(x) + \varepsilon'(x) + \varepsilon'' \tag{5.11}$$

where $m(x)$ is a deterministic function describing the structural component of Z at x, $\varepsilon'(x)$ is the term denoting the stochastic, locally varying spatially dependent residuals from $m(x)$, and ε'' is a residual, spatially independent Gaussian noise term having zero mean and variance σ^2.

The first step in kriging is to decide on a suitable function for $m(x)$. In the simplest case, where no trend or drift is present, $m(x)$ equals the mean value in the sampling area and the average or expected difference between any two places x and $x + h$, separated by a distance vector h, will be zero and

$$E[Z(x) - Z(x + h)] = 0 \tag{5.12}$$

Also, it is assumed that the variance of differences depends only on the distance between sites, h, so that

$$E\{[Z(x) - Z(x + h)]^2\} = \mathrm{E}\{[\varepsilon'(x)\varepsilon - (x + h)]^2\} = 2\gamma(h) \tag{5.13}$$

where $\gamma(h)$ is a function known as the semivariance.

The two conditions, stationarity of difference and variance of differences, define the requirements for the intrinsic hypothesis of regionalized variable theory. This means that once structural effects have been accounted for the remaining variation is homogeneous in its variation so that differences between sites are merely a function of the distance between them. As noted above, a graph of variance (or semivariance) with distance is known as the *variogram* (or semivariogram).

If the conditions specified by the intrinsic hypothesis are fulfilled, the semivariance can be estimated from sample data:

$$\Upsilon(h) = \frac{1}{2n} \sum_{i=1}^{n} [z(x_i) - z(x_i + h)]^2 \tag{5.14}$$

where n is the number of pairs of sample points separated by distance h (otherwise known as the *lag*). The variogram is an essential step in determining optimal weights for interpolation from point samples. Also, it can assist in optimizing sample networks (Burgess, et al., 1981; McBratney and Webster, 1983a, 1983b; Webster and Burgess, 1984). The variogram is used in interpolation to estimate the weights that should be assigned to point data values as a function of their distance from the site whose attribute value is being predicted. The prediction is *optimal* in the sense that it has the lowest possible variance, given the variogram and the data points available. Properly designed sampling schemes can be used to determine the main scales at which spatial variation occurs (Oliver and Webster, 1986), so sampling can be optimized. The recognition that spatial variation occurs at certain discrete scales, rather than varying over a continuum, is of course counter to the simple Brownian fractal idea.

Practical experience has shown that the experimental variograms estimated from mineralogical, soil chemical, or soil physical data can take many forms. The monotonically increasing form of the variogram of the Brownian fractal function is not common for many environmental data, particularly those relating to static properties of the landscape, such as soil and geological properties. It is not always an easy, totally objective exercise to fit a model to an experimentally observed variogram for the purposes of extracting the weights for interpolation, although guidance is available for determining the best procedures to be followed (Cressie, 1985; Cressie and Hawkins, 1980; McBratney and Webster, 1986). In general, there are two major classes of variograms, transitive and unbounded. Transitive variograms are those in which the variance increases steadily to a particular level (known as the *sill*) at a certain sample spacing or range (Figure 5.9a). Thereafter, the variance does not increase, and in certain circumstances it may slightly decrease or oscillate if there are regular fluctuations in the data. Transitive variograms suggest that spatial variation is caused by patterns no more complex than major structural control, together with local correlated, irregular variation with possible additions of measurement error (otherwise termed *noise*). In the author's experience, transitive variograms are commonly found for data taken from small sample areas where spatial variation is governed by a single, dominant process.

In contrast to transitive variograms, unbounded variograms display no sill, or sign of reaching a level at which spatial variation stops increasing with distance. The variogram of the Brownian fractal function epitomizes these unbounded variograms, which is not to say that all unbounded variograms result from a series or surface that has been generated by a Brownian fractal function. An unbounded variogram does suggest, however, that variation is occurring over a continuous range of scales in a manner similar to that of the Brownian fractals.

In Figure 5.3 the variograms (semivariance = one-half the variance of increments) of Brownian fractal processes (Equations 5.3 and 5.4) for both standard and logarithmic plots were shown. When D is large (1.95), the semivariance reaches its maximum value just after the origin. Variograms (or semivariograms) of this form are commonly encoun-

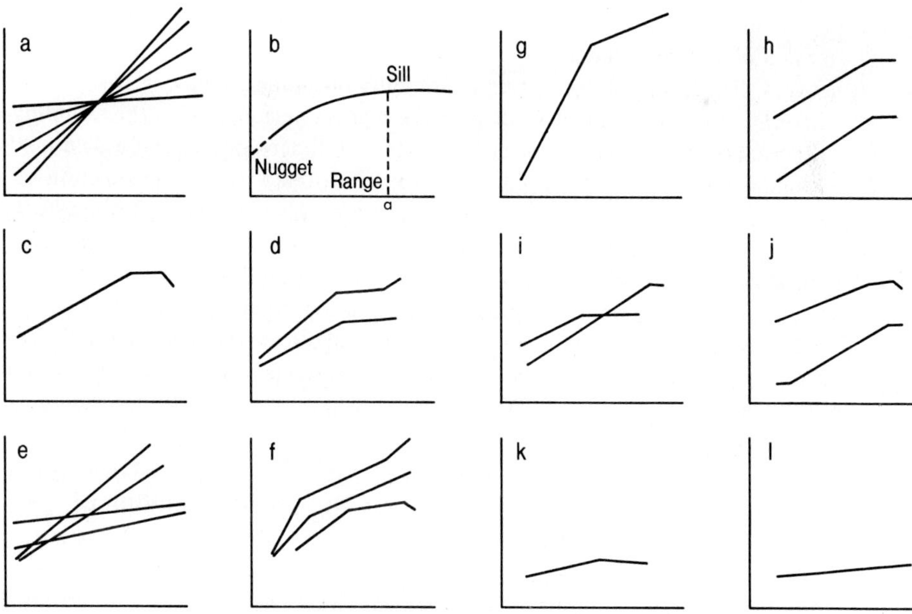

Figure 5.9 Schematized log–log variograms of actual soil chemical and landform measurements. (a) Brownian fractal (nontransitive); (b) typical transitive variogram with nugget, range, and sill found for many soil attributes; (c) soil pH (Culling, 1986); (d) and (e) soil surface microtopography and penetrometer measurements (Armstrong, 1986); (f) and (g) digital elevation models of North America (Mark and Aronson, 1984); (*h*),(*i*), and (j) percent clay, percent silt, and pH from surface and subsurface soil samples from a one-dimensional transect at Sandford in the English Midlands (Burrough, 1983a); (k) and (l) thicknesses of depositional layers in the Dutch Flevopolders (Burrough, 1983a). Only rarely do experimental variograms show double log linearity over the whole of the sampled range - transitive forms reaching a constant variance seem to be more common.

tered in soil variation (Campbell, 1978; McBratney and Webster, 1981), and opinion is divided as to whether they are the result of simple Gaussian noise (that is, homogeneous normally distributed variation resulting from measurement errors and truly homogeneous natural variation) or of a fractal process. Variograms, such as those of the Brownian fractal process at a D value of approximately 1.05 to1.2, approach the origin parabolically. Similar variograms can be obtained when the original data both vary smoothly and show a trend, commonly termed *drift*. Matheron (1973) has argued that interpolating data with trends can best be achieved by using intrinsic random functions of order $k \geq 1$ (the variogram is an intrinsic random function of order $k = 0$) to model the variation. Because the definition of what is a trend and what is a part of local variation depends greatly on the scale of the investigation, it can be argued that trends, too, are a usual part of spatial variation.

Lovejoy et al. (1986) have raised an interesting aspect of the problem of spatial sampling, that of the requirement for not only a match in the spatial resolution between a sample network and the phenomenon under study, but also for a dimension match. In many sampling networks, the distribution of data points can be highly irregular (they cite the world distribution of meteorological stations as an example). They point out that the surface being sampled (in their case the earth's surface) may have a *larger* fractal dimension (termed E) and hence be more irregular than the effective fractal dimension of the sample network D_m. A sparsely occurring or irregular phenomenon having fractal dimension D_p where $D_p < E - D_m$, cannot be detected or mapped adequately by the sampling network. Because, according to Lovejoy et al., the fractal dimension of geophysical fields (that is, all kinds of regionalized variables) is expected to be a decreasing function of sampling intensity, the most intense and sparse phenomena cannot be detected when E-D_m exceeds D_p. In a way, this tells us what we already know from experience: sparse sampling will only detect long-range variations when the surface being sampled is free from short-range noise. The contribution of Lovejoy et al. is that the expected detection success of a sampling network can be quantified in terms of fractal dimension. The idea could be useful for making better estimates about how much information is lost when interpolating from a set of irregular sampling points to a regular grid, as occurs in many geographical information systems (Burrough, 1986).

Geostatistical Evidence for Fractality in the Spatial Variation of Properties of Soil and Rocks

Soil data collected in Arizona by Gajem et al. (1981) over four orders of magnitude of length (0.2, 2, 20, and 200 m) within a total area of 2000 m give linear variograms when plotted on a double logarithm scale. An early study by Smith (1938) also presents many data on wheat yields that appear to demonstrate a power-law relationship between yields and size of plot. The distribution of gold ore in South African mines also shows log–log linearity (Journel and Huijbregts, 1978). All these cases seem to indicate that the Brownian fractal might be a useful model. In most cases, the fractal dimension D of these soil and geological properties appears to lie between 1.6 and 1.9 (Burrough, 1981, 1983a; see also Table 5.2).

Figure 5.9 presents schematized results of several published studies that used variogram analysis to estimate the fractal dimensions of surfaces and of the spatial distribution of soil physical and chemical properties (Armstrong, 1986; Burrough, 1983a; Culling, 1986; Mark and Aronson, 1984; Roy et al., 1987). In all cases the fractal dimensions were estimated from the slopes of log–log plots of the estimated semivariance versus sampling interval.

While it is clear that for some properties the log–log variograms are linear over the range of spatial scales sampled, thereby suggesting self-similarity, many plots show clear breaks of slope that correspond to the geostatistical idea of a *sill*, or a transition from one pattern of variation at one scale to another pattern at a larger scale. Some sample variograms appear to show more than one break of slope.

There are probably several contributing factors for these observed results. First, there is the problem of sampling. With the exception of the studies by Mark and Aronson (1984) and Roy et al. (1987), the estimated variograms are based on linear transects that are at most a few hundred points long (see Power et al., 1987, for further discussion on the problem of sampling). Burrough (1983b) and Taylor and Burrough (1986) have shown how the effective degrees of freedom for a variogram estimated from linear transects rapidly decrease with increasing lag, and hence the confidence limits of the estimates broaden rapidly. This is because the estimates of semivariance at the different lags are strongly correlated, being computed from a single set of data. Simulation studies confirm these findings. Simple simulations of linear Brownian fractal series also show that for short runs (up to 600 sample points) a given simulation can display pseudo "structural" features that a geologist or soil scientist would "explain" in terms of a major physical feature, such as a hill or an intrusion. Even when sampling variation does not produce pseudo scale-dependent structures, different simulations generated by the same model can easily give rise to estimates of D for a simple linear Brownian fractal trail that can vary as much as 1.50 ± 0.15.

The conclusion thus drawn is that when data are not abundant it may be difficult to get good estimates of D. This situation is probably very common in geochemical studies where large amounts of quantitative data are very expensive to gather.

The second causal factor of the breaks of slope in the log–log variograms could be that the sampled population does indeed contain variation at two or more distinct scales. This seems to be the case with Mark and Aronson's study and the rock and fault surface roughness studies. Burrough (1983b) suggested that a model of nested variation at discrete scales fitted the observed variations of soil along the Sandford linear transect well (see Figure 5.9 *h, i, j*, and below) and it may be that a similar model may be appropriate to model the fractal in homogeneity of rock surfaces and landform. McBratney et al. (1982) came to a similar conclusion when mapping trace metal concentrations in Scottish soils, and they associated each scale with a known kind of spatial variation in the landscape. Our knowledge of the geological, geomorphological, soil-forming, and land-use processes in a landscape gives us good reason to think that certain processes do operate only at certain discrete scales (for example, glaciation and the formation of volcanoes), while others (for example, wind and water erosion) operate over a wide range of scales. Because many landscapes are the result of many processes operating both simultaneously and in historical sequence, it is not surprising that patterns of variation on the earth's surface are so complex.

The departures from the ideal isotropic Brownian fractal model led Burrough (1983b) to propose the nested model of fractal functions presented above (see the section on Other Models of Stochastic Fractals). Burrough (1981, 1983a and b, 1984, 1985) examined data from several soil transects in order to see if the fBn was a sensible model. One series was a 3200-m-long transect, sampled at 10-m intervals across 14 different geological parent materials in Oxfordshire in the English Midlands (the Sandford transect). Other transects were sampled in the Flevopolders in the Netherlands and on the river terraces of the Maas (Grubbenvorst). The variograms of both the Sandford and the Grubbenvorst data showed strong evidence of the control exerted by the geological and geomorphological structure of the terrain; in other words, the condition of self-similarity was not satisfied even though variation clearly occurred over a wide range of scales.

It seems clear that the variation of many properties of soil and geology occurs over many different scales. Serra (1968) noted this and termed it the *gigogne* effect. The causes of variation over different scales lie in the chemical, biological, geomorphological, and geological processes that shape the earth's surface. Processes such as the movement of clay in a soil profile work over distances of a few hundredths of a millimeter to a few centimeters, the biological actions of plants and small animals such as ants may affect areas from a few centimeters to a few meters, river erosion and sedimentation occurs over distances from a few meters to several tens of kilometers, and the mountain-building processes of plate tectonics act at continental scales. The value of a chemical property of the soil at any given point is the sum of all these different effects acting in concert.

When a single process dominates, the simple kriging model is adequate to describe nonsmooth, continuous variation. When variations at all scales grade easily into each other, the result can be modeled well by sets of variograms that describe the variation at each scale. These variograms provide a good theoretical basis for estimating reserves and for interpolation. There seems to be no practical reason for using a fractal approach for interpolating regionalized variables.

DISCUSSION AND CONCLUSION

Most fractal studies of the variation of landscape properties seem to have had only the estimation of the fractal dimension as their goal. These studies have shown that fractal dimensions vary with process and altitude (Aviles et al., 1987; Brown et al and Scholz, 1985; Goodchild, 1982; Mark and Aronson, 1984; Power, et al. 1987; Roy et al., 1987), and, for soil surface properties at least, the estimates of fractal dimension appear to vary with the time of sampling (Armstrong, 1986). While it is interesting to know how rough or noisy a distribution is compared to an ideal Brownian fractal model, knowledge of fractal dimension is of little use unless it can be applied to comparative studies relating the characteristics of a spatial phenomenon to a given process. Applications of fractal dimension for analytic ends seem to have been few in the geosciences. Woronow (1981) used estimates of fractal dimension to distinguish between different crater types on Mars, and

Elliot (1989) studied surface roughness of forelands in Norway. Most interest in knowing what the appropriate D values for landscapes has come from cartographers who wish to simulate or to preserve the natural roughness of terrain in maps and digital terrain models (Armstrong and Hopkins, 1983; Dutton, 1981; Goodchild, 1982; Goodchild and Dubuc, 1987; Clarke, 1986, 1987; Muller, 1987; Shelberg et al., 1982). On the other hand the recent developments in the analysis of drainage basins suggest that fractals can provide a good basis for understanding the development of drainage networks (Stark, 1991).

In conclusion, (1) there are few estimates of D available for geochemical (soil chemical) properties based on more than a few hundred points. (2) Linear transects consisting of only a few tens or hundreds of points can suffer from sampling variation that can seriously affect estimates of D and may include "pseudo structures." (3) Real data contain information about several interacting processes that may operate at widely separated scales. Some researchers suggest associating a fractal dimension with every scale, but as this gets far away from the original idea of self-similarity and scaling, it is questionable whether such an approach is helpful unless it can be used to assist analytical distinctions that cannot be done in any other way. (4) The number of real applications of fractal theory in geochemical studies is small at present, and there is much work to be done to adapt the simple, elegant theory of the Brownian fractals to the complex real world. Useful applications such as scaling in porous media seem to have been achieved without explicit use of the fractal concept. (5) The advantage of the fractal concept is that it has enlightened many scientists to the fact that the world is not made up of ideal surfaces or of perfectly smooth varying distributions. On the other hand, the dangers of the fractal approach are those inherent in any methodology that attempts to reduce complex phenomena to a single number or model (Medawar, 1984). (6) Because natural phenomena are often the end result of a complex set of processes operating jointly over a wide range of scales, there is often no a priori reason why the resulting spatial or temporal variation should be self-similar and scaling. (7) Finally, the fractal approach has yet to provide practical tools for mineral estimation and interpolation, and it is not yet clear that fractal methods could improve on the established methods of geostatistics. The geostatistical approach allows for both deterministic and stochastic variation over several scales to be taken into account and permits scale-dependent variation to be related to processes that can be explained in physical terms. There may indeed be some natural processes that operate within a single medium over a wide range of scales, such as geological faulting (Aviles et al., 1987; Power et al., 1987),clouds (Lovejoy, 1982), windblown sands (Burrough, 1984, 1985), caves (Curl, 1986), or stream networks (Stark, 1991) giving rise to true fractals. These areas clearly require more research. Practical tools for mapping, however, are provided by geostatistical methods.

ACKNOWLEDGMENTS

This chapter is a revised and expanded version of a chapter entitled *"Fractals and Geochemistry"* that was first published in *The Fractal Approach to Heterogeneous Chemistry*, edited by David Avnir and published by John Wiley and Sons, 1989. Thanks are due to both David Avnir and to John Wiley Limited for granting permission to reuse this material. Thanks are also due to Oxford University Press for permission to adapt Figure 5.1 from *Principles of Geographical Information Systems for Land Resources Assessment* (Burrough, 1986), Michael Goodchild for supplying the Random Island study, the Netherlands Ministry for Public Health and Environment for the soil pollution data in Figure 5.8, and David Avnir and Olivier Dubrule for constructive comments on an earlier draft.

REFERENCES

Agterberg, F. D. 1982. Recent developments in geomathematics. *Geo-Processing* 2:1–32.

Armstrong, A. C. 1986. On the fractal dimension of some transient soil properties. *Journal of Soil Science* 37:641–652.

———, and Hopkins, L.D. 1983. Fractal enhancement for thematic display of topologically stored data. In *Proceedings, Sixth International Symposium on Automated Cartography (Auto-Carto 6)*, Vol. 2, Ottawa, Canada, pp. 309–318.

Ausloos, M., and Berman, D.H. 1985. A multivariate Weierstrass–Mandelbrot function. *Proceedings of the Royal Society of London A* 400:331–350.

Aviles, C. A., Scholz, C. H., and Boatwright, J. 1987. Fractal analysis applied to characteristic segments of the San Andreas fault. *Journal of Geophysical Research* 92:331–344.

Avnir, D. 1986. Fractal aspects of surface science—an interim report. In Better Ceramics through Chemistry II, C. J. Brinker, D. E. Clark, and D. R. Ulrich, eds. *Materials Research Society Symposium Proceedings* 73:321.

———, Farin, D., and Pfeifer, P. 1985. Surface geometric irregularity of particulate materials: The Fractal approach. *Journal of Colloid and Interface Science* 103:112–123.

Band, L. E. 1986. Topographic partition of watersheds with digital elevation models. *Water Resources Research* 22:15–24.

Beckett, P. H. T., and Webster, R. 1971. Soil variability—a review. *Soils and Fertilizers* 34:1–15.

Berry, M. V., and Lewis, Z. V. 1980. On the Weierstrass–Mandelbrot function. *Proceedings of the Royal Society of London* A 370:459–484.

Bradbury, R. H., and Reichelt, R. E. 1983. Fractal dimension of a coral reef at ecological scales. *Marine Ecology Progress Series* 10:169–171.

Brown, S. R., and Scholz, C. H. 1985. Broad bandwidth study of the topography of natural rock surfaces. *Journal of Geophysical Research* 90:12575– 12582.

Burgess, T. M., and Webster, R. 1984. Optimal sampling strategies for mapping soil types. I. Distribution of boundary spacings. *Journal of Soil Science* 35:641–654.

———, Webster, R., and McBratney, A. B. 1981. Optimal interpolation and isarithmic mapping of soil properties. IV. Sampling Strategy. *Journal of Soil Science* 32:643–659.

Burrough, P. A. 1981. The fractal dimensions of landscapes and other environmental data. *Nature* 294:240–242.

———, 1983a. Multiscale sources of spatial variation in soil. I. The application of fractal concepts to nested levels of soil variation. *Journal of Soil Science* 34:577–597.

———, 1983b. Multiscale sources of spatial variation in soil. II. A non–Brownian fractal model and its application in soil survey. *Journal of Soil Science* 34:599–620.

———, 1984. The application of fractal ideas to geophysical phenomena. *Bulletin of the Institute of Mathematics and Its Applications* 20(3/4):36–42.

———, 1985. Fakes, facsimiles and fractals: fractal models of geophysical phenomena. In *Science and Uncertainty*, S. Nash. Northwood, ed. England: IBM UK Ltd/Science Reviews, pp. 151–170.

———, 1986. *Principles of geographical information systems for land resources assessment.* Oxford, U.K.: Oxford University Press.

Campbell, J. 1978. Spatial variation of sand content and pH within single contiguous delineations of two soil mapping units. *Journal of the Soil Science Society of America 42:460–464.*

Chelidze, T. L. 1982. Percolation and fracture. *Physics of the Earth and Planetary Interiors* 28:93–101.

Clarke, K. C. 1986. Computation of the fractal dimension of topographic surfaces using the triangular prism surface area method. *Computers and Geosciences* 12:713–722.

———, 1987. Scale-based simulation of topography. In *Proceedings, Eighth International Symposium on Automated Cartography (Auto-Carto 8)*, Baltimore, pp. 680–688.

Cressie, N. 1985. Variogram estimation by weighted least squares. *Mathematical Geology* 17:563–586.

———, and Hawkins, D. M. 1980. Robust estimation of the variogram. *Mathematical Geology* 12:115–125.

Culling, W. E. H. 1986. Highly erratic spatial variability of soil pH on Iping Common, West Sussex. *Catena* 13:81–98.

———, The characterization of regular/irregular surfaces in the soil-covered landscape by Gaussian random fields. *Computers and Geosciences* 15:219–226.

———, and Datko, M. 1987. The fractal geometry of the soil-covered landscape. *Earth Surface Processes and Landforms* 12:369–385.

Curl, R. L. 1986. Fractal dimensions and geometries of caves. *Mathematical Geology* 18:65–781.

David, M. 1977. *Geostatistical Ore Reserve Estimation.* Developments in Geomathematics 2. Amsterdam, Netherlands: Elsevier.

Davis, J. C. 1986. *Statistics and data analysis in geology.* 2nd ed. New York: John Wiley & Sons.

Dawe, R. A., Wheat, M. R., and Wright, R. J. 1985. Effects of heterogeneities on miscible flow processes in EOR. *Proceedings of the Third European Meeting on Improved Oil Recovery.* Rome, April 1985. AGIP Rome.

Delhomme, J. P. 1978. Kriging in the hydrosciences. *Advances in Water Resources* 1(5):251–256.

Dicke, M., and Burrough, P. A. 1988. Using fractal dimensions for characterizing the tortuosity of animal trails. *Physiological Entomology* 13:393–398.

Dubrule, O. 1983. Two methods with different objectives: Splines and kriging. *Mathematical Geology* 15:245–255.

. 1984. Comparing splines and kriging. *Computers and Geosciences* 10:327–338.

Dutton, G. H. 1981. Fractal enhancement of cartographic line detail. *American Cartographer* 8(1):23–40.

Elliot, J. K. 1989. An investigation of the change in surface roughness through time on the foreland of Austre Okstindbreen, North Norway. *Computers and Geosciences* 15:209–218.

Farin, D., and Avnir, D. 1987a. Reactive fractal surfaces. *Journal of Physical Chemistry* 91:5517–5521.

———.———. 1987b. The fractal nature of molecule-surface chemical activities and physical interactions in porous materials. *Proceedings of IUPAC Symposium. Characterization of Porous Solids.* In Studies in Surface Science and Catalysis, K. K. Unger, D. Behrens, and H. Kral. eds. Amsterdam, Netherlands: Elsevier.

Folinsbee, R. E. 1977. World's view—from Alph to Zipf. *Bulletin of the Geological Society of America* 88:897–907.

Fournier, A., Fussell, D., and Carpenter, L. 1982. Computer rendering of stochastic models. *Communications of the Association of Computing Machinery* 25:371–384.

Gajem, Y. M., Warrick, A. W., and Myers, D. E. 1981. Spatial dependence of a typic torrifluvent soil. *Journal of the Soil Science Society of America* 45:709–715.

Geffen, Y., Aharony, A., and Alexander, S. 1983. Anomalous diffusion on percolating clusters. *Physical Review Letters* 50:77–80.

Goodchild, M. F. 1980. Fractals and the accuracy of geographical measures. *Mathematical Geology* 12:85–98.

———. 1982. The fractional Brownian process as a terrain simulation model. *Modeling and Simulation* 13:1133–1137. Proceedings, 13th Annual Pittsburgh Conference on Modeling and Simulation.

———, and Dubuc, O. 1987. A model of error for choropleth maps, with applications to geographic information Systems. In *Proceedings, Eighth International Symposium on Automated Cartography (Auto-Carto 8)*, Baltimore, pp. 165–169.

———, and Mark, D. M. 1987. The fractal nature of geographic phenomena. *Annals of the Association of American Geographers* 77(2):265–278.

Gray, D. M. 1961. Interrelationships of watershed characteristics. *Journal of Geophysical Research* 66:1215–1223.

Håkanson, L. 1978. The length of closed geomorphic lines. *Mathematical Geology* 10:141–167.

Hayward, J., Orford, J. D., and Whalley, W. B. 1989. Three implementations of fractal analysis of particle outlines. *Computers and Geosciences* 15:199–208.

Henley, S. 1987. Kriging—blue or pink. *Mathematical Geology* 19:155–158.

Howrath, R. J., White, C. M., and Koch, Jr., G. S. 1980. On Zipf's law applied to resource prediction. *Transactions of the Institution of Mining and Metallurgy. Section B. Applied Earth Science* 89:B182–190.

Jensen, S. K., and Domingue, J. O. 1988. Extracting topographic structure from digital elevation data for geographic information system analysis. *Photogrammetric Engineering and Remote Sensing* 54:1593–1600.

Jones, J. G., Thomas, R. W., and Earwicker, P. G. 1989. Fractal properties of computer-generated and natural geophysical data. *Computers and Geosciences* 15:227–235.

Journel, A. J., and Huijbregts, C. J. 1978. *Mining Geostatistics.* New York: Academic Press.

Katz, A. J., and Thompson, A. H. 1985. Fractal sandstone pores: Implications for conductivity and pore formation. *Physical Review Letters* 54:1325–1328.

Kennedy, S. K., and Lin, W.-H. 1986. Fract—A FORTRAN subroutine to calculate the variables necessary to determine the fractal dimensions of closed forms. *Computers and Geosciences* 12:705–712.

Koplik, J., Wilkinson, D., and Willemsen, J. F. 1983. Percolation and capillary fluid displacement. In *The Mathematics and Physics of Disordered Media.* Lecture Notes in Mathematics No. 1035, B. D. Hughes and B. W. Ninham, eds., pp. 169–183. New York: Springer-Verlag.

Lam, N. S. 1983. Spatial interpolation methods: a review. *The American Cartographer* 10:129–149.

Laverty, M. 1987. Fractals in Karst. *Earth Surface Processes and Landforms* 12:475–480.

Lenormand, R., and Zarconne, C. 1985. Invasion percolation in an etched network: Measurement of a fractal dimension. *Physical Review Letters* 54:2226–2229.

Longley, P. A., and Batty, M. 1989. Fractal measurement and line generalization. *Computers and Geosciences* 15:167–184.

Lovejoy, S. 1982. Area-perimeter relation for rain and cloud areas. *Science* 216:185–187.

———, Schertzer, D., and Ladoy, P. 1986. Fractal characterization of inhomogeneous geophysical measuring networks. *Nature* 319:43–44.

———, Schertzer, D., and Tsonis, A. A. 1987. Functional box counting and multiple elliptical dimensions in rain. *Science* 235:1036–1038.

Mandelbrot, B. B. 1975. Stochastic models for the Earth's relief, the shape and the fractal dimension of the coastlines, and the number-area rule for islands. *Proceedings of the National Academy of Sciences* 72(10):3825–3828.

———, 1977. *Fractals, Form, Chance and Dimension.* San Francisco: W.H. Freeman & Co.

———, 1982. *The Fractal Geometry of Nature.* San Francisco: W.H. Freeman & Co.

———, and Wallis, J. R. 1969. Computer experiments with fractional Gaussian noises. Part I. Averages and variances. *Water Resources Research* 5:228–241.

Mark, D. M. 1984. Fractal dimension of a coral reef at ecological scales: a discussion. *Marine Ecology Progress Series* 14:293–94.

———, and Aronson, P. B. 1984. Scale-dependent fractal dimensions of topographic surfaces: An empirical investigation with application in geomorphology and computer mapping. *Mathematical Geology* 16:671–683.

Matheron, G. 1971. *The Theory of Regionalized Variables and Its Applications.* Paris: Les Cahiers du centre de morphologie mathematique de Fontainbleu, Ecole Nationale Superieur des Mines de Paris.

———, 1973. The intrinsic random functions and their applications. *Advances in Applied Probability* 5:439–468.

McBratney, A. B., and Webster, R. 1981. Spatial dependence and classification of the soil along a transect in northeast Scotland. *Geoderma* 26:63–82.

———, and ———. 1983a. Optimal interpolation and isarithmic mapping of soil properties. V. Coregionalization and multiple sampling strategy. *Journal of Soil Science* 34:137–162.

———, and ———. 1983b. How many observations are needed for regional estimation of soil properties? *Soil Science* 135:177–183.

———, and ———. 1986. Choosing functions for semivariograms of soil properties and fitting them to sample estimates. *Journal of Soil Science* 37:617–640.

———, et al. 1982. Regional variation of extractable copper and cobalt in the topsoil of south-east Scotland. *Agronomie* 2:969–982.

Medawar, P. 1984. *Plato's Republic.* New York: Oxford University Press.

Miller, E. E. 1980. Similitude and scaling of soil-water phenomena. In *Applications of Soil Physics*, D. Hillel, ed. New York: Academic Press.

Muller, J. C. 1987. Optimum point density and compaction rates for the representation of geographic lines. In *Proceedings, Eighth International Symposium on Automated Cartography (Auto–Carto 8)*, Baltimore, pp. 221–230.

Nielsen, D. R., and Bouma, J. 1985. *Spatial Analysis of Soil Data.* PUDOC, Wageningen, The Netherlands.

Oliver, M. A., and Webster, R. 1986. Combining nested and linear sampling for determining the scale and form of spatial variation of regionalized variables. *Geographical Analysis* 18:227–242.

Paliwal, H. V., Bhatnagar, S. N., and Haldar, S. K. 1986. Lead-zinc resource prediction in India: an application of Zipf's law. *Mathematical Geology* 18:539–549.

Peitgen, H. O., and Richter, P. H. 1986. *The Beauty of Fractals.* New York: Springer-Verlag.

———, and Saupe, D. 1988. *The Science of Fractal Images.* New York: Springer-Verlag.

Power, W. L., et al. 1987. Roughness of natural fault surfaces. *Geophysical Research Letters* 14:29–32.

Richardson, L. F. 1961. The problem of contiguity: An appendix of statistics of deadly quarrels. *General Systems Yearbook* 6:139–187.

Ripley, B. 1981. *Spatial Statistics.* New York: John Wiley and Sons.

Rivier, N., Guyon, E., and Charlaix, E. 1985. A geometrical approach to percolation through random fractured rocks. *Geological Magazine* 122:157–162.

Robinson, P. C. 1983. Connectivity of fracture systems—a percolation theory approach. *Journal of Physics A* 16:605–614.

Rowland, N. S., and Sampey, D. 1977. Zipf's law—an aid to resource inventory prediction in partially explored areas. *Mathematical Geology* 9:383–391.

Roy, A. G., Gravel, G., and Gauthier, C. 1987. Measuring the dimension of surfaces: a review and appraisal of different methods. In *Proceedings, Eighth International Symposium on Computer-assisted Cartography (Auto-Carto 8)*, Baltimore, pp. 68–77.

Rudenno, V. 1981. The probability of economic success in exploring for the tin deposits. *Bulletin of the Canadian Institute of Mining and Metallurgy* 74(828):99–101.

Sen, P. N., Scala, C., and Cohen, M. H. 1982. Stochastic modeling of two-phase flow in porous media. *Society of Petrological Engineering Paper No. 11128.*

Serra, J. 1968. Les structures gigognes: morphologie mathématique et interprétation métallogénique. *Mineralium Deposita* 3:135–154.

Scheidegger, A. E. 1970. Stochastic models in hydrology. *Water Resources Research* 6:750–755.

Shelberg, M. C., Moellering, H., and Lam, N. 1982. Measuring the fractal dimensions of empirical cartographic curves. In *Proceedings, Fifth International Symposium on Computer-Assisted Cartography (Auto-Carto 5)*, Washington D.C., pp. 481–490.

Shepard, D. 1968. A two-dimensional interpolation function for irregularly spaced data. In *Proceedings of the 23rd National Conference of the Association for Computing Machinery*, pp. 517–523. Princeton, NJ: Brandon/Systems Press Inc.

Smith, H. F. 1938. An empirical law describing heterogeneity in the yield of agricultural crops. *Journal of Agricultural Science* 28:1–23.

Stanley, H. E. 1984. Application of fractal concepts to polymer statistics and to anomalous transport in randomly porous media. *Journal of Statistical Physics* 36:843–860.

Stark, C. P. 1991. An invasion percolation model of drainage network evolution. *Nature* 352:423–425.

Tapp, B. A., and Moniuszko, V. 1978. Application of Zipf's law to mineral distribution patterns in the northern Australian orogenic provinces. *Eleventh Commonwealth Mining and Metallurgical Congress*, Institute of Mining and Metallurgy. Paper 35, 8 pp. Australia.

Taylor, C. C., and Burrough, P. A. 1986. Multiscale sources of spatial variation in soil. III. Improved methods for fitting the nested model to one-dimensional semivariograms. *Mathematical Geology* 18:811–821.

Thornes, J. 1990. Big rills have little rills.... *Nature* 345:764–765.

Turcotte, D. L. 1986. Fractals and fragmentation. *Journal of Geophysical Research* 91(B2):1921–1926.

———. 1989. Fractals in geology and geophysics. *Pure and Applied Geophysics* 131:171–196.

Unwin, D. 1989. Fractals and the geosciences. Introduction to the Special Issue Fractals and the Geosciences. *Computers and Geosciences* 15:163– 166.

Verly, G., David, et al., eds. 1984. *Geostatistics for Natural Resources Characterization.* Parts 1 and 2. Doordrecht: the Netherlands: D. Reidel Publishing Co.

Voss, R. F. 1985. Random fractal forgeries: From mountains to IBM music. In *Science and Uncertainty*, S. Nash, ed., pp. 69–88. Northwood, England: IBM UK Ltd. Science Reviews.

Webster, R. 1984. Elucidation and characterization of spatial variation in soil using regionalized variable theory. In *Geostatistics for Natural Resources Characterization*, G. Verly, M. David, A. G. Journel, and A. Marechal, eds. Part 2, pp. 903–913. Doordrecht, the Netherlands D. Reidel Publishing Co.

Webster, R. 1985. *Quantitative Spatial Analysis of Soil in the Field.* Advances in Soil Science, Volume 3. New York: Springer–Verlag.

———, and Burgess, T. M. 1984. Sampling and bulking strategies for estimating soil properties in small regions. *Journal of Soil Science* 35:127–140.

———, and Oliver, M. 1990. *Statistical Methods in Soil and Land Resources Survey.* Oxford, U.K.: Oxford University Press.

Whalley, W. B., and Orford, J. D. 1989. The use of fractals and pseudo-fractals in the analysis of two-dimensional outlines: Review and further exploration. *Computers and Geosciences* 15:185–198.

Wilke, S., Guyon, E., and de Marsilly, G. 1985. Water penetration through fractured rocks: Test of a tridimensional percolation description. *Mathematical Geology* 17:17–27.

Williams, J. K., and Dawe, R. A. 1986. Fractals—an overview of potential applications to transport in porous media. *Transport in Porous Media* 1:201–209.

Woronow, A. 1981. Morphometric consistency with the Hausdorff-Besicovitch dimension. *Mathematical Geology* 13:201–216.

Zonneveld, J. I. S. 1973. Convergentie en luchtfotointerpretatie. *Geografisch Tijdschrift* (Koninklijk Nederlands Aardrijkskundig Genootschap) 7(1):38–48.

CHAPTER 6

STATISTICS OF CHANNEL NETWORKS ON FRACTIONAL BROWNIAN SURFACES

Michael F.Goodchild

National Center for Geographic Information and Analysis, University of California, Santa Barbara

Brian Klinkenberg

Department of Geography, University of British Columbia, Vancouver, Canada

ABSTRACT

Many natural forms can be successfully simulated using fractals. We argue that fractals can therefore provide useful models of natural forms in the absence of specific effects and constraints: in effect, null hypotheses for natural forms. The fractional Brownian process is a useful simulation for terrain and offers a method of determining the characteristics of channel networks formed by random processes. We simulate random channel networks and tabulate the statistical distributions of various parameters. The chapter demonstrates that recently observed departures from the random topology model of Shreve are attributable to the geometric constraints imposed by packing networks onto a topographic surface, and not to geologic and geomorphic controls, as previously supposed.

INTRODUCTION

Hypothesis testing is a powerful strategy in science, particularly in areas where data are subject to substantial uncertainty. Suppose that observations suggest the presence of an effect E. If E is not the only effect present, then it is not capable of accounting fully for the observed variation in the data. The general strategy is to consider what would have happened in the absence of effect *E*. By comparing this hypothetical outcome with the real one, we may succeed in proving the presence or absence of the effect. We term the hypothetical process in the absence of effect *E* the null hypothesis or H_0. The strategy has its limitations: it does not evaluate the strength of effect *E*, merely its existence; if the effect is weak, it may be masked by other sources of variation (termed a Type II statistical error); and other sources of variation may cause us to falsely conclude the presence of the effect (Type I statistical error). In addition, it may be difficult to formulate an appropriate null hypothesis.

Suppose, for example, that we wish to investigate the effect of age on voting behavior. Two samples, one of Republican voters and the other of Democrats, would each show variation in age, but the ages in the two sets would not be the same, and the means would likely be somewhat different. In the absence of the effect of interest (that is, under the null hypothesis), both samples would be drawn from the same population. But if the effect is real, the population of Democrats would be expected to have a different distribution of ages from the population of Republicans.

Many common forms of hypothesis testing involve very simple null hypotheses and standard testing procedures. However, in some cases the null hypothesis is sufficiently complex or mathematically difficult that, while it is easy enough to state, the only way of investigating its outcomes is by simulation. In such cases the general form of the approach is as follows. An effect is suspected, and data are collected. A null hypothesis is stated, and simulation is used to generate a sample of its outcomes. These are compared to the data. If the data are found to be a possible outcome under the null hypothesis, then the effect is declared to be absent. On the other hand, if the data are inconsistent with the outcomes of the simulation, then the effect is declared to be confirmed.

In this chapter, we examine the use of this strategy with a fractal null hypothesis and simulation process to test an important issue concerning stream networks. The complex patterns formed by branching streams fall into the general class of graphs known as trees. Horton (1945) introduced an ordering and numbering scheme for stream networks that was later refined by Strahler (1952), as follows. Each tributary stream (the leafs of the tree network) is order 1; the joining of two order 1 streams forms an order 2; and in general the joining of two order i streams forms an order $i + 1$. Joining of an order i stream with an order j stream ($i \neq j$) produces a stream of order equal to the larger of i and j. The order of the stream forming the root of the tree becomes the order of the basin. There is now a large literature on the analysis of ordered channel networks, much of it prompted by the early work of Horton. Of particular concern here is what is now known as the *Horton law of stream numbers*: if the logarithm of the number of streams of given order is plotted against the order for a network or collection of networks, the points fall close to a straight line.

Although the effect represented by the Horton law is dramatic, we might nevertheless wish to confirm it by a hypothesis test. For bivariate plots, the conventional form of the null hypothesis proposes that the observed values of one variable have no effect on the corresponding values of the other. Although we would almost certainly reject the null hypothesis in this case, confirming the existence of the Horton law, this particular null hypothesis is clearly inappropriate, as it is impossible for the number of streams to be unrelated to order, given the system of ordering: the number of streams of order i can be no more than one-half the number of order i - 1. We can therefore reject this particular null hypothesis out of hand.

The objective of this chapter is to propose a more appropriate null hypothesis for channel networks and to test its outcomes using fractal simulation. The model is a significant departure from one proposed by Shreve (1966, 1967), described below, which has been the focus of much research in channel networks over the past two decades. Our model accounts for several consistent and puzzling deviations from the Shreve model observed in stream networks.

More generally, the chapter illustrates the value of fractal models as null hypotheses for a whole range of natural systems (Goodchild, 1988). Mandelbrot (1982a) has argued very convincingly that fractals are a useful model of natural systems. But many geomorphologists find the fractal simulations of terrain in Mandelbrot's book too artificial (Mark and Aronson, 1984). Self-affine landscape simulations lack the scale-specific effects of geology and landscape-modifying processes such as wind, rain, and glaciation that provide visual cues as to scale. However, Goodchild and Mark (1987) argue that this lack can be an advantage, in the sense that fractal simulations provide a test-bed for geomorphic processes—an initial or boundary condition landscape on which to simulate the effects of real processes. In general, fractal simulations are a powerful way of visualizing and analyzing what would be expected in the absence of real effects or constraints on natural form.

THE SHREVE MODEL

In the Shreve model of channel network topology (Shreve 1966, 1967), all networks are assumed to be part of an infinite topologically random channel network (TRCN) in which all possible alternative channel network topologies occur with equal probability. There are, for example, five topologically distinct ways in which four first-order streams can combine, four yielding second-order basins and one a third-order basin (Figure 6.1). According to the Shreve model, the five arrangements should be found with equal likelihood in real stream basins. The model has been remarkably successful at explaining the Horton law of stream number, as it predicts not only the observation of straight-line plots, but also the slopes (the ratio of the number of streams of order i - 1 to the number of order i is known as the *bifurcation ratio*, and is constant for a straight-line plot). It has also been extended with success to certain geometrical properties of channel networks, such as link lengths and basin areas (Shreve, 1975).

In the hypothesis testing tradition of statistics, the Shreve model is a null hypothesis, reflecting what would be expected to occur under a simple random process. Thus, if empirical data are found to be consistent with the model, the normal inference would be that the basin or basins from which the data were obtained evolved in a way that was largely free of geological or other constraints on the random combination of streams. On the other hand, deviations from the model's predictions would be interpreted as indicative of the operation of such effects.

Unfortunately, standard hypothesis-testing procedures provide only a weak test of the model, since confirmation of a null hypothesis amounts to no more than a failure to reject. The likelihood of a Type II statistical error, or acceptance of the null hypothesis when in fact it is false, depends both on sample size and on the strength of the effect. In other words, large amounts of data are needed to demonstrate small deviations from the model: the smaller the amount of data available, the more likely is the confirmation of the null hypothesis. Understandably, then, early small-scale empirical tests largely confirmed the predictions of the infinite TRCN model.

A recent paper by Abrahams (1984) undertook a comprehensive review of the field and identified a growing consensus on systematic empirical departures from the model. For large basins (magnitude $M > 50$, where magnitude is defined as the number of first-order streams), there is a tendency for bifurcation ratios (and thus slopes of Horton law plots) to exceed predictions, particularly in areas of high relief. First-order streams can be classified as *S* (source) or *TS* (tributary source) depending on whether they join to other streams of order 1 or to streams of order greater than 1, respectively (in Figure 6.1, the four second-order basins have two *S* and two *TS* links, while the third-order basin has four *S* links and no *TS* links). In theory, the proportion of *TS* links should tend to 50% in very large basins, but in real networks the proportion appears to be higher than expected (networks with a high proportion of *TS* links are said to have *fishbone* topologies). Abrahams (1977) found a strong correlation between the amount of deviation from the model and the relative relief of the topography.

Abrahams (1984, p.164) finds a pattern emerging in empirical results, that "small channel networks fit the model better than large ones." Under the hypothesis-testing framework in which a failure to reject the null hypothesis is often equated with confirmation of the stochastic model, we are tempted to look for explanations of this pattern in erosion processes. But a simpler explanation, suggested by the comments above, is that the null hypothesis is not precisely correct and has been accepted for small data sets because of the greater probability of Type II errors.

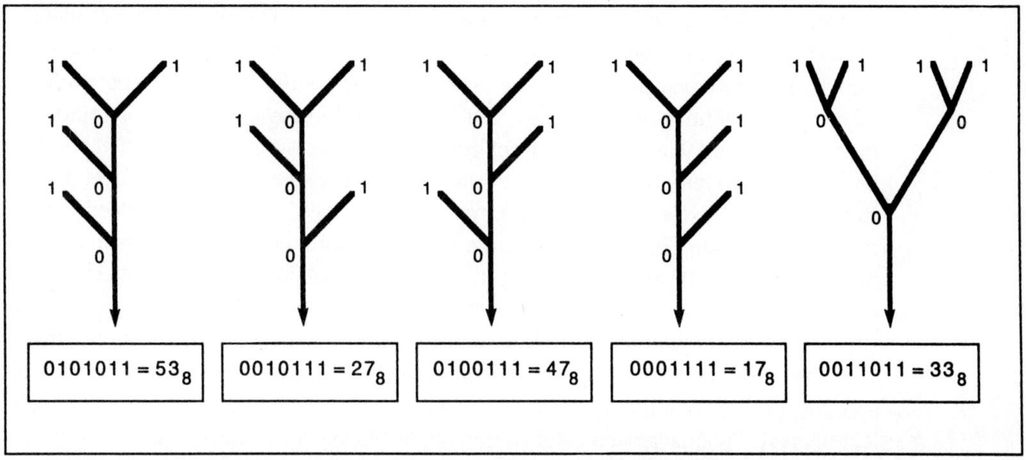

Figure 6.1 The five alternative (topologically distinct) channel networks generated by four first-order streams. The notation denotes the coding schemes for networks and ambilateral classes.

Although much early research supported the random topology model, it is quite possible, given the remarks above, that a reexamination of earlier data would reveal much more widespread disagreement. For example, Smart (1969, pp.1766–1767) analyzed a total of 1157 basins of magnitude 4 (four first-order streams, five possible topologies, see Figure 6.1) and 629 of magnitude 6 (six first-order streams, 42 possible topologies) in a study of the topologies of U.S. networks, separated into eastern and western samples. Although the Shreve random topology model was accepted as a null hypothesis at the 0.05 level in three out of the four analyses, in all cases the data show a bias consistent with those discussed by Abrahams and analyzed later in this paper, which if real would presumably have led to rejection of the null hypothesis given a sufficiently large sample. Smart regarded the results of his analysis as inconsistent, because the null hypothesis was accepted in only three cases, and looked for an explanation in terms of structural controls. Given the work summarized by Abrahams, the alternative explanation of a real bias and three Type II errors seems much more acceptable.

The more straightforward and conventional implication of these results is that there are processes operating on the landscape, in the form perhaps of geological controls, that prevent or bias the random joining of streams, assigning unequal probabilities to certain alternative topologies and resulting in deviations from the model. Clearly, this is an appropriate inference if the model is itself an appropriate null hypothesis and describes correctly the result of random, uninhibited channel network development. However, this is not necessarily the case. Streams developing on a real terrain are subject to constraints of a purely geometrical origin, which will undoubtedly affect the relative likelihoods of alternative topologies. First, basins must pack onto the surface, leading to dependencies between the geometries of adjacent basins, which will in turn affect their topologies. A second class of geometrical constraints is introduced by the requirement of continuity in the surface, which forces dependencies between flow directions of neighboring streams. In principle, then, since all topologies cannot be equally likely if networks are forced to pack onto a surface, the Shreve null hypothesis cannot be true and we can reject it out of hand.

We term a null hypothesis appropriate if it differs from reality only in the absence of the effect of interest, in this case geologic or other control on channel network development, so the implications of rejection are clear and unambiguous. An inappropriate null hypothesis is one that differs from reality in several additional ways, so that rejection may be due either to the effect of interest or to one or more of the additional, irrelevant effects. In this sense, the Shreve model is an inappropriate null hypothesis if indeed the development of an infinite TRCN on real terrain would be inhibited by constraints of a purely geometrical nature. A null hypothesis that failed to include such constraints would be inappropriate, as it would not allow us to separate out the effects of geometrical constraints, which are uninteresting from a geological or geomorphological perspective.

The effect of geometrical constraints on the random topology model is unknown, and it is unlikely that a more appropriate null hypothesis would be amenable to mathematical analysis. The approach used in this paper has been to simulate the random development of stream networks on surfaces under a broad range of conditions. If agreement is found with the infinite TRCN model, then the implication will be that real deviations are due to additional, geological or other forms of control over network development. On the other hand, deviations observed in the simulations will tend to confirm the inappropriateness of the Shreve model as a channel network null hypothesis. Finally, if the form of the simulated deviation agrees with observed deviations in real networks, the implication will be that the origin of such deviations lies in geometrical constraints rather than in physical processes. Earlier results from this study appeared in Goodchild et al. (1985).

FRACTIONAL BROWNIAN MOTION

There have been several previous simulations of random channel development on surfaces. Leopold and Langbein (1962), Schenck (1963), and others allowed streams to wander randomly over a square lattice, generating a move at each lattice cell from a random integer corresponding to one of four move directions (we refer to four move directions as the rook's case, to distinguish it from the queen's case of eight move directions, including four diagonal moves). Special means had to be devised to avoid topological inconsistencies such as closed loops and spirals. The packing constraint is imposed in these models in the form of a uniform density of one stream move per cell: the elimination of loops and spirals has the effect of a weak surface continuity constraint. Seginer (1969) imposed a stronger continuity constraint by first generating a random surface and then extracting the stream network from it, thus ensuring topological consistency. The initial random surface consisted of a square grid of independent elevations, characterized by frequent pits and no general trend. A study by Craig (1980) of simulated erosion processes on surfaces extracted stream networks and compared them to the random topology model as a means of validation of the simulation process, a dubious procedure given the known departures of real networks from the model.

A much more appropriate means of generating suitable random surfaces is available in a class of stochastic processes known as fractional Brownian motion (fBm). Their use for terrain simulation was first explored by Mandelbrot (1977, 1982a) and has attracted considerable attention, due at least partly to striking computer-generated illustrations. More significantly, fBm surfaces have several attractive features that make them suitable for the purposes of this study. The variogram of an fBm surface has the form

$$E\,[z(x) - z(x + d)]^2 = k|d|^{2H} \tag{6.1}$$

where x denotes a location, d is a displacement from this location, z is the elevation of the surface at location x, E is the statistical expectation, $|d|$ is the length of the displacement d, k is a constant, and H is a parameter in the range 0 to 1. $H = 0$ corresponds to independent elevations and thus an infinitely rugged surface with no trend, while as H tends to 1 surfaces become locally smoother with stronger general trend. $H = 0.5$ corresponds to Brownian motion, in which the difference in elevation between adjacent points has a probability distribution that is symmetrical about zero.

Because of the power law form of the variogram, fBm surfaces are scale free and termed self-affine: a small piece of the surface if sufficiently enlarged is statistically indistinguishable from the surface as a whole. Since most if not all geomorphological processes are scale dependent, having different amounts of influence on the landscape at different scales, fBm surfaces appear to lack evidence of erosion. So, by extracting drainage networks from fBm surfaces of varying *H*, we can simulate network development under a broad range of surface conditions and subject to packing and continuity constraints. Kirkby (1985) used a fractal surface as the random, erosion-free starting point for simulations of slope and stream evolution.

There have been several comparisons between the fBm model and real terrain. Goodchild (1982) examined the topography of Random Island, Newfoundland, and found deviations from self-similarity that could be interpreted in terms of the varying influence of geomorphological processes. Mark and Aronson (1984) reached similar conclusions in their analyses of areas in the eastern United States. fBm surfaces have equal abundances of peaks and pits; rugged surfaces of low *H*, with high peak and pit frequencies, are consequently more suggestive of karst or dead ice topographies than real, eroded terrain.

The surfaces used in the study (Figure 6.2) consisted of square arrays of 256 by 256 elevations, generated by the method described by Mandelbrot (1977) in which an initially flat surface is faulted along randomly located, straight lines. For $H=0.5$, the surface on either side of each fault is displaced upward or downward by a constant amount; for other values of *H* a nonuniform displacement is used. This method was adopted in preference to the more commonly used algorithm of Fournier et al. (1982) because of Mandelbrot's theoretical objections to the latter (Mandelbrot, 1982b). A total of 1000 faults was applied to each surface to ensure a high probability that at least one fault occurred between every pair of adjacent elevations. No scaling was applied to the elevations, since both the self-similarity properties and the process of network extraction are invariant under linear transformation. Surfaces were generated with values of *H* ranging from 0.3 to 0.7 in steps of 0.1. Three replications of the 0.7 surface were generated because this value of *H* was thought by Mandelbrot to give the closest resemblance to certain real terrains. Figure 6.2 shows examples of fBm surfaces for values of *H* ranging from 0.3 to 0.9.

DRAINAGE NETWORK EXTRACTION

fBm surfaces have no eroded stream channels, so the derivation of channel networks must be based on local terrain slopes. O'Callaghan and Mark (1984) have described the extraction of networks under similar conditions from real digital elevation models, and see also Jensen (1985). In their algorithm, each cell may flow to one of eight queen's case neighbors; in this study moves were restricted to the four rook's case neighbors because of the problems that would be introduced by allowing up to seven streams to join at once.

Each cell was assigned an integer between 0 and 4 according to the following rules. If any of the four neighboring cells were strictly lower, flow was to the lowest neighbor, designated by an integer between 1 and 4. Otherwise, the cell was designated a sink and assigned 0. This is termed the deterministic or *D* assignment.

Several objections can be raised to this simple approach. First, it assigns precisely one channel flow direction to each cell, whereas real channels are observed to develop only after the accumulation of a sufficient amount of nonchannel flow. We allow for this by the truncated or *T* assignment, which modifies the networks obtained under assignment

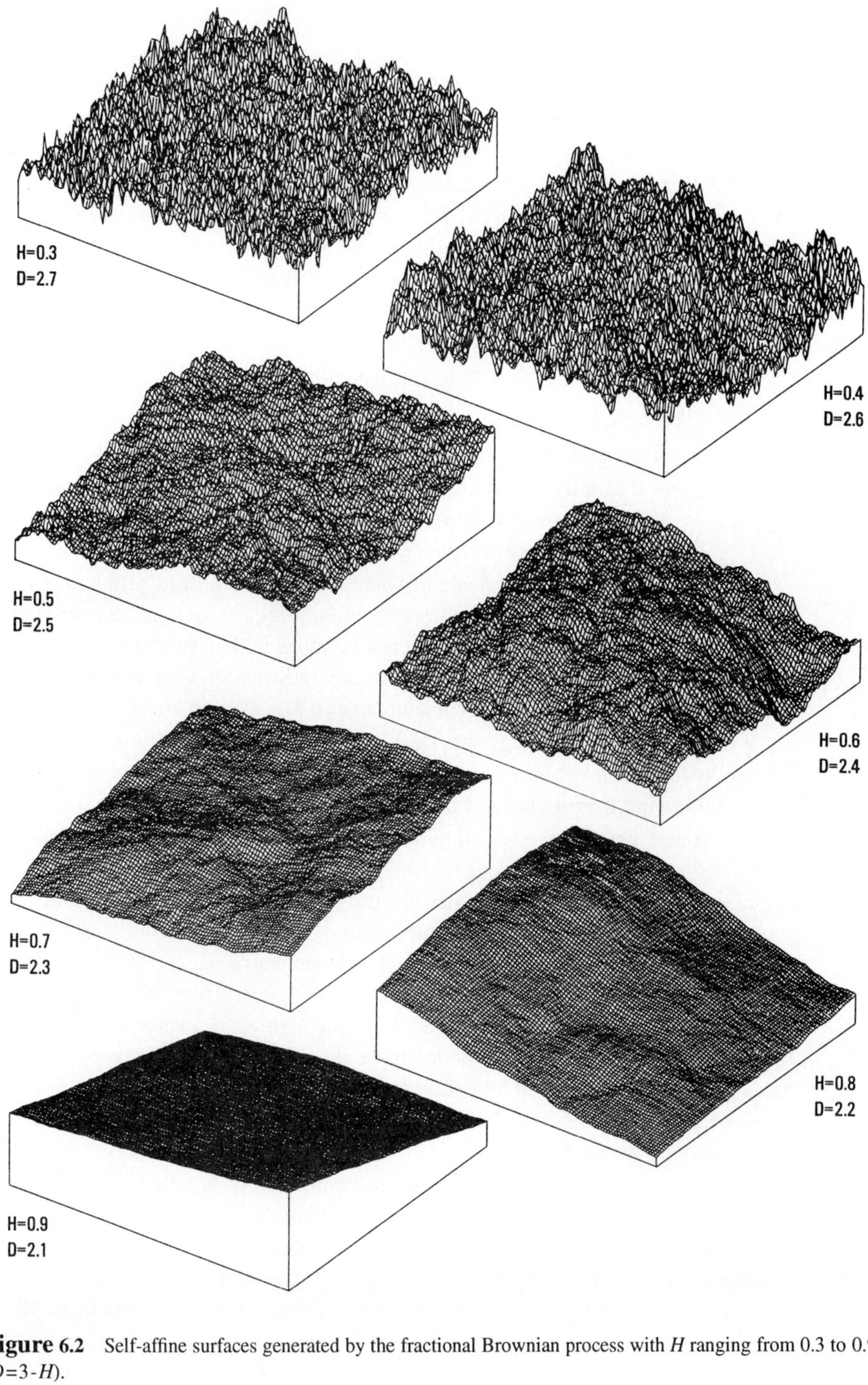

Figure 6.2 Self-affine surfaces generated by the fractional Brownian process with *H* ranging from 0.3 to 0.9 (*D*=3-*H*).

D above by deleting downstream from each first-order stream head until a certain minimum flow level is reached, measured by the number of upstream cells (see also Mark, 1983; Hugus and Mark, 1984; O'Callaghan and Mark, 1984). Thus, the T assignment assumes a certain threshold-contributing upstream area before a channel can develop. This process can result in the deletion of junctions and redesignation of stream orders.

Second, the use of a limited set of move directions introduces a potential bias in flow direction. Consider an area of uniform slope. In reality all streams would flow in the direction of maximum gradient without joining. Under the D assignment, all streams would flow without joining in the permitted move direction closest to the direction of maximum gradient, introducing a consistent bias of up to 45°. The random or R assignment allows for this through the following rules:

> If no neighbor is strictly lower, the cell is a pit.
>
> Else if exactly one is lower, flow is in that direction,
>
> > or if exactly two are lower and in opposite directions, flow is to the lowest,
> >
> > or if exactly two are lower and in neighboring directions, or more than two are lower, assign direction based on probabilities proportional to the differences in elevation between the cell and the lower neighbors.

The R assignment also deals to some extent with a third problem. Because of the self-similarity property, fBm surfaces vary within-cells as well as between them, although the use of a discrete array has the effect of deleting within-cell variation. The consequent straightening of derived channels within cells reduces the probability of joining. The R rules compensate for this by allowing streams to join even in areas of uniform slope.

The problem of the edge was dealt with in all cases by adding a border of cells with very high elevations. This ensured that flow either sank at the edge of the 256 by 256 array or flowed inward from it, but never outward.

Channel networks were obtained from the integer arrays in the form of binary strings (Shreve, 1967; Scheidegger, 1967). In our implementation, the binary string representation of each basin is computed by traversing the tree beginning and ending at the root, or outlet stream, in a clockwise or left to right direction, turning left at every intersection and reversing direction at every leaf or source stream. A 0 is encoded every time a junction is encountered from downstream and a 1 every time a leaf or source stream is reached. This binary notation provides a unique coding for every possible network topology, in which the number of 1s is one more than the number of 0s and that always terminates in a 1. Furthermore, at any point before the end of the string the number of 1s is less than or equal to the number of 0s. Figure 6.1 demonstrates the coding scheme with the five topologies of magnitude $M = 4$.

The literature on channel networks has consistently assumed all junctions to be three valent, consisting of one outlet and two inlet streams, whereas the method of network extraction described above will frequently produce four-valent junctions where three inflowing streams merge to form one outflowing stream. The extraction algorithm allowed for this by breaking all four-valent junctions into two three-valent junctions plus a dummy link. To avoid favoring certain topologies, breaks were made randomly, the two alternatives being given equal probabilities. This is illustrated in Figure 6.3. Suppose the elevations predict that flow is into the central cell from the left, from above and from the right, and that outflow is downward, thus creating a four-valent junction (this could be the result if the cells to the left, above, and to the right had higher elevations, and the cell below had a lower elevation). The letters *a, b,* and *c* in Figure 6.3 denote the binary strings describing the three subbasins upstream of the junction. Figure 6.3 shows the two equally likely networks created by randomly breaking the junction of four streams into two junctions of three streams each. The results reported earlier (Goodchild et al., 1985) did not randomize the breaking of four-valent junctions and were limited to the *D* method of flow assignment.

Channel-length information was encoded in the form of a second string of the same length, giving the length of the link downstream of each junction for each 0 in the binary string and the length of the source link for each 1.

All analyses of the basins were carried out by manipulations of these pairs of strings. Stream order was defined using the Strahler (1952) system and obtained with the following algorithm (see also Liao and Scheidegger, 1968). Replace all instances of 0*xx* in the string by $x+1$ (*x* not equal to 0), and increment a counter for the number of streams of order $x+1$ by 1. Replace all instances of 0*xy*, *x* not equal to *y* and neither *x* nor *y* equal to 0, by the greater of *x* and *y*. This process must terminate with a single digit equal to the order of the basin.

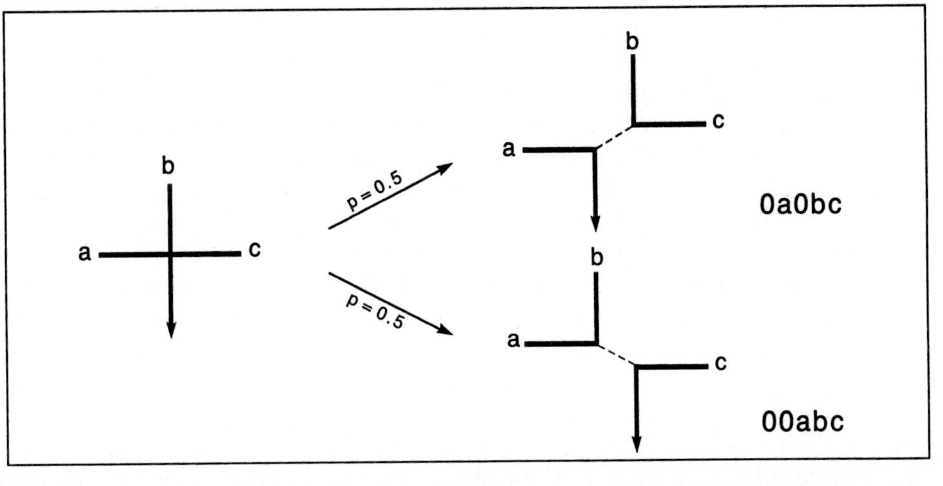

Figure 6.3 A four-valent junction and the two equiprobable pairs of three-valent junctions derived from it.

ANALYSIS: *D* ASSIGNMENT

Under the extraction rules described above, a basin is defined as a tree network rooted or terminating in a sink. This has no direct equivalent in the infinite TRCN model, since all links are part of one infinitely large, singly rooted tree. A randomly chosen basin in the infinite TRCN model is best defined as the subbasin rooted in and upstream of a randomly chosen link. However, we might assume that the basins in the simulation have arisen by a process in which any link in the model is equally likely to become a sink. This would allow us to compare the abundances of basins by order in the simulation with the known probabilities that a randomly chosen basin or subbasin in the model is of order w. The relative abundances of basins by order under the D assignment rules are shown in Table 6.1.

TABLE 6.1 Abundances of Basins by Order

H	1	2	3	4	5	Total
0.3	14,363	4,949	642	1	0	19,955
0.4	10,646	4,502	806	5	0	15,959
0.5	7,524	3,589	827	16	0	11,956
0.6	3,900	2,130	751	40	1	6,822
0.7a	621	380	178	50	9	1,238
0.7b	1,543	808	383	56	4	2,794
0.7c	1,801	1,009	458	68	0	3,336

The infinite TRCN model predicts that the number of basins of order i be half the number of order $i - 1$. This is clearly more nearly true of the smoother surfaces of high H than of the more rugged, low H surfaces with much more frequent pits. Although basins on the smoother surfaces reach higher orders, there is clearly severe truncation, which may be due to limitations on the abundances of high-magnitude basins or to the relative infrequency of basins of high order for a given magnitude. The latter interpretation is more likely for the high H surfaces where pits are much less common.

The analysis of relative abundances of streams and links on the simulated surfaces is similarly affected by the presence of pits, since the infinite TRCN model is again concerned with a randomly chosen feature in an infinite network. The simulations show similar overabundances of low-order streams and links, particularly on the more rugged surfaces, and truncation at higher orders.

In the infinite TRCN model, all possible topologies of basins of a given magnitude are equally likely. Table 6.2 shows the relative abundances of the five possible topologies of magnitude 4 basins (basins with four first-order streams; see Figure 6.1). Each basin topology is identified by the octal representation of its binary string; thus, the string 0001111 becomes basin 17. Note that in this and subsequent analyses in this paper, a basin is defined as terminating in a sink; subbasins of a given magnitude that might exist embedded in basins of larger magnitude are not included in the sample.

TABLE 6.2 Abundances of Topologies for Magnitude 4 Basins

H	17	27	33	47	53	Total
0.3	110	150	178	126	163	727
0.4	128	157	170	158	137	750
0.5	118	131	115	119	113	596
0.6	66	87	63	77	64	357
0.7a	10	17	7	7	14	55
0.7b	37	30	20	26	29	142
0.7c	31	33	26	30	43	163

The clearest trend in the table is that shown by basin 33, or 0011011, which is overabundant for the rugged $H = 0.3$ and $H = 0.4$ surfaces and underabundant for the remainder, although only the $H = 0.3$ results are significantly different from the model predictions at the 0.05 level on a chi-squared test. Basin 33 is unique in the magnitude 4 basins as the only one of third-order. As noted earlier, the second-order basins, 17, 27, 47, and 53, have been loosely referred to as fishbone topologies in the literature, as a high proportion of first-order streams, in these cases 50%, join streams of a higher order (*TS* or tributary source links). In the case of basin 33, all first-order streams join other first-order streams (*S* or source links). We will informally refer to the degree to which a basin approximates a fishbone form as its *linearity* and use the proportion of *TS* links as a more objective measure.

Two topologies are said to be of the same ambilateral class (Smart, 1969) if one can be obtained from the other by exchanging the two subbasins incident at one or more junctions, in effect removing the distinction between left and right. In Figure 6.1, basins 17, 27, 47, and 53 are all of the same ambilateral class, but cannot be transformed into basin 33 by mirroring at junctions. Smart first suggested aggregation to ambilateral classes as a means of reducing the number of alternative topologies in high-magnitude basins, while retaining the significant hydrological differences between classes (Smart, 1969, p.1761). In addition, all members of one ambilateral class clearly share the same proportions of *S* and *TS* links, and therefore the same degree of linearity, or similarity to a fishbone form.

To identify a basin's ambilateral class, each 0, or junction, in the basin's binary string is examined. The two subbasins incident at the junction are described by two consecutive blocks of binary code immediately following the 0 and can be identified from the rule that any subbasin must have exactly one more 1 than it has 0s. The binary representations of the two subbasins are compared numerically and, if necessary, switched so that the first has a numerical value less than or equal to the second. When all junctions have been examined in this way, the binary string will have been reduced to that of a common ambilateral class. For example, consider the basin 00111 (basin 7, magnitude 3). The first junction has incident subbasins 011 and 1. Since the first is greater in value than the second, they are reversed, giving the string 01011. The second 0 has symmetrical incident subbasins, 1 and 1, so no reversal is necessary, and the ambilateral class is identified as 13, the octal representation of 01011.

The magnitude 5 basins, for which there are 14 possible topologies and 3 ambilateral classes, are shown tabulated by ambilateral class in Table 6.3. The most linear topology, in this case class 253, which has 60% *TS* links, is overabundant for all surfaces and significantly so in all but one.

TABLE 6.3 Abundances of Basins by Ambilateral Class, M = 5

H	233	153	253	Chi squared
0.3	26	129[a]	252[a]	20.8[b]
0.4	33	124[a]	221[a]	10.7[b]
0.5	28	93	246[a]	19.0[b]
0.6	19	64	194[a]	21.6[b]
0.7a	2	7	32[a]	7.6
0.7b	3	21	63[a]	11.3[b]
0.7c	5	30	83[a]	12.3[b]
Model probability	2/14	4/14	8/14	
Proportion *TS* links	20%	20%	60%	

a. Denotes observed abundance exceeds model predictions.
b. Denotes significant at the 0.05 level

The pattern for magnitude 6 basins, shown in Table 6.4, is similar, as is the analysis of magnitude 7 basins. In both cases there is increasing bias toward the most linear topologies, 1253 and 5253, respectively, with increasing *H*. In Table 6.4, it is evident that 553 is also overabundant on all surfaces but one. Since 553 has the same proportion of *TS* links as three other ambilateral classes, it is clear that the proportion of *TS* links is not a completely satisfactory predictor. Basin 553 is symmetrical, with two linear subbasins (01011) meeting at the top of the root link.

In general, basin order and proportion of *TS* links are inversely related; as Table 6.4 shows, the most linear basins are also those of lowest order. The overabundance of linear basins thus has the effect of truncating the distribution of basin orders so that fewer high-order basins are found than expected under the infinite TRCN model.

TABLE 6.4 Abundances of Basins by Ambilateral Class, $M = 6$

H	633	1233	553	1153	653	1253	Chi squared
0.3	4	15	18[a]	26	36[a]	82[a]	7.4
0.4	6	20	33[a]	37	54[a]	102[a]	10.7
0.5	2	14	25[a]	28	36	123[a]	32.3[b]
0.6	0	7	18[a]	26	20	105[a]	42.8[b]
0.7a	1	2	3	4	2	20[a]	[c]
0.7b	0	6	9[a]	11	8	53[a]	23.0[b]
0.7c	2	5	8[a]	7	12	49[a]	17.4[b]
Model probability	2/42	4/42	4/42	8/42	8/42	16/42	
Proportion *TS* links	0/6	2/6	2/6	2/6	2/6	4/6	
Basin order	3	3	3	3	3	2	

a. Denotes overabundant.
b. Denotes significant at the 0.05 level.
c. Denotes chi squared not computable due to small numbers.

Table 6.5 shows the proportions of *TS* links observed in basins of magnitude 4 through 20, compared to the proportions expected under the model, for three of the surfaces. It is easy to identify *TS* links in a basin's binary string since the two 1s in any sequence 011 are *S* links and all other 1s are *TS*. The expected proportions in the model are given by $(M-2)/(2M-3)$, which is asymptotic to 0.5 as M tends to infinity (Mock, 1971). In the limit, the probability that a randomly chosen bit is a 1 in the infinitely long basin string is 0.5, and each bit is independent of its neighbors. There are eight equiprobable sequences of the two bits preceding and the one bit following the 1, four of which dictate that the 1 denote an *S* link and four a *TS* link:

0010, *TS* 0011, *S* 0110, *S* 0111, *S*

1010, *TS* 1011, *S* 1110, *TS* 1111, *TS*

Table 6.5 shows that the observed proportion of *TS* links is generally higher than expected, and increasingly so with increasing basin magnitude and increasing *H*.

Finally, we tabulate the bifurcation ratios for these three surfaces and for basin magnitudes 4 through 20 in Table 6.6. The ratios are calculated in two ways. *B*0 is the (i - 1)th root of the number of first-order streams, where i denotes the basin order, while *B*1 is the ratio of first-order to second-order streams. For given magnitude, the model predicts bifurcation ratios close to 4.0. Instead, observed ratios tend to be higher than 4.0, particularly on surfaces of high *H*, *B1* is consistently higher than *B0* and both tend to increase with increasing magnitude.

TABLE 6.5 Observed and Expected Proportions of *TS* Links

		$H = 0.3$		$H = 0.5$		$H = 0.7$a	
M	Expected	n	Observed	n	Observed	n	Observed
4	0.400	727	37.8	596	40.4	55	43.6
5	0.429	407	44.8	367	46.8	41	51.2
6	0.444	181	47.7	228	51.0	32	53.1
7	0.455	120	51.4	194	51.7	32	61.6
8	0.462	69	46.7	125	53.6	24	57.3
9	0.467	44	54.0	94	54.4	13	62.4
10	0.471	18	55.6	56	55.0	13	66.2
11	0.474	13	58.0	46	56.9	12	63.6
12	0.476	3	55.6	28	53.0	7	73.8
13	0.478	4	50.0	29	53.8	14	62.6
14	0.480	2	35.7	14	56.1	7	75.5
15	0.482	0		17	53.7	7	63.8
16	0.483	1	62.5	11	54.5	11	67.0
17	0.484	0		4	58.8	11	65.8
18	0.485	2	55.6	3	51.9	4	69.4
19	0.486	0		5	60.0	8	68.4
20	0.487	0		2	60.0	4	72.5

ANALYSIS: *T* AND *R* ASSIGNMENTS

The above analyses were repeated using the *T* assignment, truncating each stream downstream from the head until the flow from a prescribed number of cells *F* had been accumulated. This minimum number of cells was set to 1, 2, and then 5, and results were compared in each case to those obtained under the *D* assignment.

Truncation tends to delete large numbers of small first-order basins and to reduce the order of larger basins. For $F = 1$, the model was rejected as a predictor of the relative abundances of the five $M = 4$ topologies (compare Table 6.2) for all surfaces except $H = 0.7$c, in all cases due to the underabundance of basin 33 and the overabundance of the linear topologies. The most linear ambilateral class was overabundant in every case where chi squared could be calculated for every surface and for each of $M = 4$, 5, 6, and 7. Observed proportions of *TS* links tended to be higher for given magnitudes than those shown in Table 6.5, and bifurcation ratios higher than those in Table 6.6. Although sample sizes were substantially lower, the same trends were observed for increasing truncation at $F = 2$ and $F = 5$. In summary, truncation produces increasing divergence from the model.

TABLE 6.6 Bifurcation Ratios by Basin Magnitude

	$H = 0.3$			$H = 0.5$			$H = 0.7$a		
M	*n*	*B0*	*B1*	*n*	*B0*	*B1*	*n*	*B0*	*B1*
4	727	3.51	3.51	596	3.61	3.61	55	3.75	3.75
5	407	3.95	4.05	367	4.09	4.18	41	4.39	4.45
6	181	4.06	4.34	228	4.36	4.61	32	4.67	4.84
7	120	4.17	4.67	194	4.24	4.71	32	5.50	5.80
8	69	3.43	4.14	125	4.15	4.87	24	4.98	5.50
9	44	4.36	5.13	94	3.89	4.87	13	5.31	6.12
10	18	4.68	5.42	56	4.51	5.27	13	6.32	7.05
11	13	3.91	5.29	46	3.96	5.20	12	4.60	6.11
12	3	3.46	4.67	28	3.42	4.57	7	8.34	9.14
13	4	3.29	4.44	29	4.17	5.17	14	4.28	5.88
14	2	3.74	3.15	14	3.55	4.83	7	6.67	9.00
15	0			17	3.54	4.69	7	3.67	5.89
16	1	4.00	5.33	11	3.73	4.70	11	4.00	6.59
17	0			4	4.12	5.31	11	4.12	6.31
18	2	4.24	4.80	3	4.24	4.50	4	4.24	7.13
19	0			5	4.36	5.19	8	4.36	6.53
20	0			2	4.47	5.83	4	4.47	7.92

By increasing the probability of streams joining, the *R* assignment tends to increase basin order. Larger numbers of order 5 basins were obtained compared to Table 6.1, and the $H = 0.7$b and $H = 0.7$c surfaces produced order 6 basins. For the magnitude 4 ambilateral classes, surfaces $H = 0.6$, $H = 0.7$a, and $H = 0.7$b showed overabundance of the linear class, but the remaining four surfaces showed underabundance. For magnitude 5, the most linear class, 253, was overabundant for all surfaces except $H = 0.3$, and similarly for magnitude 6, the linear classes were overabundant for all except $H = 0.3$ and $H = 0.4$. Proportions of *TS* links and bifurcations ratios tended to be higher than expected, although again not to the same degree as with the *D* and *T* assignments, and the differences again tended to increase with increasing magnitude. As we might expect, then, by increasing the probability of joining, randomization tends to reduce the linearity of the extracted basins, but not sufficiently to produce consistency with the infinite TRCN model.

DISCUSSION

The simulated networks derived from fBm surfaces under a range of assignment rules show significant deviations from the infinite TRCN model. These deviations agree in many ways with those identified by previous empirical research and summarized by Abrahams (1984). Furthermore, there is evidence that a reexamination of data previously interpreted as supporting the random model would also show deviations of the same type, although not sufficiently strong or based on sufficiently large sample sizes to cause rejection of the model.

More specifically, the analysis has identified the following grounds for rejection of the random model:

1. Ambilateral classes for magnitude 4 through 7 basins do not occur with predicted frequencies: instead most surfaces show a bias toward classes with linear or fishbone form, low order, and high proportions of *TS* links. The bias is strongest with smoother surfaces of high *H*.
2. The proportion of *TS* links is higher than expected for basins of magnitude 4 through 20, and the deviation increases with magnitude and with the parameter *H*.
3. There is a tendency for both evaluations of the bifurcation ratio to exceed model predictions. *B1* tends to be higher than *B0*, and the bias increases with basin magnitude and with *H*.

There are other forms of observed deviation from the model, such as asymmetry and the distributions of *cis* and *trans* links, whose interpretation is quite different and that are not therefore relevant to this paper.

In his 1966 paper, Shreve proposed the hypothesis that "in the absence of geologic controls a natural population of channel networks will be topologically random" (Shreve, 1966, p.27). Subsequent research has tended to follow the same paradigm, interpreting deviation from the model as due to the influence of geologic factors and acceptance of the null hypothesis as confirming their absence. Abrahams (1984, p.162) identifies the first of the underlying postulates of the random topology model as "In the absence of environmental controls, channel networks are topologically random."

We have argued in this paper that the packing of channel networks onto a continuous topographical surface imposes additional constraints of a purely geometrical nature, raising the possibility of a second basis for rejection of the model. Given this alternative interpretation, the model becomes an inappropriate null hypothesis for testing for the presence or absence of geologic controls. Many results that have previously been taken as confirming the model can be reinterpreted as failing to reject it simply because of inadequate sample size.

It would be very difficult to devise a more appropriate null hypothesis because of the wide range of possible packing and continuity conditions. The scale-free fBm surfaces used in this study provide continuity conditions ranging from extreme ruggedness to local smoothness and thus possess a reasonable degree of generality. Quite different constraints might be imposed by surfaces with strong scale-dependent or periodic variation, such as glaciated landscapes, or surfaces with regional variation in smoothness. It is evident from the results obtained above that much stronger biases are imposed by packing basins onto relatively smooth surfaces of high H than onto rugged ones.

Another source of difficulty lies in the presence of pits on the fBm surfaces. At low H, with high pit frequencies, we suspect that removing downstream drains from the surface, as in a karst landscape, has the effect of reducing the severity of packing constraints, which may account for the smaller divergences observed from the random model on these surfaces. The assignment rules used in these experiments to simulate channel development may also be unrealistic. Real channels modify their own landscapes, perhaps in ways that affect the operation of packing and continuity constraints. In summary, although they are useful for illustrative purposes, fBm surfaces are not sufficiently general that we would argue for their use in a more appropriate null hypothesis.

We would agree therefore with the conclusion reached by Abrahams (1984, p.185) that, despite commonly observed deviations, "the random model will continue to serve as a standard against which to compare natural channel networks." However, it is not necessarily true that deviations from the model indicate the operation of geologic constraints. Furthermore, it is demonstrable by simulation that channel networks simulated on a wide range of random surfaces are constrained so as to deviate significantly from the random model, irrespective of geologic or environmental constraints. It is therefore difficult to see how acceptance of the model or, more correctly, failure to reject can be interpreted as anything other than a Type II statistical error; equally, it would be difficult to place an unequivocal interpretation on a rejection.

Finally, we hope this chapter has demonstrated the value of fractal simulations, not only in providing interesting visual effects, but also in providing standards and models for investigating a wide range of physical effects. There are numerous instances in science where it is necessary to compare reality to some hypothetical version in which a given effect is absent, but often we have no clear notion of what that hypothetical version might be. Fractal simulations provide precisely that, particularly, as we have demonstrated in this chapter, in the case of topography. fBm is useful both as a starting point for simulating the effects of landform processes and also as a null hypothesis in investigating the effects produced by those processes.

REFERENCES

Abrahams, A. D. 1977. The factor of relief in the evolution of channel networks in mature drainage basins. *American Journal of Science* 277:626–645.

———, 1984. Channel networks: a geomorphological perspective. *Water Resources Research* 20:161–188.

Craig, R. G. 1980. A computer program for the simulation of landform erosion. *Computers and Geosciences* 6:111–142.

Fournier, A., Fussell, D., and Carpenter, L. 1982. Computer rendering of stochastic models. *Communications of the Association for Computing Machinery* 25:371–384.

Goodchild, M. F. 1982. The fractional Brownian process as a terrain simulation model. *Modeling and Simulation* 13:1133–1137. Proceedings of the Thirteenth Annual Pittsburgh Conference on Modeling and Simulation.

———, 1988. Lakes on fractal surfaces: a null hypothesis for lake-rich landscapes. *Mathematical Geology* 20:615–630.

———, and Mark, D. M. 1987. The fractal nature of geographic phenomena. *Annals of the Association of American Geographers* 77:265–278.

———, et al. 1985. Statistics of hydrologic networks on fractional Brownian surfaces. *Modeling and Simulation* 16:317–323. Proceedings of the Sixteenth Pittsburgh Conference on Modeling and Simulation.

Horton, R. E. 1945. Erosional development of streams and their drainage basins: Hydrophysical approach to quantitative morphology. *Bulletin of the Geological Society of America* 56:275–370.

Hugus, M. K.,and Mark, D. M. 1984. Spatial data processing for digital simulation of erosion. In *Proceedings of the Fall ASP–ACSM Convention*, San Antonio, TX, pp.683–693.

Jensen, S. K. 1985. Automated derivation of hydrologic basin characteristics from digital elevation model data.In *Proceedings, Seventh International Symposium on Automated Cartography (Auto-Carto 7)*, Washington, DC, pp.301–310.

Kirkby, M. J. 1985. A two-dimensional simulation model for slope and stream evolution. Paper presented at the Binghamton Conference, State University of New York at Buffalo.

Leopold, L. B., and Langbein, W. B.1962. The concept of entropy in landscape evolution. *U.S. Geological Survey Professional Paper* 500A. Reston, VA: U. S. Geological Survey.

Liao, K. H., and Scheidegger, A. E. 1968. A computer model for some branching-type phenomena. *Bulletin of the International Association for Scientific Hydrology*13:5–13.

Mandelbrot, B. B. 1977. *Fractals: Form, Chance and Dimension*. San Francisco: W.H. Freeman & Co.

———, 1982a. *The Fractal Geometry of Nature*. San Francisco: W.H. Freeman & Co.

———, 1982b. Comment on computer rendering of stochastic models. *Communications of the Association for Computing Machinery* 25:581–583.

Mark, D. M. 1983. Automated detection of drainage networks from digital elevation models. In *Proceedings of the Sixth International Symposium on Automated Cartography (Auto-Carto 6)*, Vol.2, Ottawa, Canada, pp.288–298.

———,and Aronson, P. B. 1984. Scale-dependent fractal dimensions of topographic surfaces: An empirical investigation, with applications in geomorphology and computer mapping. *Mathematical Geology*16:671–684.

Mock, S. J. 1971. A classification of channel links in stream networks. *Water Resources Research* 7:1558–1566.

O'Callaghan, J. F., and Mark, D. M. 1984. The extraction of drainage networks from digital elevation data. *Computer Vision, Graphics, and Image Processing* 28:323–344.

Scheidegger, A. E. 1967. On the topology of river nets. *Water Resources Research* 3:103–106.

Schenck, H. S. Jr. 1963. Simulation of the evolution of drainage basin networks with a digital computer. *Journal of Geophysical Research* 68:5739–5745.

Seginer, I. 1969. Random walk and random roughness models of drainage networks. *Water Resources Research* 5:591–607.

Shreve, R. L. 1966. Statistical law of stream numbers. *Journal of Geology* 74:17–37.

———, 1967. Infinite topologically random channel networks. *Journal of Geology*75:178–186.

———, 1975. The probabilistic-topologic approach to drainage basin geomorphology. *Geology* 3:527–529.

Smart, J. S. 1969. Topological properties of channel networks. *Bulletin of the Geological Society of America* 80:1757–1774.

Strahler, A. N. 1952. Hypsometric analysis of erosional topography. *Bulletin of the Geological Society of America* 63:1117–1142.

CHAPTER 7

INTERPRETING THE FRACTAL DIMENSION OF RIVER NETWORKS

Jonathan D. Phillips
Department of Geography and Planning, East Carolina University, Greenville, NC

ABSTRACT

Topological representations of river channel networks are fractals. The fractal dimension (*D*) of networks has been linked to classical fluvial morphometry. Methods for estimating *D* based on traditional morphometric parameters have been devised, but estimates of *D* based on bifurcation and stream-length ratios are shown to be unreliable. Although the concept of a space-filling network is inconsistent with knowledge of basin hydrology, *D* values for networks have been interpreted by others as representing the space-filling properties of drainage systems. It is more accurate to interpret *D* as an indication of the degree of geological constraints on network evolution over a particular range of spatial scales. The fractal dimension of river systems is similar to the random model of stream network topology in that both provide hypothetical reference standards to which real networks can be compared, and that both are limited in terms of providing insight into process–response mechanics. It is important to remember that the fractal dimension, like any other morphometric parameter, reflects forms resulting from geomorphic processes, but is not an inherent property of fluvial systems.

INTRODUCTION

Fractals can successfully describe complex geomorphic features, and there are visual similarities between fractal-simulated terrain and representations of real landscapes. This has led to considerable speculation that a wide variety of landforms are fractals (that is, their Hausdorff-Besicovitch dimension is greater than their topological dimension). River channel networks have been described with fractal models by several authors (Mandelbrot, 1982; Seiler, 1986; Hjelmfelt, 1988; Tarboton et al., 1988; La Barbera and Rosso, 1989).

The interpretation of the fractal dimension (D) of the topological representation of channel networks has generated information and hypotheses very similar to those generated by classical morphometric relationships and by the random model of network topology (Tarboton et al., 1988; La Barbera and Rosso 1989; Robert and Roy, 1990). The fractal dimensions of networks have also been linked to classical fluvial morphometry, and efficient means for estimating D from standard morphometric relationships have been developed (Seiler, 1986; Hjelmfelt, 1988; Tarboton et al., 1988; La Barbera and Rosso, 1989; Robert and Roy, 1990). The latter are potentially very useful because direct measurements of D for networks can be extremely laborious and expensive.

The purpose of this paper is to place the fractal dimension of channel networks in the context of both classical fluvial morphometric studies and contemporary thinking in fluvial geomorphology. The goal is to assess the validity of fractal estimations based on standard morphometric relationships and to refine the interpretation of the fractal dimension so that it is consistent with contemporary thinking in fluvial geomorphology and hydrology. Of special interest are the assumptions underlying estimates of fractal dimensions from Horton stream length and bifurcation ratios and the validity of the concept of a space-filling drainage network.

LITERATURE REVIEW

There is a well-known power-function relationship between mainstream length (L) and drainage basin area (A):

$$L = aA^{m} \tag{7.1}$$

where a and m are empirical parameters. Hack (1957) introduced this relationship and reported a value of $m = 0.6$. Others reported different values, but generally greater than 0.5 (see the review by Robert and Roy, 1990). Because dimensional analysis suggests that an isometric basin where shape is independent of size or length should have $m = 0.5$, the values of the exponent have been interpreted as representing allometric trends (size- or scale-related changes in shape or morphology), such as the trend toward basin elongation (Church and Mark, 1980; Hack, 1957).

Mandelbrot (1982) suggested that a value of $m > 0.5$ in Equation 7.1 arose from the fractal nature of river networks, which cause measurements to vary with spatial (map) scale. He maintained that the length of a stream channel should increase as the power $D/2$ of its drainage area (that is, $m = D/2$), and based on earlier empirical results using Equation 7.1, he speculated that rivers are fractals with $D = 1.1$ to 1.2. Mandelbrot (1982) further describes some fractal geometric patterns resembling river networks where the fractal dimension of individual lines is 1.1, but where the complete network pattern is spacefilling with $D = 2$. Mandelbrot's interpretation of the exponent in the length–area relationship has led hydrologists and geomorphologists to review their interpretations of stream network morphometry.

A relationship similar to Equation 7.1 exists between total stream length (ΣL) in a basin and drainage area, where

$$\Sigma L = cA^b \tag{7.2}$$

and identical conclusions with regard to the relationship between the exponent and the fractal dimension hold (that is, $b = D/2$), except that in this case D represents the fractal dimension of the entire network, rather than a single stream line, and that Mandelbrot suggests that $b \approx 1$.

Seiler (1986) used fractal arguments and Mandelbrot's interpretation of Equations 7.1 and 7.2 to argue that river networks are self-similar. This allowed an estimation of upstream drainage area to be determined for any point along a river solely from distance downstream, which Seiler used to develop estimates of water-pollutant dilution factors.

Hjelmfelt (1988) examined data from eight basins to test Mandelbrot's (1982) hypothesis that the D of the mainstream channel as estimated from Equation 7.1 is in the range of 1.1 to 1.2. He found a mean of $D = 1.158$ and supported the hypothesis that m values greater than 0.5 reflect the fractal nature of river channels or networks. Robert and Roy (1990), however, concluded that D could not be inferred from the exponent of the length–area relationship. They found that the value of the exponent may reflect scaling relationships and the effects of cartographic generalization, as well as fractal characteristics of the network.

Tarboton et al. (1988) measured D from digital elevation models of stream channel networks using two different methods. Both methods were based on the slope of plots of (1) apparent length (that is, estimated ΣL) versus size of the "ruler" used to measure length, or (2) the number of cells intersected by a channel against cell (pixel) size. Two distinct regions of self-similarity were found. In the self-similarity region where these slopes were approximately unity for the length measurements or about 2 for the box-counting method, Tarboton et al. (1988) determined that $D = 2$ for the network, as Mandelbrot speculated.

La Barbera and Rosso (1989) estimated D from morphometric data and found fractal dimensions to vary from 1.5 to 2, with typical values of about 1.6 or 1.7. These results were interpreted in the context of Mandelbrot's (1982) claim that a drainage network must penetrate everywhere to drain an area thoroughly, and that following both banks of the network would, in the limit, be a plane-filling motion. La Barbera and Rosso (1989) consider this a limiting case, with most actual values to be somewhat less than the space-filling value of $D = 2$, while Tarboton et al. (1988) contend that networks are spacefilling.

RELATIONSHIPS TO CLASSICAL FLUVIAL MORPHOMETRY

All the studies cited above relate the fractal dimension of the mainstream channel or the channel network to classical fluvial morphometry. The purpose of this linkage has been to provide theoretical support for the implications of the fractal measurements (Tarboton et al., 1988; La Barbera and Rosso, 1989; Hjelmfelt 1988) or to develop and test methods for estimating D for channel networks (Seiler, 1986; La Barbera and Rosso, 1989; Robert and Roy, 1990). The relationships between river network fractal dimensions and traditional fluvial morphometry have involved the relationships between mainstream or total channel length and basin area as described above and Horton's (1945) bifurcation and length ratios. Robert and Roy (1990) have already shown that the exponent of the length–area relationship is not a reliable means for estimating D. Similar problems exist in those methods based on the Horton ratios, as addressed below.

Estimating D from Morphometric Relationships

The most common and direct way to measure the fractal dimension of channel networks or other linear features is to measure their length at least twice using two separate measurement increments or unit "rulers." If statistical self-similarity is assumed then D can be estimated from

$$D = \frac{\ln\left(\frac{n'}{n''}\right)}{\ln\left(\frac{e''}{e'}\right)} \tag{7.3}$$

where e' and e'' are two separate sampling intervals or ruler lengths and n' and n'' the number of steps or increments needed to measure the entire length, such that length at each measurement scale is $n'e'$ or $n''e''$. Variations of Equation 7.3 are used to estimate D, depending on the measurement techniques.

Tarboton et al. (1988) directly measured D from digital elevation models. Unless a digital representation of the network and the appropriate software are available, calculation of D can be quite tedious. Robert and Roy (1990), for example, had to rely on a curvimeter, overlay grid, and manual cartometry to derive their data. Thus, methods for estimating D from more easily obtained and readily available parameters such as Horton ratios can considerably simplify the estimation of the fractal dimension of channel networks. Both Tarboton et al. (1988) and La Barbera and Rosso (1989) estimate D using a method based on the bifurcation and stream-length ratios. La Barbera and Rosso (1989) derive the fractal estimate by introducing the concept of a measurement function relating the log-transformed values of Z, the measured length of the network, to the unit e used for measurement.

The bifurcation ratio R_b and stream-length ratio R_l (after Horton, 1945, and using the ordering system of Strahler, 1952) are given by

$$R_b = \frac{n(u)}{n(u+1)} \tag{7.4}$$

and

$$R_l = \frac{l(u+1)}{l(u)} \tag{7.5}$$

where $n(u + 1)$ and $n(u)$ are the number of stream segments of order $u + 1$ and u, and $l(u + 1)$ and $l(u)$ are the mean lengths of segments of order $u + 1$ and u. In deriving estimates of D based on the Horton ratios, R_b and R_l are taken as characteristic values for an entire network and are thus assumed constant over all stream orders.

By assuming that the Horton ratios hold exactly at all scales and that the measure of the total length of streams in the network is

$$L_t \sim e^{*\,1 - (\log R_b \log R_l)} \tag{7.6}$$

where e^* is the resolution with which first-order streams are detected. Both La Barbera and Rosso (1989) and Tarboton et al. (1988) independently derived

$$D = \frac{\log R_b}{\log R_l}, \quad \text{if } R_b > R_l \tag{7.7a}$$

and

$$D = 1, \quad \text{if } R_b \le R_l \tag{7.7b}$$

La Barbera and Rosso (1989) also note that Equations 7.7a and b can be derived from intuitive arguments under the assumption that ordered drainage networks described by Horton's laws are self-similar. Tarboton et al. (1990) later extended the argument by assuming that e^*, a characteristic scale length, is itself fractal. This gives $\log R_b/\log R_l = D/d$, where d is the dimension of e^*. However, La Barbera and Rosso (1990) show that introduction of a fractal measure for e^* is inconsistent with the assumption of constant Horton ratios, and the d values proposed by Tarboton et al. (1990) are in any case close to unity. Estimated D values from Equation 7.7a for nine networks analyzed by Tarboton et al. (1988) ranged from 1.7 to 2.5. La Barbera and Rosso (1989) estimated D values ranging from about 1.2 to slightly more than 2, with typical values of 1.6 to 1.7. These results greater than 2 are problematic, because 2 is the upper limit for the fractal dimension of linear features. These values can be explained by the faulty assumption that the bifurcation and length ratios hold over all scales.

The assumption of constant bifurcation and length ratios implies (in a manner consistent with Horton, 1945) that the mean length of streams l_u of order u increases as a geometric series such that the individual stream-length ratios λ_u are given by

$$\lambda_u = \frac{l_{u+1}}{l_u} \approx R_l, \quad \text{for } u \ge 1 \tag{7.8}$$

Note that λ_u applies only to the length ratio between two specific orders, while R_l is an average that applies over the entire network to any pair of orders u and $u + 1$. If $\log R_l \le \log R_b/2$, then $D \le 2$, as should be the case. Otherwise, D estimated by Equations 7.7 will exceed the theoretical maximum. If channel networks are topologically random and lengths of interior channel links are independent random variables drawn from a single population, it can be deduced that, for a sufficiently large number of stream segments, $\lambda_u = B_u/2$, where B_u is the bifurcation ratio between two specific orders u and $u + 1$ (as opposed to R_b, the network mean bifurcation ratio) (Smart, 1972). A number of studies reviewed by Abrahams (1984) have examined the distribution of stream link lengths with order, with variable results. Some have substantiated the independence of link lengths, other studies have shown that lengths increase with order, and still others have found that lengths decrease with order (Abrahams, 1984). For the case where l_u increases with increasing u, l_{u+1} would be much larger than l_u, resulting in $\log R_l > \log R_b/2$ and unrealistically high estimates of D.

Equations 7.7a and b thus provide a means for interpreting the fractal dimension of stream networks in the context of classical fluvial morphometry. However, Equations 7.7a and b are not a reliable means for estimating D unless the distribution of stream link lengths is known to be independent of stream order. It should also be understood that the Horton ratios themselves have little physical significance. They are mathematical abstractions that would apply to any branching network subject to the ordering scheme and have not been found to be clearly correlated to geomorphological or hydrological phenomena (Richards, 1982, p. 35).

Application to Southern Appalachian Basins

An example of the problems associated with estimating D from the Horton ratios and the difficulties in interpreting the results is provided using data from 50 third-order drainage basins along a 560-km transect from the Piedmont of North Carolina to the Highland Rim of Tennessee (McCluskey, 1980; see Figure 7.1). One third-order basin was selected from each U.S. Geological Survey 1 : 24,000 topographic map along the transect. Basins were thus sampled in four lithologically and topographically distinct physiographic provinces: the Piedmont (19 basins), Blue Ridge Mountains (11 basins), Ridge and Valley (12 basins), and the Appalachian Plateau (8 basins) (Figure 7.2). These physiographic provinces and the characteristics of the drainage systems within them are described by Fenneman (1938). The specific study basins are described by McCluskey (1980), and a typical study basin is shown in Figure 7.3.

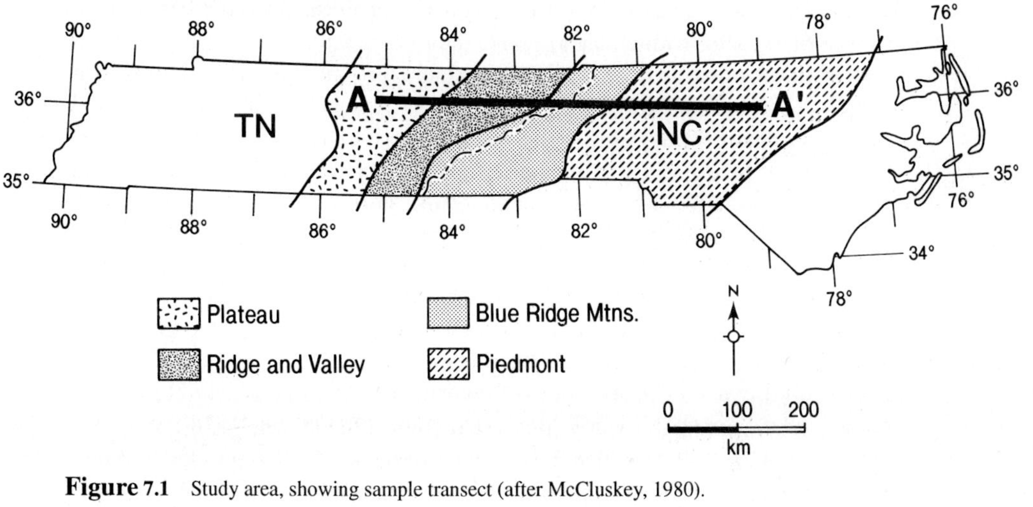

Figure 7.1 Study area, showing sample transect (after McCluskey, 1980).

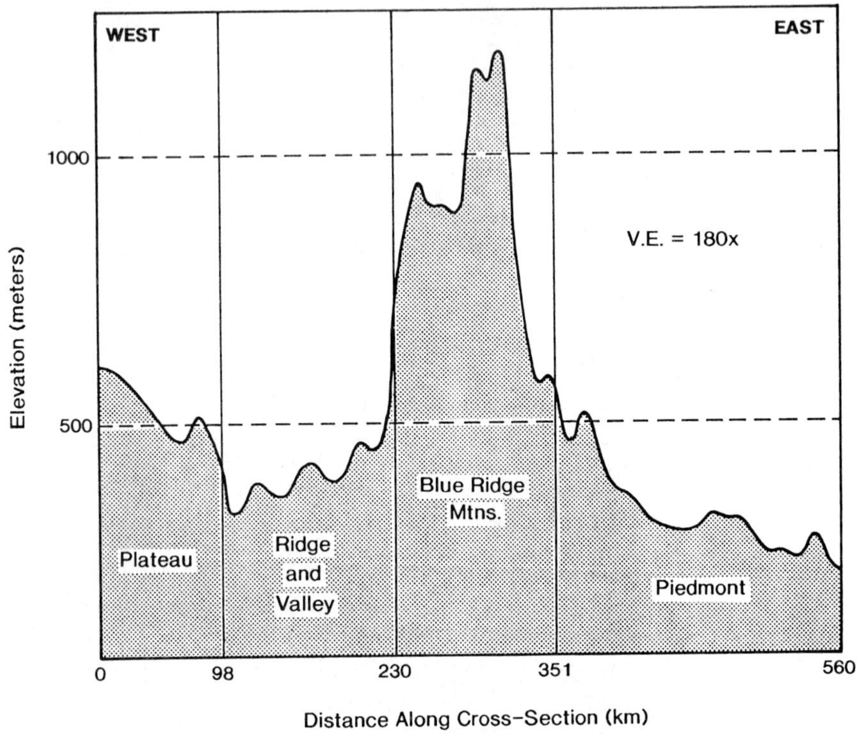

Figure 7.2 Topographic cross section along sample transect with 180× vertical exaggeration (after McCluskey, 1980).

Morphometric data were measured from the topographic maps to derive a number of standard parameters, including the network bifurcation and stream-length ratios. Measurement methods were standard procedures of map analysis (see both description and criticism by Mark, 1983) which clearly have their shortcomings, but which are typical of the type of information that is or would be available for estimating fractal dimensions of river networks. D was estimated for each basin using Equation 7.7a, as $R_b > R_l$ in every case. Because Equation 7.7a yields negative D values in some basins, the mean D for each physiographic province was computed using the mean bifurcation and length ratios for the basins, rather than by taking the average of the individual D values. The negative values were obtained because length ratios, although normally greater than unity, were less than 1 for some individual basins. Results are shown in Table 7.1, while the calculations of related morphometric variables are included in Appendix 7.1.

Mean fractal dimensions for the four provinces were greater than the theoretical maximum of 2. As a value approaching 2 implies an unconstrained network (see below), these values are not reasonable, even allowing for measurement error. Structural control of drainage networks in the study area is common (Fenneman, 1938), suggesting that D should be significantly less than 2. But in the Ridge and Valley province, where structural control in the form of trellis network patterns and subsequent streams is especially evident, half the basins had $D > 2$! In theory, and in the aggregate in real basins, stream-length ratios are greater than unity. But occasionally, and in nearly one-third of the study basins, $R_l < 1$. This creates problems for estimating D, as it produces meaningless and uninterpretable negative fractal dimensions. This was the case for 30% of the study basins. Overall, estimated D values were within the theoretical range of $1 < D < 2$ for 36% of the basins and greater than 2 for 34%.

Results in this application illustrate the problems associated with estimating D from the Horton ratios.

TABLE 7.1 Estimated Fractal Dimension (D) Values in the Southern Appalachians

	Physiographic Province			
	Piedmont	Blue Ridge Mountains	Ridge and Valley	Appalachian Plateau
Mean R_b	3.79	4.23	4.13	4.39
Mean R_l	1.68	2.01	1.85	1.33
Mean D	2.57	2.07	2.31	5.19
Number of basins	19	11	12	8
Basins where $D < 0$	5	3	2	5
Basins where $1 < D < 2$	6	6	4	2
Basins where $D > 2$	8	2	6	1

Source: McCluskey, 1980; author.

PHYSICAL INTERPRETATIONS OF THE FRACTAL DIMENSION

Mathematically, the fractal dimension of linear features such as a stream channel network is limited to the range $1 < D < 2$. These theoretical bounds have a physical interpretation. If a network completely drains a basin and if every drainage path could be followed, then (in theory) walking both banks would be a plane-filling action and $D = 2$. Conversely, if the channels and flow paths are confined to a single rectifiable linear pattern, $D = 1$. Thus, the more thoroughly the drainage pattern penetrates the area, the closer D is to 2. The more limited the drainage pattern is by geological controls, the closer the value is to 1.

Recalling from Equation 7.2 the relationship between total stream length and drainage area, $\Sigma L = cA^b = cA^{D/2}$. Numerous empirical studies have confirmed that total stream length varies as a power function of drainage area, although the nature and cause of the relationship have been widely debated (Pethick,1975, 1978; Gardiner et al., 1977; Church and Mark, 1980). Typical b values have been found to fall in the 0.6 to 0.8 range.

A theoretical range for b can be established from consideration of the role of environmental constraints on drainage density. At one extreme, consider the case where drainage density is unconstrained and the network has had sufficient time to attain an equilibrium. In this case, drainage density is constant and $b = 1$. At the opposite extreme, consider a case where drainage density and stream length are constrained by environmental controls (lithology and climate) and L is isometrically related to A. In this case, dimensional analysis gives $b = 0.5$. Any allometry associated with variation of environmental constraints with basin size would be positive, as observed in the trend toward basin elongation, giving $b \geq 0.5$. In this way the theoretical range of b is established: $0.5 < b < 1.0$. Church and Mark (1980) presented a similar argument. They also showed that a limiting value of $b = 1$ can be derived from the random hypothesis of network topology. The random hypothesis or random topology model holds that all topologically distinct channel networks with the same number of first-order streams are equally likely, and that in the absence of environmental constraints, a natural population of channel networks will be topologically random (Shreve, 1966).

Using the theoretical range of b to derive the theoretical range of D, we obtain $1 \leq D \leq 2$. This suggests that the interpretation of the fractal dimension of channel networks is

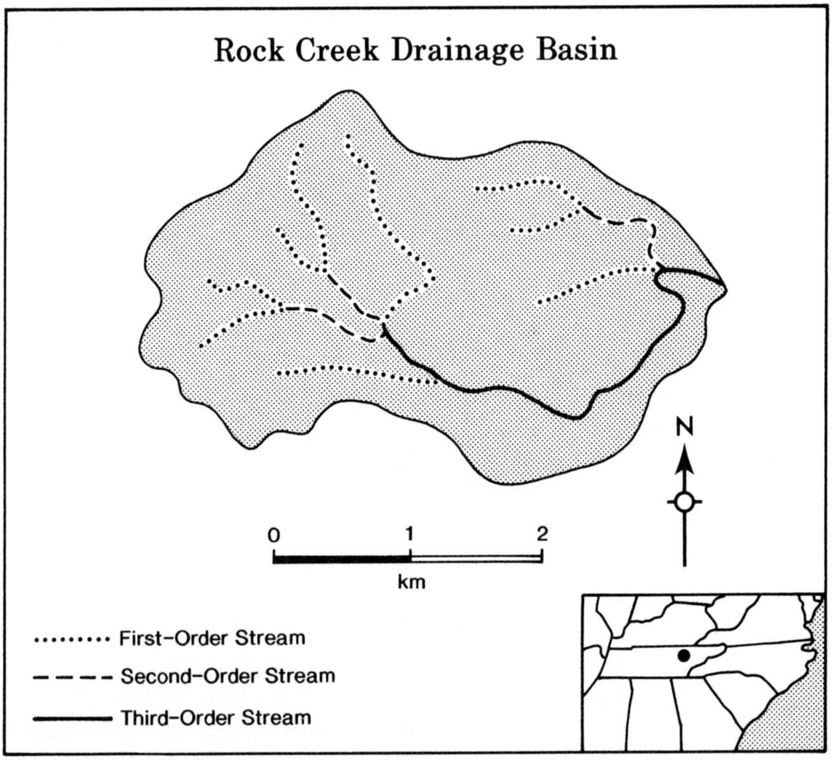

Figure 7.3 Typical study basin, indicating Strahler's method of stream ordering (after McCluskey, 1980). The Horton length ratio for this basin is 0.85, resulting in a negative D value estimated from Equation 7.7a.

generally consistent with classical morphometric relationships. However, the derivation of the range of D from the total stream length–drainage area exponent also raises the question of the distinction between space-filling and unconstrained networks.

If self-similarity prevails at all scales and D is constant at all scales, $D = 2$ must imply a space-filling network, as delineation of the channels and flow paths in greater and greater detail would eventually fill a planar surface. However, it is known that self-similarity does not prevail over all scales (as recognized by Mandelbrot) and that D is not stable over a range of spatial scales for a variety of geomorphic phenomena (Anderle and Abrahams, 1989; Burrough, 1983; Mark and Aronson, 1984; Tarboton et al., 1988, 1989; Robert and Roy, 1990). In fact, Tarboton et al. (1988) data show D approaching 1 at more detailed spatial scales, indicating that $D \approx 2$ does not hold in the region where it would be expected to if the network were indeed space filling.

The concept of a space-filling drainage network is also inconsistent with existing knowledge of hydrologic processes. It is increasingly recognized that water flow at scales ranging from the largest basins to individual soil pores follows preferential paths and that not all areas within a watershed or other hydrologic unit contribute runoff to stream systems (Beven 1987). These general principles imply that space-filling drainage systems are rare, if they exist at all.

The space-filling network is antithetical to the even more widely accepted variable source area concept in hydrology (reviewed by Troendle, 1985). This concept recognizes the fact that not all areas within a watershed contribute runoff to the stream system, that contributing areas do not produce runoff at equal rates or quantities, and that the runoff-contributing areas expand and contract temporally in response to subsurface moisture fluxes. The argument that networks must be space filling if an area is to be completely drained (Tarboton et al., 1988; Mandelbrot, 1982) is true as an abstraction, but of little meaning in reality since few areas are "completely drained" by the surface hydrologic system.

For these reasons, it is recommended that a reliably measured D value near 2 for channel networks should be interpreted as an indication of a geologically unconstrained pattern, rather than a space-filling one, recognizing that unconstrained drainage is not necessarily (or even usually) space filling. It follows that the interpretation of D between the range of 1 and 2 should be based on the degree of geological constraints at a particular range of spatial scales, rather than on the space-filling properties of the network.

Fractals and the Random Model

Conceptual models of channel networks occupy a continuum from a simple, completely determined system that is entirely predictable and has no elements of randomness to the completely stochastic random model (Abrahams, 1984). The random model of network topology holds that, in the absence of climatic, structural, or topographic controls, network topology and geometry evolve at random (Shreve, 1966). Because the random model gives a limiting value for the total stream length versus area exponent of $b = 1$ (Church and Mark, 1980), the random hypothesis also implies that $D = 2$.

Thus, the analysis of D for stream networks raises the same questions and provides the same information as tests of the random hypothesis (see review by Abrahams, 1984). Both the random hypothesis and $D = 2$ provide reference standards against which real networks can be judged. Because the random model of network typology has a long history of successful use (despite its shortcomings) and well-developed theory, it would seem in one sense to be a superior reference standard against which real networks can be judged. On the other hand, the increasing generation of fractal measurements of terrain from maps, digital elevation data, and remotely sensed data provides not only a ready source of morphometric information, but a link between fluvial geomorphology and contemporary quantitative spatial analysis. It would seem that the choice between the traditional conceptualization of the random model of network topology and the fractal dimension as a reference standard is mostly a matter of taste.

As with any morphometric parameter, we should be careful as to the extent to which the fractal dimension's value is ascribed to fundamental physical properties (that is, factors that control force–resistance relationships and fluxes of energy and matter). The fractal dimension can be said to derive from fundamental physical properties of fluvial systems only to the extent that randomness is an inherent component of fluvial process–response relationships. Topological or geometric morphometric parameters do not necessarily have direct interpretation in terms of processes or mechanics.

CONCLUSION

River networks have fractal characteristics in that their topology is a spatially complex pattern where the Hausdorff–Besicovitch dimension exceeds the topological dimension. Fractals are linked in theory to elements of classical fluvial geomorphology: the power-function relationships between stream length and basin area and the Horton bifurcation and stream-length ratios.

Because direct measurements of D can be difficult and tedious, estimates based on length–area exponents and Horton ratios have been developed. Robert and Roy (1990) showed that D cannot be reliably estimated from stream length–drainage area relationships. It is shown here that D cannot be reliably estimated from Horton ratios, either, unless detailed information on the distribution of stream lengths over a range of stream orders is available.

It has been suggested that the fractal dimension of stream networks represents their space-filling properties and that many networks are space filling, implying complete drainage. This is inconsistent with knowledge of hydrologic processes and also depends on assumptions of self-similarity and stable D values over all scales, which often do not hold. A more reasonable interpretation of D is that it reflects the degree of geological constraints on network evolution over a particular range of spatial scales.

Fractal dimensions of river networks provide identical information and produce identical questions to those produced by the venerable random model of network topology. Fractals have a much smaller body of supporting geomorphological theory and literature, but much greater utility than the random model beyond fluvial geomorphology. The choice of the classical random model or the fractal dimension as a conceptual standard for comparison to real networks is primarily a matter of personal taste or the goals of a specific research endeavor.

This analysis supports Goodchild and Mark's (1987) conclusion that fractals are useful primarily as cartographic tools and morphometric parameters, but do not (any more or less than morphometric parameters in general) have physical significance. Fractals are a relatively new and elegant concept and have proven utility for quantifying the spatial complexity of landforms, modeling terrain, and identifying scale thresholds. Because of this, there may be a tendency for geomorphologists to view the fractal dimension as itself somehow being an inherent property of geomorphic systems, rather than as a reflection of the forms that result from the process–response relationships of the system. It is therefore prudent to be mindful of the limitations of fractals (or other morphometric parameters) and of the need to link forms with processes.

ACKNOWLEDGMENTS

Thanks are extended to Jim McCluskey for allowing use of his data from the Southern Appalachians. Critical reviews of an earlier draft by the editors resulted in significant improvements in this paper.

REFERENCES

Abrahams, A. D. 1984. Channel networks: A geomorphological perspective. *Water Resources Research* 20:161–188.

Anderle, R., and Abrahams, A. D. 1989. Fractal techniques and the surface roughness of talus slopes. *Earth Surface Processes and Landforms* 14:197–209.

Beven, K. 1987. Towards a new paradigm in hydrology. In *Water for the future: Hydrology in perspective*, J. C. Rodda and N. C. Matalas, eds. pp. 393–403. Wallingford, UK: International Association of Hydrological Sciences Publication 164.

Burrough, P. A. 1983. Multiscale sources of spatial variation in soil. I. The application of fractal concepts to nested levels of soil variation. II. A non-Brownian fractal model and its application to soil survey. *Journal of Soil Science* 34:599–620.

Church, M., and Mark, D. M. 1980. On size and scale in geomorphology. *Progress in Physical Geography* 4:342–390.

Fenneman, N. M. 1938. *Physiography of the Eastern United States*. New York: McGraw-Hill Book Co.

Gardiner, V., Gregory, K. J., and Walling, D. E. 1977. Further notes of the drainage density-basin area relationships. *Area* 9:117–121.

Goodchild, M. F., and Mark, D. M. 1987. The fractal nature of geographic phenomena. *Annals of the Association of American Geographers* 77:265–278.

Hack, J. T. 1957. Studies of longitudinal stream profiles in Virginia and Maryland. *U.S. Geological Survey Professional Paper* 294-B:45–94.

Hjelmfelt, A. T. 1988. Fractals and the river length–catchment area ratio. *Water Resources Bulletin* 24:455–459.

Horton, R. E. 1945. Erosional development of streams and their drainage basins: A hydrophysical approach to quantitative morphology. *Geological Society of America Bulletin* 56:275–370.

La Barbera, P., and Rosso, R. 1989. On the fractal dimension of river networks. *Water Resources Research* 25:735–741.

———, and ———. 1990. Reply. *Water Resources Research* 26:2245–2248.

Mandelbrot, B. B. 1982. *The Fractal Geometry of Nature*. San Francisco: W.H. Freeman & Co.

Mark, D. M. 1983. Relations between field-surveyed channel networks and map-based geomorphometric measures, Inez, Kentucky. *Annals of the Association of American Geographers* 73:358–372.

———, and Aronson, P. B. 1984. Scale-dependent fractal dimensions of topographic surfaces: An empirical investigation, with applications in geomorphology and computer mapping. *Mathematical Geology* 16:671–683.

McCluskey, J. M. 1980. Selected morphometric characteristics and their relationship to the genetic provinces of the Southern Appalachian Highlands. Greenville, N.C.: M.A. thesis, Department of Geography and Planning, East Carolina University.

Pethick, J. S. 1975. A note on the drainage density–basin area relationship. *Area* 7:217–222.

———. 1978. Comment: Drainage density–basin area relationship. *Area* 10:349–350.

Richards, K. S. 1982. *Rivers: Form and Process in Alluvial Channels*. London: Methuen.

Robert, A., and Roy, A. 1990. On the fractal interpretation of the mainstream length drainage area relationship. *Water Resources Research* 26:839–842.

Seiler, F. A. 1986. Use of fractals to estimate environmental dilution factors in river basins. *Risk Analysis* 6:15–25.

Shreve, R. L. 1966. Statistical law of stream numbers. *Journal of Geology* 74:17–37.

Smart, J. S. 1972. Quantitative characterization of channel network structure. *Water Resources Research* 8:1487–1496.

Strahler, A. N. 1952. Hypsometric (area–altitude) analysis of erosional topography. *Geological Society of America Bulletin* 63:1117–1142.

Tarboton, D.G., Bras, R.L., and Rodriguez-Iturbe, I. 1988. The fractal nature of river networks. *Water Resources Research* 24:1317–1322.

———,———, and ———. 1989. Scaling and elevation in river networks. *Water Resources Research* 25:2037–2051.

———, ———, and ———. 1990. Comment on "On the fractal dimension of stream networks." *Water Resources Research* 26:2243–2244.

Troendle, C.A. 1985. Variable source area models. In *Hydrological Forecasting*, M. G. Anderson, and T. P. Burt., eds. pp. 347–403. New York: John Wiley & Sons.

APPENDIX 7.1 Summary of Morphometric Variables

Quadrangle	Drainage Basin	R_b	R_l	R_e	D_d	H	E	Θ_g	Area (km^2)
Piedmont:									
Hillsborough	Little Creek	3.50	1.98	0.90	2.53	55	0.67	5.22	3.52
Efland	No name	6.00	2.45	0.94	3.19	52	0.54	4.52	4.86
Mebane	No name	2.00	1.39	0.76	4.86	58	0.60	6.08	7.82
Burlington	No name	3.00	2.24	0.77	2.04	49	0.54	6.36	3.28
Gibsonville	No name	3.00	1.55	0.77	1.46	40	0.42	5.68	2.35
McLeansville	No name	4.75	1.62	0.86	6.91	55	0.62	5.83	11.12
Greensboro	No name	5.50	4.63	0.97	3.94	55	0.50	8.24	6.34
Guilford	No name	4.00	2.98	0.76	3.49	58	0.59	6.38	5.02
Kernersville	No name	4.50	0.85	0.85	3.33	46	0.49	7.92	5.36
Winston-Salem East	Sokas Creek	5.00	0.28	0.61	4.83	64	0.51	9.12	7.77
Winston-Salem West	Side Branch	2.50	0.63	0.79	1.02	55	0.48	9.17	1.62
Clemmons	Ellison Branch	4.33	1.39	0.79	5.11	83	0.46	11.04	8.22
Farmington	No name	6.33	2.69	0.70	7.87	80	0.45	8.02	12.66
Lone Hickory	Fisher Creek	3.00	0.92	0.70	3.02	82	0.55	13.37	4.86
Brooks Crossroads	Walkers Branch	2.50	1.40	0.71	2.49	85	0.55	10.06	4.01
Osbornville	Osborn Creek	3.25	1.24	0.84	5.29	158	0.37	9.34	8.51
Guilreath	No name	3.00	1.09	0.83	4.00	347	0.53	31.90	6.44
Moravian Falls	Moravian Creek	2.60	0.76	0.81	6.35	469	0.32	31.58	10.22
Boomer	Little Warrier Creek	3.25	1.74	0.85	6.00	246	0.29	20.07	9.65
Mean x		3.79	1.68	0.80	4.09	112	0.50	11.05	
s		1.26	1.01	0.09	1.87	117	0.10	8.10	

Quadrangle	Drainage Basin	R_b	R_l	R_e	D_d	H	E	Θ_g	Area (km^2)
Blue Ridge Mountains:									
Grandin	No name	4.50	1.81	0.74	1.78	472	0.48	37.98	9.70
Buffalo Cove	Green Park Branch	2.00	1.49	0.85	1.95	427	0.57	47.50	3.11
Globe	Johns River	6.50	2.84	0.84	2.62	658	0.43	45.74	7.21
Grandfather Mtn.	Wilson Creek	2.50	0.69	0.77	2.19	963	0.47	41.60	5.91
Newland	No name	3.50	0.94	0.94	1.50	372	0.42	20.93	6.45
Carvers Gap	Hanson Creek	3.00	0.73	0.92	2.01	785	0.39	41.28	8.74
Bakersville	Sweet Creek	4.50	2.21	0.98	2.26	399	0.33	33.12	5.21
Huntdale	Brummett Creek	6.33	2.91	0.82	4.28	561	0.38	65.42	3.23
Chestoa	No name	5.00	3.44	0.74	5.42	732	0.50	21.88	1.71
Flag Pond	Big Branch	5.20	1.62	0.74	5.28	764	0.47	50.00	3.09
Greystone	No name	3.50	3.39	0.70	3.77	524	0.52	50.00	1.51
Mean x		4.23	2.01	0.82	3.01	605	0.45	42.31	
s		1.48	1.02	0.09	1.43	190	0.07	11.59	
Ridge and Valley:									
Davy Crockett Lake	Flag Branch	4.50	3.58	0.78	3.21	259	0.27	17.71	2.74
Cedar Creek	Gregg Creek	7.00	1.71	0.82	3.45	116	0.29	11.41	3.77
Rankin	McCowan Creek	6.00	3.15	0.95	3.67	125	0.35	32.22	1.75
Parrotsville	Sinking Creek	3.17	1.20	0.84	5.10	88	0.53	41.25	4.26
White Pine	Leadvale Creek	3.50	1.82	0.89	2.24	69	0.46	10.42	3.67
Jefferson City	Goose Creek	3.00	1.23	0.88	2.12	139	0.52	15.31	4.05
New Market	Dance Branch	2.00	1.27	0.74	3.31	163	0.43	11.50	2.40
Mascsot	Clift Creek	7.00	1.86	0.70	2.45	126	0.23	17.04	6.31
John Sevier	Legg Creek	3.00	0.68	0.87	2.37	200	0.36	15.36	5.20
Fountain City	Cox Creek	5.00	3.24	0.66	1.71	152	0.30	14.90	5.97
Powell	Patt Branch	2.33	1.60	0.95	2.89	119	0.30	13.27	2.65

Quadrangle	Drainage Basin	R_b	R_l	R_e	D_d	H	E	θ_g	Area (km^2)
Clinton	No name	3.00	0.86	0.89	1.95	119	0.34	14.79	5.28
Mean x		4.13	1.85	0.83	2.87	140	0.37	17.93	
s		1.75	0.96	0.09	0.95	51	0.10	9.26	
Appalachian Plateau:									
Windrock	Hoskins Creek	5.00	0.45	0.75	3.03	600	0.27	45.63	5.28
Petros	Middle Creek	2.00	0.71	0.80	2.92	506	0.44	37.08	2.44
Camp Austin	Hall Branch	6.00	0.91	0.95	3.70	521	0.35	32.05	4.41
Lancing	Rock Creek	3.00	0.85	0.85	2.93	323	0.54	27.38	4.97
Hebbertsburg	Hawn Spring Branch	2.50	2.35	0.94	2.46	119	0.66	20.00	1.77
Fox Creek	South Fork	6.00	2.46	0.71	2.69	110	0.58	9.80	5.26
Isoline	Scott Creek	6.00	2.07	0.86	2.09	165	0.75	12.03	12.08
Campbell Junction	Clear Creek	4.60	0.85	0.80	2.90	84	0.64	11.13	10.15
Mean x		4.39	1.33	0.83	2.84	304	0.53	24.39	
s		1.67	0.82	0.08	0.47	212	0.16	13.31	

R_b = bifuncation ratio
R_l = length ratio
R_e = elongation ratio
D_d = drainage density
H = basin relief
E = elevation/relief ratio
θ_g = mean ground slope (%)

CHAPTER 8

NONLINEAR VARIABILITY OF LANDSCAPE TOPOGRAPHY: MULTIFRACTAL ANALYSIS AND SIMULATION

Daniel Lavallée
Earth Science Research Group, University of California, Santa Barbara

Shaun Lovejoy
Physics Department, McGill University, Montreal, Canada

Daniel Schertzer
Laboratoire de Météorologie Dynamique, Université Pierre et Marie Curie, Paris, France

Philippe Ladoy
Météorologie Nationale, Paris, France

ABSTRACT

We argue that geographic and geophysical fields are generally multifractal (characterized by an infinite hierarchy of fractal dimensions) and that inconsistencies are inevitable when they are forced into narrow geometric frameworks involving single fractal dimensions. As an example, we show how commonly used monofractal relationships between fractal dimensions and spectral and variogram exponents break down for multifractals. We then review some results on multifractal processes showing how they lead to universal multifractals with generators characterized by the three basic parameters H, C_1, and α. The parameter H measures the degree of nonstationarity of the process, C_1 is the codimension that characterizes the sparseness–inhomogeneity of the mean of the process, and α characterizes the degree of multifractality; $\alpha = 0$ is monofractal, $\alpha = 2$ is the maximum.

We then describe a new Double Trace Moments (DTM) technique, which is the first data-analysis technique specifically designed to estimate these parameters. We apply the technique to digital elevation maps of Deadman's Butte (50-m resolution) and French topography (1-km resolution), finding roughly compatible estimates of the multifractal index $\alpha = 1.9 \pm 0.1$ and 1.7 ± 0.1, respectively (close to its maximum). Interestingly, $C_1 \approx 0.045 \pm 0.005$ and $C_1 \approx 0.075 \pm 0.005$, respectively, which indicate that the mean of the process is not far from the space-filling value 0. We also find $H \approx 1/2$, and we use dimensional analysis to suggest possible physical explanations. The low values of C_1 show why monofractal models such as fractional Brownian motion may give reasonable results for the mean of the topography, but are poor models for the extremes; multifractal effects (that is, large sparse fluctuations) become important rapidly for high-order moments and for regions at very high (or very low) altitudes. Finally, we produce simulations of topography with the observed multifractal parameters.

INTRODUCTION

Scale Invariance and Nonlinear Variability of Geographical Data

Geophysical and geographical systems are characterized by extreme spatial and temporal variability, fractal structures spanning wide ranges of scale, and nonlinear dynamics. This ubiquitous nonlinear variability is increasingly being recognized as a central geophysical and geographical problem (see, for example, the preface and various contributions to Schertzer and Lovejoy, 1991a). The explosion of interest in chaos, fractals, and scaling, and, more recently, multifractals and multiscaling, has led to rapid advances in understanding this variability. The most promising of the new developments for geophysics and geography has undoubtedly been in scaling ideas that are now helping to provide a unified picture of atmospheric and other geophysical systems. Indeed, for some time we have proposed that scale invariance is a symmetry principle capable of unifying the earth sciences.

Unfortunately, the current honeymoon between scaling and geophysics has been dominated by restrictive monofractal ideas associated with fractal sets and the obsession with unique scaling exponents (especially the fractal dimension). Furthermore, only very special types of scale-changing operations have been considered; only self-similar or occasionally self-affine transformations have been employed. In the former case, the small scales are reduced carbon copies of the large, and in the latter case, "squashing" or "compression" along coordinate axes is also permitted. With the development of multifractals and generalized scale invariance (GSI) (Schertzer and Lovejoy, 1983, 1985), it is now clear that the types of scaling can be much broader. In particular, scale-invariant systems can now include fields rather than sets (that is, multifractal measures), and the types of possible scale transformations encompass not only differential stratification (self-affinity) but also rotation and anisotropy which varies from one place to another. A further step away from geometry toward a dynamical formalism of scale invariance was the recognition that multifractals possess stable and attractive universal generators (Schertzer and Lovejoy, 1987, 1989a, 1989b; Schertzer et al., 1988, 1991). This greatly simplifies the analysis and modeling of multifractals by replacing the infinite number of multifractal exponents by a small number of dynamically significant parameters. In the following, we illustrate some of these ideas by estimating the fundamental multifractal index α using a digital elevation model and a new data analysis technique called *double trace moments* (DTM) (Lavallée, 1991). We then use the estimated parameters to produce multifractal simulations of topography. Our results help explain both the successes and limitations of the monofractal models: although we find a nearly maximum degree of multifractality, the mean of the process is not very sparse; hence multifractal effects will not be too noticeable for low-order statistics; they will, however, dominate the behavior of the extremes, such as the high-order moments or very high/low-altitude regions.

Scaling, Fractals, and Topography

The problem of adequately conceptualizing and representing the earth's topography (for example, coastlines and rivers) has a long history. It includes Perrin's (1913) discussion of the "tangentless" coastline of Brittany and Steinhaus's (1954) discussion of the nonrectifiable nature of rivers.[1] The scaling nature of the topography has also been known for some time. In a paper entitled "A Remarkable Feature of the Earth's Topography," Venig-Meinesz (1951) found a power-law spectrum over wide ranges and correctly recognized its fundamental significance. Bell (1975) showed that Venig-Meinisz's spectral scaling of the altitude of the topography extends from planetary scales at least down to 0.15 km; using satellite data, Bills and Kobrick (1985) showed that Venus, Mars, and the moon have very similar scaling exponents to the earth. Similarly, using yardsticks of varying lengths, Richardson (1961) found various coastlines to be scaling over wide ranges and determined the relevant exponents. Mandelbrot (1967) related Richardson's exponents to fractal dimensions.

A new impetus for fractal analyses of topography came from fractional Brownian motion models of terrain (Mandelbrot 1975, 1977; Voss, 1983; Fournier et al., 1982; Clarke, 1988), which produced strikingly realistic looking models of mountains. However, in spite of the now mushrooming interest in the field, direct estimates of fractal dimensions of various topographic sets (for example, lines of constant altitude) have not been numerous. The main examples of which we are aware are Goodchild (1980), Aviles et al. (1987), Okubo and Aki (1987), and Turcotte (1989) (who also pointed out some apparent contradictions in the monofractal analyses).[2] See also Klinkenberg and Goodchild (1992) and Klinkenberg and Clarke (1992) for reviews and discussions. Other relevant direct

[1]"...The left bank of the Vistula, when measured with increased precision would furnish lengths ten, hundred, and even a thousand times as great as the length read off a school map. A statement nearly adequate to reality would be to call most arcs encountered in nature not rectifiable. This statement is contrary to the belief that not rectifiable arcs are an invention of mathematicians and that natural arcs are rectifiable: it is the opposite that is true..." (Steinhaus, 1954). In a popular book (Steinhaus, 1962), the argument is repeated and as an illustration of the point, a (fractal) Peano curve is shown.

[2]We specifically exclude variogram methods such as those reported in Mark and Aronson (1984) and Rees and Muller (1990) or spectral exponent methods such as those reported in Gilbert (1989) and Mareschal (1989), since both methods measure different scaling exponents, neither of which are fractal dimensions. They could have been used to infer fractal dimensions only if the topography had been monofractal rather than multifractal.

empirical estimates of fractal dimensions of surface features are given in De Cola (1989, 1990) (the former also includes a short discussion of multifractals). The multifractal analyses of topography are more recent and are found in Ladoy et al. (1990) and Lovejoy and Schertzer (1990a), and the corresponding multifractal simulation studies are in Wilson et al. (1991) and Sarma et al. (1990) (see Figures 8.1a and b and 8.2a and b). Other relevant empirical multifractal analyses in geophysics are the rainfield, clouds, wind and temperature, networks, and rain measurement (Lovejoy et al., 1987; Tessier et al., 1993; Schmitt et al., 1992a, b; Gabriel et al., 1988). For reviews, see Lovejoy and Schertzer (1988a, b, 1990b, 1991) and Schertzer and Lovejoy (1988, 1989a, b). Below, we show how a new multifractal analysis technique called the double trace moment (the DTM technique), when carefully applied to topographic data, enables us to estimate directly the fundamental universal multifractal parameters of topography.

Multifractals: Beyond the Geometry of Sets

A fractal is a geometrical set of points; a multifractal is a mathematical measure. When integrated–averaged over a resolution scale, this will be a field with intensity values

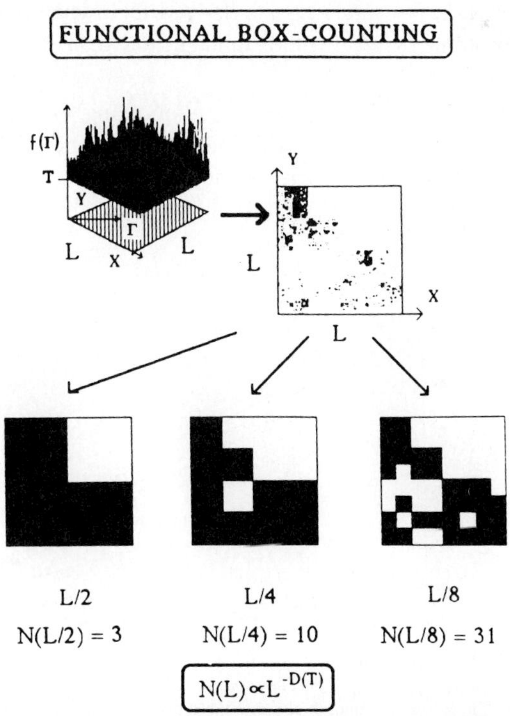

Figure 8.1(a) Schema showing how functional box counting can be used to estimate the fractal dimensions at various thresholds *T*.

defined everywhere. In geophysics and often in geography, we are rarely directly interested in sets of points. Indeed, a significant part of a map maker's skill is to find an adequate way of representing various fields such as population density, altitude, depth of ocean, temperature, and so on. To obtain sets, we typically establish thresholds and define the set of interest (for example, an *exceedance* set) to comprise all the points above a threshold, delineating them with *isolines*. Closer inspection shows that the relationship between the real world and various cartographic sets is rather indirect: the remote sensor has an intrinsic spatial–temporal resolution typically much larger than the smallest scale of the phenomenon we seek to represent. Therefore, the very first step of analysis, as we impose an inner scale to obtain a finite resolution *field* (pixel world), already involves a fundamental symmetry breaking through averaging, smoothing, or sampling (Schertzer and Lovejoy, 1991b). On the contrary, most geographers and geophysicists have forgotten this important first step, and the monofractal approach exemplified by Mandelbrot (1982) remains the general approach (see the special issue of *Pure and Applied Geophysics* on fractals in *Physics*, Vol. 132, 1989). They have taken for granted the representation of nature by geometrical fractal sets and by various mathematical functions that yield fractal sets in simple ways (for example, the exceedance sets mentioned above). In contrast, the study of nonlinear variability (in particular, that arising in turbulent cascade processes) has shown that scaling processes generically give rise to multifractal measures with very singular small-scale limits; many results derived for fractal sets and monofractal functions will not apply (This is much more fundamental than the problems discussed by Fox, 1989).

The reason that the initial averaging, smoothing, or sampling operation is a nontrivial step is that multifractals are not defined point by point, but rather they are defined on small neighborhoods of points (that is, they are nonlocal). Furthermore, they are also statistical; their variability can be "hard" in the sense that the spatial–temporal averaging

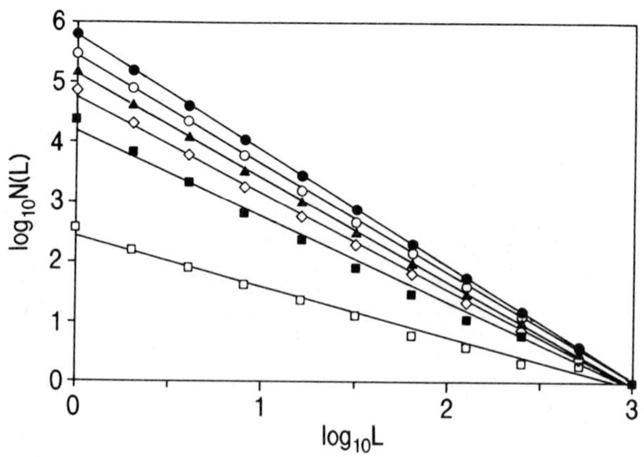

Figure 8.1(b) The results of functional box counting when applied to 1024 × 1024km topographic map of France at 1-km resolution. The lines (bottom to top) are the box-counting results for altitude thresholds decreasing by factors of 2 from 3600 m above sea level. The corresponding dimensions increase from 0.84 (at 3600 m) to 1.92 (at 28 m) (From Lovejoy and Schertzer, 1990a).

may not be enough to tame (that is, smooth out) some of the more violent fluctuations (leading for example to the phenomenon of divergence of high-order statistical moments; the "hard" behavior discussed in Schertzer and Lovejoy, 1992). Limiting our attention to fractal sets and monofractal analysis techniques is partly a result of the over emphasis on geometry at the expense of the more abstract but more powerful statistical notions of scale

Figure 8.2(a) This figure and the next illustrate the zooming technique applied in the case of the continuous multiplicative cascade with Gaussian generator ($\alpha = 2$ and $C_1 = 0.5$). The 256 × 256 window, chosen exactly in the center of this 512 × 512 resolution image, is zoomed by a factor of 2 in Figure 8.2b. The insertion of small-scale detail is done in a multiplicative way by developing the cascade processes to the appropriate scale length. These figures are reproduced from Wilson, et al., 1991.

invariance needed to deal with multifractals. Specifically, it has led to frustrating attempts to square simplistic monofractal analysis methods with multifractal data. In the following, using an example of topographic data, we wish to clear up some of these issues and show how multifractals can explain some of the apparently contradictory results reported in the literature.

Figure 8.2(b) The resulting images where the large structures are well reproduced (for example, the sharp peak in the center), but new small-scale details have appeared, characterized by the same universal scaling parameters.

FRACTAL SETS, MONOFRACTAL FUNCTIONS, AND MULTIFRACTALS

Fractal Sets, Dimensions, and Codimensions

The simplest illustration of scaling and scale invariance is to consider the dimension of a set of points (S). The intuitive definition is that the *contents* (or "volume") of the set $n(l)$ at scale l is given by

$$n(l) \propto l^{D(S)} \tag{8.1}$$

where $D(S)$ is the dimension (for example, the length, or the "volume," of a line $\propto l$, the area of a plane $\propto l^2$, and so on). The volume (that is, the measure of the set) is therefore a simple scaling (power law) function, and the dimension is important precisely because it is scale invariant (independent of l) (See Lovejoy et al., 1986a, b for the example of geophysical measuring networks). Alternatively, rather than looking at larger regions of the set [as in $n(l)$], we may consider $N(l)$, the number of boxes of size l, needed to cover it. The covering could be regarded as a spatially degraded version of the original set of scale l. Since the number of points of S covered by each box is $n(l)$, we clearly have $N(l)n(l) \approx$ constant; hence $N(l) \propto l^{-D(S)}$. The $D(S)$ obtained by this box-counting procedure can be considered as an *estimate* of the Hausdorff dimension of the set; it is also called the fractal dimension, box-counting dimension, and, earlier, capacity dimension (Kolmogorov and Tihomirov, 1959). In practice, the Hausorff dimension is often too difficult to deal with directly, even for sets studied by mathematicians. In physical applications, the primary complications that can arise if box-counting dimensions are used as approximations to Hausdorff dimensions are various log-corrections in $N(l)$. However, in specially concocted mathematical sets, significantly different dimensions can be obtained; see Falconer (1990) for a discussion.

Although a given set may be fractal, we are usually more interested in its distribution, which is generally multifractal and involves codimensions rather than dimensions. The codimension $C(S)$ of S is the scaling exponent of the fraction of the space occupied by the set, which is also equal to the probability $P(l)$ of finding the set in a given l sized box:

$$P(l) \propto l^{C(S)} \tag{8.2}$$

Since $C(S)$ determines the scaling of the probability, it is a statistical exponent that is defined *intrinsically*. This means that it is determined by the underlying (usually stochastic) generating process and requires no reference to the dimension of the space on which it is observed. In contrast, the fractal dimension depends on the intersection of the process with an observing space of dimension D, where $N(l)$ is given by the probability $P(l)$ multiplied by the total number of boxes on the observing set l^{-D}. Hence, if the observing set has a large enough dimension [$D > C(S)$] for intersection to occur (and hence for observation to be possible), then $D(S) = D - C(S)$. Since we will be generally interested in stochastic processes defined on probability spaces with infinite dimension ($D \to \infty$), the use of codimensions (which will still be finite and still characterize the sparseness of the probability space) is obligatory. The use of codimensions avoids the paradoxes of *latent* (that is, negative) fractal dimensions arising when a multifractal process is studied on low-dimensional cuts with $D < C(S)$.

Monofractal Functions, Exceedance Sets, and Isolines

Since geographers and geophysicists are interested in fields, models for scaling phenomena have usually involved scaling functions. When the latter are linear (for example, produced by summing either correlated or independent random noises), then the functions produced can generally be characterized by a single scaling exponent, and the associated sets will have unique fractal dimensions. In contrast, multifractals are produced by the (nonlinear) multiplication of elementary random noises and involve multiple scaling exponents and multiple fractal dimensions. Furthermore, they are generally not defined at mathematical points; they are not functions but rather measures. Below, we first discuss monofractal functions emphasizing their limitations.

Consider the function $h(\underline{x})$, which represents the height at a point $\underline{x}$. In this subsection, we consider the special case where h is a random scaling function with the following properties:

$$\Delta h_{l/\lambda} \stackrel{d}{=} \lambda^{-H} \Delta h_l \tag{8.3}$$

where the small-scale altitude difference is $\Delta h_{l/\lambda} = h(\underline{x}_1 + \lambda^{-1}\, \Delta\underline{x}) - h(\underline{x}_1)$ and the large-scale difference is $\Delta h_l = h(\underline{x}_2 + \Delta x) - h(\underline{x}_2)$ where $\underline{x}_1$, $\underline{x}_2$ are arbitrary, $\lambda(>1)$ is a reduction ratio, $l = |\Delta\underline{x}|$ is the separation distance, and H is the unique scaling parameter that characterizes the degree of nonstationarity of the topography. The equality $\stackrel{d}{=}$ means equality in probability distributions, $a \stackrel{d}{=} b$ if and only if $\Pr(a > q) = \Pr(b > q)$ for all q, where Pr indicates probability. This type of simple scaling[3] could also be called *scaling of the increments*. A special case in which the probability distributions of the altitude increments Δh are Gaussian is called *fractional Brownian motion* and can be obtained, for example, by passing Gaussian white noise through a power-law filter (that is, by fractional integration; normal Brownian motion is obtained by the usual nonfractional integration). This notion was introduced by Kolmogorov (1940); the expression was coined by Mandelbrot and Van Ness (1968); these are the very special processes used to make the now familiar monofractal mountains.

Given a function $h(\underline{x})$, sets can be defined in many different ways. First, consider the exceedance set $S_{\geq T}$ as the set of all points x satisfying $h(x) \geq T$ (that is, those regions whose altitude equals or exceeds the threshold T). In general, if $T_1 > T_2$, then $S_{\geq T1}$ will be a subset of $S_{\geq T2}$ and hence $D(S_{\geq T1}) \leq D(S_{\geq T2})$ (see Figures 8.1a and b for empirical confirmation of this multifractal behavior). However, for the special case of simple scaling, $S_{\geq T}$ will not be a fractal set al all, and $D(S_{\geq T}) = D$ irrespective of T. Similarly, the *graph of h, G*, is defined as the set of points $(x, h(x))$ in a $(D + 1)$-dimensional space, for example, when $D = 2$, on a

[3]This type of scaling was first introduced by Lamperti (1962) under the name "semistable." It was called "self-similarity" by Mandelbrot and Van Ness (1968). However, this name turned out to be a misnomer since the actual functions f were not self-similar but self-affine, and self similarity is a much wider concept anyway.

topographic surface. Again, in the special case of simple scaling, there is a unique codimension (and dimension) associated with G, where $C(G) = H$, and we obtain the well-known result $D(G) = 3 - H$ for Voss's (1983) and Mandelbrot's (1982) celebrated monofractal landscapes.

We can also define the *perimeter set* P_T associated with $S_{\geq T}$. P_T is the *border* set of $S_{\geq T}$ defined as the *level set* or *T-crossing set* of G with the plane $h(\underline{x}) = T$. It is the set of points such that arbitrarily small neighborhoods of $\underline{x}$ contain some points with $h(\underline{x}) < T$ and some such that $h(\underline{x}) \geq T$. $D(P_T)$ will be the dimension of the isolines on a topographic map of $h(\underline{x})$. By definition P_T is a subset of $S_{\geq T}$; hence $D(P_T) \leq D(S_{\geq T})$. In general, it will depend on the connectedness properties of $S_{\geq T}$ and increase or decrease with T only being bounded above by the decreasing function $D(S_{\geq T})$. Finally, we may define $S_{=T}$ as the set of points x such that $h(\underline{x}) = T$. Clearly, it will be a subset of P_T; hence generally, $D(P_T) \geq D(S_{=T})$. However, in simple scaling we obtain the simple result $C(P_T) = C(S_{=T}) = H$; hence isolines in simple scaling models of topography will have dimensions $D(P_T) = D(S_{=T}) = 2 - H$ independent of T.

Variograms, Spectra, and the Distribution of Islands

The relationship $C(P_T) = H$ combined with Equation 8.1 has led to a popular way of estimating $D(P_T)$. By taking qth powers of the modulus of both sides of Equation 8.3 and taking ensemble (that is, statistical) averages (indicated by $< >$), we obtain

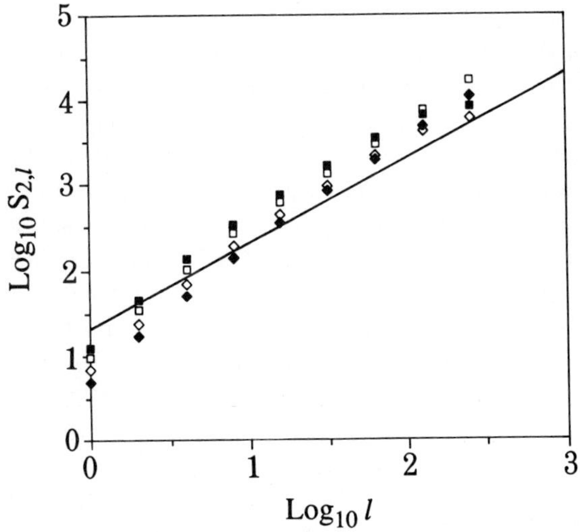

Figure 8.3 First- and second-order structure functions or variograms of the Deadman's Butte data, in both east–west (x) and north–south (y) directions, are shown along with a reference line of slope = 1 (corresponding to a spectral exponent of $\beta = 2$). The functions $S_2(\Delta x)$, $(S_1(\Delta x))^2$, $S_2(\Delta y)$, and $(S_1(\Delta y))^2$ are shown, respectively, by white squares, black diamonds, black squares, and white diamonds.

$$S_{q,l} = \langle |\Delta h_l|^q \rangle$$

$$S_{q,l/\lambda} = \lambda^{K_h(q)} S_{q,l} \tag{8.4}$$

where $K_h(q)$ is the scaling exponent of the qth-order structure function $S_{q,l}$ (also called *variogram* when $q = 2$). In the special case of simple scaling, $K_h(q)$ is linear: $K_h(q) = -Hq$; in multifractals, it will be convex (nonlinear). Unfortunately, the linearity of $K_h(q)$ is usually simply assumed and H is estimated as $K_h(2)/2$. The relation $H = C(P_T)$ is then used to estimate the fractal dimension of the isolines and graph. For an example on topography, Figure 8.3 shows the results for the Deadman's Butte digital elevation map (50-m resolution, 512 × 512 points). We already can see (Figure 8.4) that $K_h(2) \neq 2K_h(1)$; discussion of this multiple scaling is given in the following section. It involves convex functions $K_h(q)$, and exceedance sets $S_{\geq T}$ will have fractal dimensions that decrease as the threshold is increased.

Now consider the energy spectrum $E(k)$ of scaling processes.[4] Irrespective of whether or not there is simple or multiple scaling, we will have $E(k) = k^{-\beta}$, where k is the wave number. (This ignores possible log corrections or effects of anisotropic scale invariance.) This type of spectrum is relevant for the problem of surface topography since it has been known for some time (Venig-Meinesz, 1951) that $\beta \approx 2$ for the earth. Bills and Kobrick (1985) show that this behavior holds for at least five orders of magnitude in scale (from planetary scales down to at least ≈0.15 km). Figure 8.5 illustrates that it is also roughly true for the Deadman's Butte data analyzed here where we obtained $\beta \approx 1.93$.

Since the energy spectrum is the Fourier transform of a second-order moment ($q = 2$), it is easy to show that $\beta = 1 - K_h(2)$. For example, with the Deadman's Butte data (Figure 8.3a and b), we have $K_h(2) = -1.1 \pm 0.2$, $\beta = 1.93 \pm 0.03$. In the case of simple scaling, $K_h(2) = -2H$; we therefore have the two special relations $\beta = 1 + 2H = 1 + 2C(P_T)$. These relations have been widely popularized for the case of lines embedded in planes ($D = 2$), surfaces embedded in volumes ($D = 3$) in the form $D(P_T) = (5 - \beta)/2$, $D(P_T) = (7 - \beta)/2$ respectively. Since in virtually all these applications, the lines or surfaces are likely to be multifractals, these relations will not hold. This means that there is not necessarily any contradiction between estimates of fractal dimensions of lines of constant altitude and spectral or variogram exponents.

---

[4]In Fourier space, the dilation (magnification) $x \longrightarrow \lambda x$ is equivalent to $k \longrightarrow \lambda^{-1}k$. In spaces dimension 2 and higher, more general anisotropic dilations (generalized scale invariance) are required. These will give rise to the appearance of texture, which will generally give logarithmically periodic modulations of the spectra about the simple power law given here (see Pflug et al., 1993 for more details and an application to cloud texture). This could explain the small discrepancy in power law topography spectra reported by Gilbert (1989). Probably the most significant source of deviation from the expected power laws followed in the latter's spectra are due to small sample sizes, (usually, the spectrum of a single realization is used), not the ensemble averages. For precise estimates of the necessary size (the *sampling dimension*), see Lavallée et al., 1991.

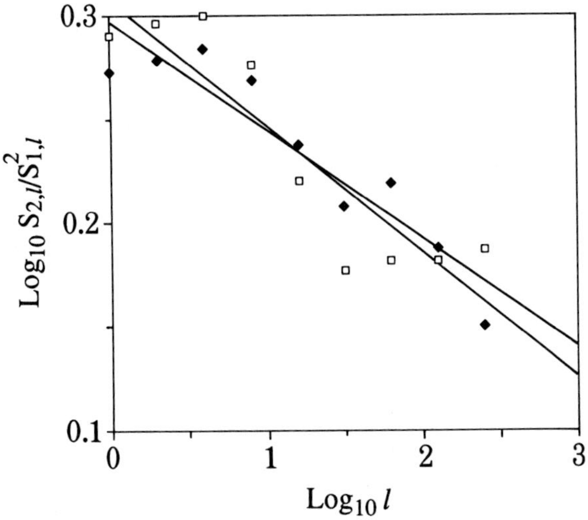

Figure 8.4 We show the log ratios of log $(S_2(\Delta x)/(S_1(\Delta x))^2)$ and log $(S_2(\Delta y)/(S_1(\Delta y))^2)$ as functions of the log lags, along with their best fit slopes, which are estimates of $K_h(2) - 2K_h(1) \approx C_1$ (see text). Although the method is not accurate due to the small sample size, the slopes are 0.055 and 0.065, which are consistent with the C_1 estimated below.

Finally, we may consider the number of simply connected areas (*islands*) that exceed the threshold T. Due to the scale invariance, we may expect the following power-law distribution:

$$Nr\,(A_T > a_T) = a_T^{-B_T}, \qquad a_T >> 1 \tag{8.5}$$

where Nr indicates the number of islands defined by the threshold T whose area A_T exceeds a fixed threshold a_T. For the earth, when T is taken as sea level, the simply connected regions are true islands; B_T has been estimated by Korcak (1938) as $B_T \approx 0.8$. Mandelbrot (1982) popularized this Korcak law and pointed out that fractional Brownian motion will yield $B_T = D(P_T)/2$ = constant (= 1 - H/2). However, as discussed in detail in Lovejoy and Schertzer (1991) for multifractals, not only is B_T a function of T, but it has no special relation to $D(P_T)$

PROPERTIES OF MULTIFRACTAL MEASURES

Multiplicative Processes and Topography

The general monofractal nature of processes produced by the addition of random noises and the general multifractal nature of the corresponding multiplicative processes have already been mentioned. The multiplicative processes that interest us here were first developed as models of turbulent cascades, initially proposed as a description of atmospheric dynamics by Richardson (1922). There are now a whole series of such models: Novikov and Stewart (1964), Yaglom (1966), Mandelbrot (1974), Frisch et al. (1978), Schertzer and Lovejoy (1983, 1987), Benzi et al. (1984), and Meneveau and Sreenivasan (1987). In these models, the energy flux ε from large to small scales is conserved; hence, it is the basic cascade quantity. Directly observable fields such as the velocity shear (Δv_l) for two points separated by distance l are related to the energy flux through dimensional arguments:

$$\Delta v_l \approx \varepsilon_l^{1/3} l^{1/3} \tag{8.6}$$

The scaling $l^{1/3}$ can be viewed as a power-law filter ($k^{-1/3}$) in Fourier space (see the section on Nonstationary processes below). In contrast, the dynamical equations responsible for the topography are not known; the best we can do at present is to speculate on the fundamental dynamical quantities. In analogy with turbulence, we may expect the observable altitude fluctuations (Δh_l) to be related to a fundamental conserved quantity φ_l and the horizontal separation of the points as

$$\Delta h_l = \varphi_l^a l^H \tag{8.7}$$

Equation 8.7 indicates that the nonstationary topography process for altitude fluctuations Δh could be related to the underlying stationary φ. In the following, we show how a highly inhomogeneous multifractal topography will generically result if orographic processes are multiplicative and we outline some of their basic properties. Indeed, elsewhere (Lovejoy, et al., 1993) we show how equation 8.7 can be viewed as a multifractal general-

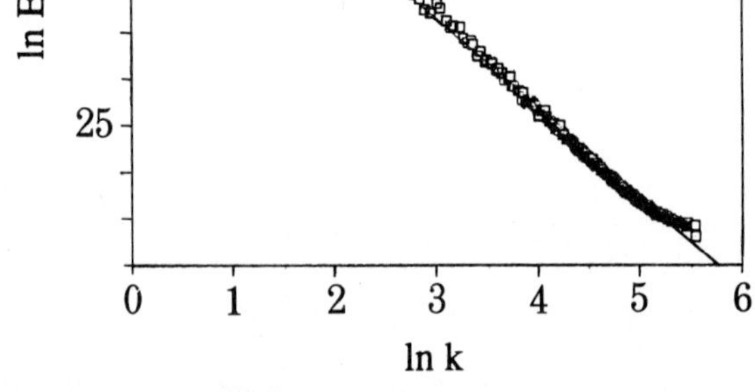

Figure 8.5 The spectral energy density as a function of the radial wave number for a grid mesh of 512 × 512, at 50-km resolution, of the Deadman's Butte data. The straight line is the best fit slope yielding exponent estimate $\beta \approx 1.93 \pm 0.03$.

ization of deterministic models of topography near ocean floor ridges, including a dimensional argument giving $H = 1/2$. Figure 8.6 illustrates an example of such a discrete multiplicative process for φ in two dimensions: a large structure of characteristic length l_0 with an initial uniform density φ_0 is broken up (via nonlinear interactions with other structures or through internal instability) into smaller substructures of characteristic length $l_1 = l_0/\lambda$ ($\lambda = 2$ is the scale ratio between two construction steps in this particular example), transferring in the process to each substructure a fraction of its density $\mu\varphi_0$. Figures 8.7 a–c show the analogous continuous cascade process for h, with the parameters corresponding to the topography as estimated below.

When the process is repeated, larger and larger values of φ appear, concentrated on smaller and smaller volumes; the result will be highly intermittent. Indeed, analysis shows that we obtain the following basic multifractal relations:

$$\varphi_\lambda \approx \lambda^\gamma, \quad \lambda > 1$$
$$Pr(\varphi_\lambda \geq \lambda^\gamma) \approx \lambda^{-c(\gamma)} \tag{8.8}$$

where φ_λ is the value of the multifractal at resolution λ, which is the ratio of the largest scale of interest to the smallest scale of the process. This indicates that the probability distribution of singularities of order higher than a value γ is related to the fraction of the space occupied by them, as determined by their codimension function $c(\gamma)$. Note that Equation 8.8 is nonlocal, that is, it is a relationship between the histograms of *incipient singularities* $\gamma_\lambda(\underline{x}) = \log \varphi_\lambda(\underline{x})/\log \lambda$ at different resolutions about the point $\underline{x}$ (it does *not* assume that a well defined limit exists at each point $\underline{x}$ when $\lambda \to \infty$), the singularities must be viewed as statistical exponents rather than point values.

Similarly, the qth order statistical moments of the field intensities will generally have a scaling exponent $K(q)$; using the definition for the statistical moments, we obtain the following relation:

$$<\varphi_\lambda^q> \propto \lambda^{K(q)} \tag{8.9}$$

(In the limit $\lambda \to \infty$, the scaling exponents are related by the following Legendre transformations: $K(q) = \max_\gamma (q\gamma - c(\gamma))$; $c(\gamma) = \max_q (q\gamma - K(q))$. The turbulent notations $c(\gamma)$ and $K(q)$ are related to the more familiar strange attractor notation (Halsey et al., 1986) by the following: $\alpha_D = (D - \gamma)$, $f_D(\alpha_D) = D - c(\gamma)$, and $\tau_D(q) = (q - 1)D - K(q)$. We have added the subscript D to emphasize that the strange attractor notation fundamentally depends on the dimension of the observing space D; it cannot be used in stochastic processes where $D \to \infty$.

The multiplicative processes discussed above are quite general; however, various other (restrictive) types of multifractals exist. Table 8.1 gives a comparison of some of their properties. Up to now, most attention has been paid to the geometric multifractals (Parisi and Frisch, 1985; these very calm multifractals are defined purely geometrically

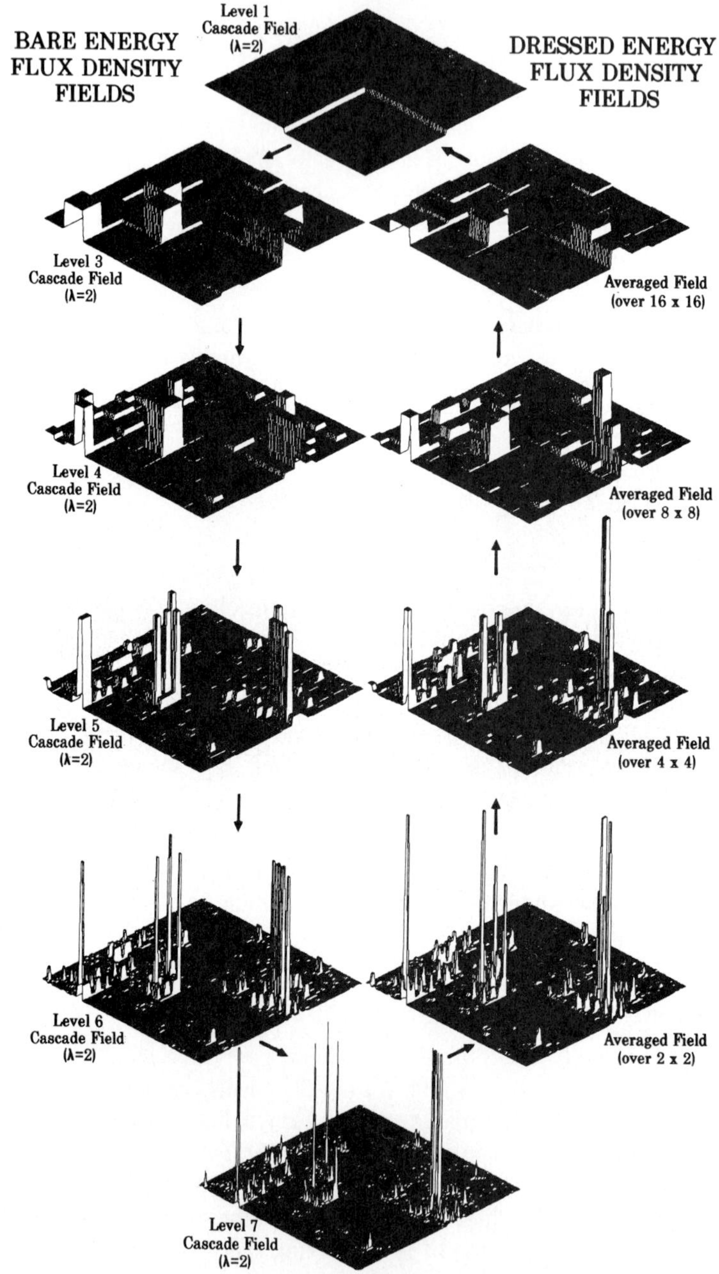

Figure 8.6 The left side shows the step by step construction of a "bare" multifractal process (the α model), starting with an initially uniform unit φ field. The vertical axis represents the density φ_λ flux to smaller scale with its ensemble average conserved ($<\varphi_\lambda> = 1$). At each step the horizontal scale is divided by 2. The developing spikes are incipient singularities of various orders (characteristic of multifractal processes). The right side shows the effect of smoothing over larger and larger scales, yielding the "dressed" quantities.

without any probability space) and to the "microcanonical" multifractals, which are also too calm for our present purposes.

Table 8.1 Classification of Multifractals According to Their Extreme Singularities

Type of Multifractal	Geometric	Microcanonical	Canonical
Types of singularities present	Calm	Calm	Calm, wild usually hard
Localized	Yes	No	No
Conservation per realization	Yes	Yes	No
Convergence of moments	All orders	All orders	Divergence of high orders

Figure 8.7 Three stages in the construction of a continuous multiplicative process for the topography field h, showing the effect of finer and finer resolution. The resolution increases from (a) 1/16th of the image size, to (b) 1/64th, to (c) 1/256th. This field is obtained by fractional integration of a continuous φ field, with parameters $H = 1/2$, $C_1 = 0.1$, and $\alpha = 1.8$.

Universal Multifractals

It has been shown that, for canonical processes, universality classes exist (for example, Schertzer and Lovejoy, 1987, 1988, 1989a; Schertzer et al., 1991; Wilson et al., 1991), which means that we may expect $K(q)$ to be given by the following functional form:

$$K(q) = \begin{cases} \dfrac{C_1}{\alpha - 1}(q^{\gamma} - q) & \alpha \neq 1 \\ C_1 q \log(q), & \alpha = 1, \quad \text{for } (\alpha < 2,\ q \geq 0) \end{cases} \tag{8.10}$$

where C_1 and α $(0 \leq \alpha \leq 2)$ are the fundamental parameters needed to characterize the processes. The Lévy index α indicates the class to which the probability distribution belongs; it tells us how far we are from monofractality: $\alpha = 0$ corresponds to the monofractal β model; $\alpha = 2$ is the maximum. The above functions are for conserved (stationary) quantities and are the multiplicative analogs of the standard central limit theorem for the addi-

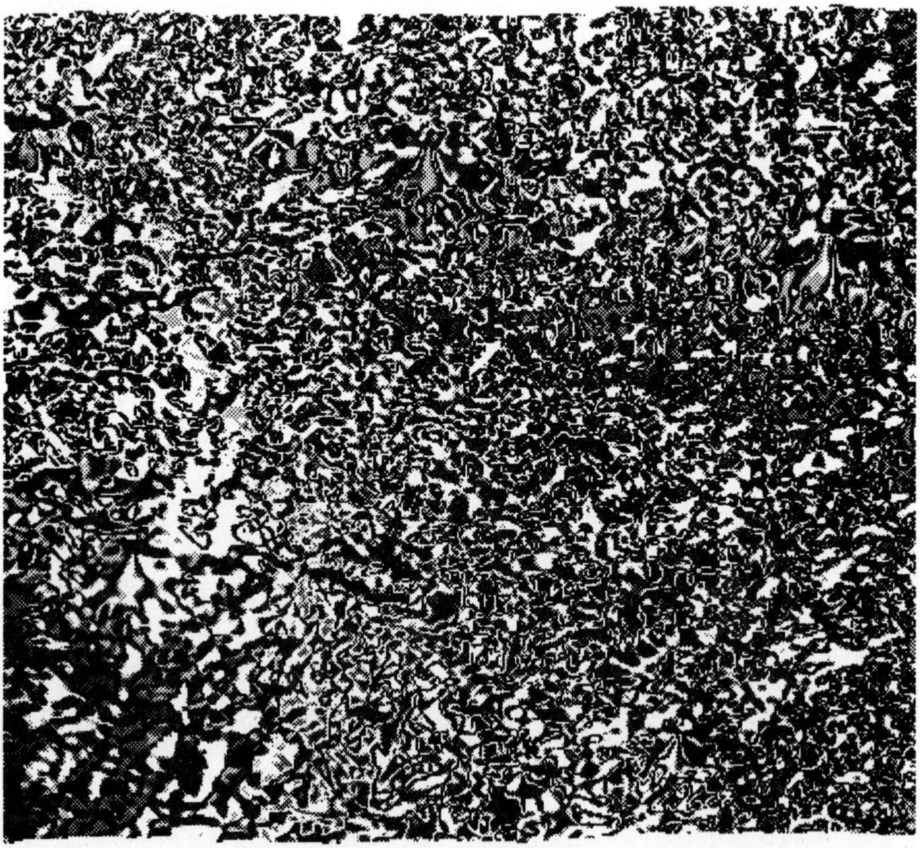

Figure 8.7(b)

tion of random variables. Closer analysis shows that there are actually five qualitatively different cases for the Lévy index α.

The case $\alpha = 2$ corresponds to log normal multifractals, $1 < \alpha < 2$ belongs to (log) Lévy processes with unbounded singularities, $\alpha = 1$ corresponds to log Cauchy multifractals, when $0 < \alpha < 1$ we have (log) Lévy processes with bounded singularities, and finally $\alpha = 0$ corresponds to the most popular and well-known monofractal β model (Novikov and Stewart, 1964; Mandelbrot, 1974; Frisch et al., 1978). A more detailed discussion of these fives cases and in particular of the generators of the Lévy variables can be found in Schertzer et al. (1988), Fan (1989), or Schertzer and Lovejoy (1989b).

The parameter C_1 is the fractal codimension of the singularities contributing to the average values of the field; it tells us about the sparsity of the average level of intensity. Furthermore, if $C_1 > D$ (the dimension of space in which the process is observed), then the multifractal is *degenerate* on the space; that is, it almost surely vanishes everywhere. When α is not much smaller than 2 (as in the topography data below), these are approximately lognormal, since with the exception of their extreme tails, these Lévy distributions

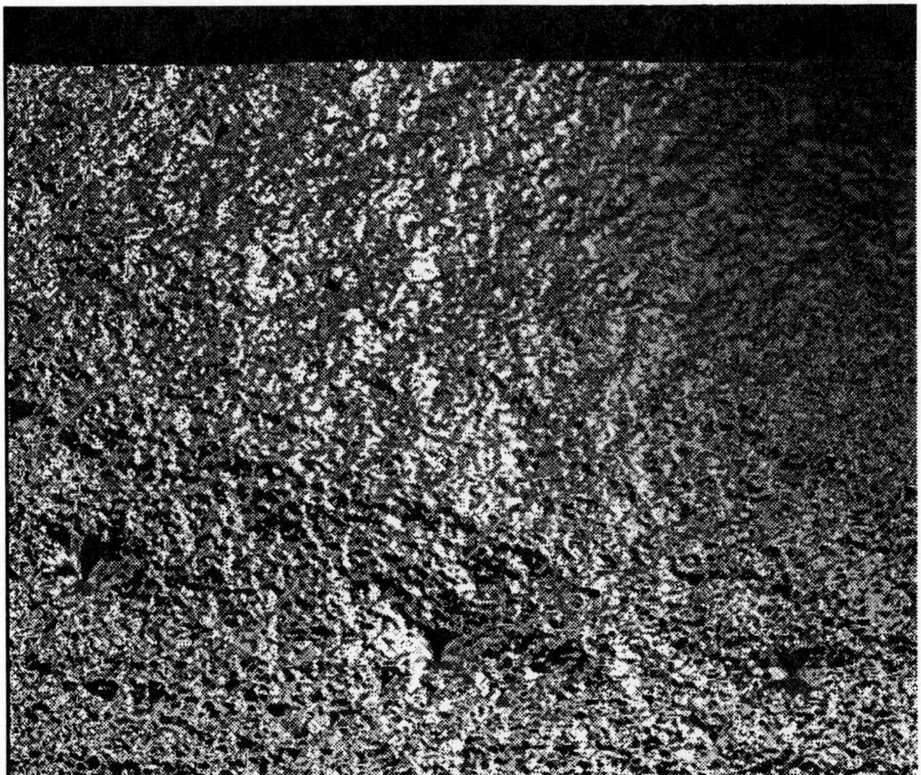

Figure 8.7(c)

are themselves nearly normal (this tail is pushed to lower and lower probability levels as $\alpha \to 2$). The widespread lognormal phenomeology is geophysics and geography is therefore an indication that the corresponding fields are actually multifractals, and are not exactly lognormal.

Nonstationary Processes

We have seen that in multiplicative processes there is always an underlying conserved quantity that has basic physical significance. In turbulence, it was the energy flux to smaller scales; in topography, it does not yet seem to be known, but we denoted it φ and related it to the altitude fluctuations via Equation 8.7. In terms of the scaling, conservation means $<\varphi_\lambda>$ = constant (independent of λ), hence $K(1) = 0$. If we consider the energy spectrum of φ_λ, it is $k^{-\beta}$ with $\beta = 1 - K(2)$; that is, the spectrum is always less steep than a $1/f$ noise.

If we consider the directly measurable altitude fluctuations Δh rather than φ, we find that $\Delta h_\lambda \approx \varphi_\lambda{}^a \lambda^{-H}$, where a is unknown (changing a is essentially the same as changing C_1; see below), and we anticipate that from the measurements below, $H = 1/2$ (if φ_λ was constant, then the altitude difference between two points would grow as the 1/2 power of their separation distance $\lambda^{-1/2}$). In Fourier space, the relation between $\varphi_\lambda{}^a$ and Δh_λ is simple: they will differ by k^{-H}; the spectra (the modulus squared of the Fourier transforms)

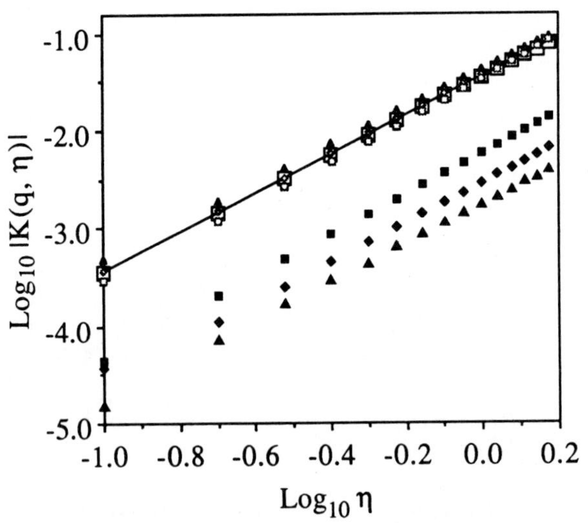

Figure 8.8. Log $|K(q, \eta)|$ versus log η, with $\alpha = 2$ (log normal), $C_1 = 0.15$, and $D = 2$, are given for $q = 0.5$. The curve of the stationary processes (big hollow square) is compared to those of the same processes after fractional differentiation (white symbols, $H = -2, -1$, and -0.5, from top to bottom). The fractional differentiation and integration do not affect the estimate of α (all the slopes are parallel), but fractional integration leads to biased estimates of C_1 (the curves with black symbols are all shifted downward compared to the theoretical stationary processes shown by the line).

will differ by k^{-2H}. Since (normal) differentiation of a function corresponds (ignoring constant factors) to multiplication by k in Fourier space and integration corresponds to division by k, we obtain the altitude fluctuations by the fractional integration (order H) of $\varphi_\lambda{}^a$. Conversely, if Δh_λ is measured, we may obtain φ_λ by fractional differentiation of the same order (see Schertzer and Lovejoy, 1991b, Appendix B.2, for more discussion of fractional derivatives and integrals).

The reason for dwelling on this is that it illustrates a basic point common to most geophysical fields, that their spectra have $\beta > 1$ and hence they cannot be stationary processes; they must be fractionally differentiated (that is, the spectra must be power law filtered) to become stationary. For topography, this means removing the λ^{-H} term and taking the a^{-1} power of the result in order to obtain the stationary φ from the nonstationary Δh_λ. The importance of this for standard data analysis has long been realized; for example, we have already seen that when $\beta > 1$ we should use variograms rather than autocorrelation functions [for example, analyze the differences (finite derivatives) of a time series rather than the series itself]. The same considerations apply to the use of the Double Trace Moment technique (below). Figure 8.8 shows the result when a simulated stationary process is fractionally integrated and differentiated by varying amounts: as long as we differentiate (filtered by k^H with $H > 0$), we obtain stable and accurate estimates of both C_1 and α; however, when we fractionally integrate ($H < 0$), we only recover α and C_1 is not accurately determined. Figure 8.8 also clearly indicates that, as long as the spectrum is less steep than the underlying stationary process [$\beta < 1 - K(2)$], we can recover C_1. After C_1, α are estimated, we can determine $K(2,1)$ from Equation 8.13 in the next section, and hence

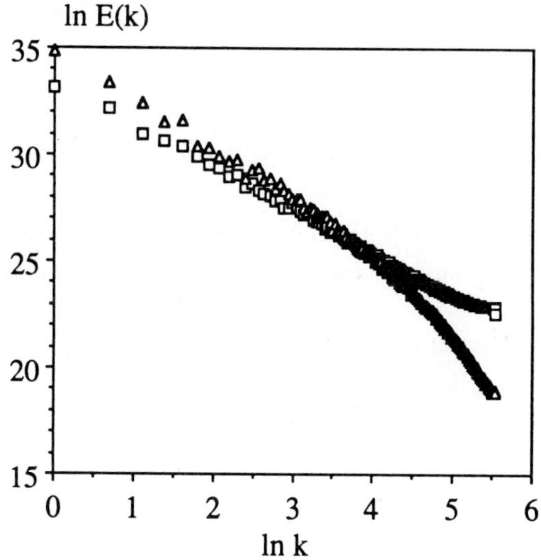

Figure 8.9 The spectrum of Deadman's Butte compared with the spectrum of the modulus of the gradient (which approximates a fractional differentiation of order 1), followed by a fractional integration of order 1. As expected, the curves are parallel for low frequencies, but at the high wave number end, they differ significantly over roughly the last factor 4.

the β of the stationary process, and infer the amount of fractional integration required to go from the underlying stationary process to the observed nonconserved process. Writing β for the spectral slope of the observed process, the order of fractional integration (the H in Equation 8.7) required to go from the stationary process to the nonstationary process is therefore given by

$$H = \frac{\beta - 1 + K(2)}{2} = \frac{\beta - 1}{2} + \frac{C_1 (2^{\alpha} - 2)}{2(\alpha - 1)} \tag{8.11}$$

This formula will be useful in the simulations described below where we produce multifractals with the same α, β, and C_1.

As a final comment before turning to the actual data analysis, we describe a shortcut that in many cases enables us to avoid the use of Fourier space. Recall that replacing the time series by its differences is essentially the same as multiplying by k in Fourier space. To generalize this to two (or more) dimensions, one possibility would be to use a finite difference Laplacian. This would multiply by $|\underline{k}|^2$ in Fourier space and hence the spectrum by $|\underline{k}|^4$; it would be quite drastic. We find that a compromise involving less smoothing that works quite well is to replace the field by the modulus of the local finite difference gradient operator. Figure 8.9 shows this explicitly on the height field data, which is discussed below. It compares the spectrum of the modulus of the gradient multiplied by k^{-1} in Fourier space with the spectrum of the original height field; the two curves are parallel as expected, except for high-wave-number deviations occurring roughly in the last factor of 4 of wave numbers. This deviation is not surprising: if a "cleaner" filtering is desired, Fourier methods should be used.

COMPARISON OF MONO AND MULTIFRACTAL EXPONENT RELATIONS

Now that we have introduced the essential difference between the nonstationary altitude h, and the stationary underlying flux φ, we summarize and compare their various properties in table 8.2.The monofractal limit ($\alpha \to 0$) is obtained when the convex curve of $K(q)$ becomes a straight line (a linear function of q). Simple scaling is the limit when $C_1 \to 0$ ($H \neq 0$).

The multifractal nature of the topography and the consequent failure of the equation $\beta = 2C(P_T) + 1$ can reconcile some of the apparently contradictory results summarized in Turcotte (1989). For example, using the value $\beta = 2$ for topography and the relation $\beta = 1 + 2C(P_T)$, we obtain $C(P_T) = 0.5$ and hence, $D(P_T) = 1.5$ for one dimensional cross-sections of the topography. This is in contradiction with various analyses of the lines of constant altitude (values of $D(P_T) \approx 1.2$–1.3 have often been reported.) However, as figure 1.1b shows, the value of $D(S \geq {}_T)$ (which is an upper bound for $D(P_T)$) is simply a function of altitude apparently tending to zero for large T; the discrepancy is simply a reflection of the multifractal nature of the topography.

Table 8.2 Summary of Various Scaling Exponents and the Simple Scaling Special Case

	General Multifractal/Multiscaling Result	Simple Scaling
Conserved flux scaling	$<\varphi_l^q> = l^{-K(q)}$, $K(q)$ convex, $\neq 0$ [except $K(0)$, $K(1) = 0$]	$<\varphi_l^q>$ independent of l, $K(q) = 0$
Altitude Structure function	$S_{q,l} = <\|\Delta h_l\|^q>$; $K_h(q) = K(q) - qH$	$K_h(q) = -qH$
Spectral exponent	$\beta = 1 - K_h(2)$	$\beta = 1 + 2H = 1 + 2C(P_T)$
Codimension of exceedance sets $S \geq _T$	$C(S \geq _T)$ = increasing function of altitude threshold T	$C(S \geq _T) = 0$
Codimension of isolines P_T	$C(PT) \geq C(S_T \geq)$	$C(P_T) = H$ = constant
Exponent for the distribution of the area of islands	$B_T > 0$, function of T	$B_T = D(P_T)/2 = 1 - H/2$ = constant, independent of T

DOUBLE TRACE MOMENT ANALYSIS TECHNIQUE

When applied to turbulent and/or geophysical data (rather than to strange attractors), existing multifractal analysis techniques have had limited accuracy for a number of reasons. Techniques based on statistical moments (for example, partition function methods) suffer from both undersampling and divergence of moments, whereas others based on histograms or wavelets (with the exception of the probability distribution multiple scaling technique, Lavallée et al., 1991) assume that multifractals are local. Finally, these techniques have attempted to estimate an infinite number of parameters: the codimension function. We now describe a simple technique that overcomes these problems by exploiting the universality to estimate C_1 and α directly; $K(q)$ is then obtained using Equation 8.10 and $c(\gamma)$ by Legendre transformation of the same equation.

Consider the stationary multifractal process $\varphi_{\lambda'}(<\varphi_{\lambda'}> = 1)$, where the resolution λ' is the ratio of the outer (or largest) scale of interest to the smallest scale of homogeneity. The basic idea is to generalize the application of statistical methods to the quantity $\varphi_{\lambda'}^{\eta}$, which is no longer conserved [$<\varphi_{\lambda'}^{\eta}> = (\lambda')^{K(\eta)}$]. This is done by taking the ηth power of φ_λ at the largest scale ratio λ and then studying the scaling behavior of the various qth moments at decreasing values of the scale ratio $\lambda' \leq \lambda$. The q and η, double trace moments at resolution λ' and λ, have the following multiple scaling behavior (Lavallée, 1991; Lavallée et al., 1992):

$$\Pi_{\lambda'}^{(\eta)}(B_\lambda) = \int_{B_\lambda} \varphi_{\lambda'}^{\eta} d^D x \tag{8.12a}$$

$$Tr_{A_\lambda}(\varphi_{\lambda'}^{\eta})^q = \langle \sum_i \left[\Pi_{\lambda'}^{(\eta)}(B_{\lambda,i})\right]^q \rangle \approx \lambda^{K(q,\eta)-(q-1)D} \tag{8.12b}$$

where B_λ is a dimensional "ball" (square in figure 8.10), $\Pi_{\lambda'}^{(\eta)}(B_\lambda)$ is the "η flux" over the ball, and A_λ is the set A at resolution λ which results from covering the set with the balls (indexed with i). The sum is over all the balls in the covering. The integration over A_λ rescales the fields and "dresses" the quantities. When $\eta = 1$, the right side of Equation 8.12 reduces to the usual trace moments (which are themselves the ensemble average of the usual partition function).

The scaling exponent $K(q, \eta)$ is related to the usual $K(q) = K(q, 1)$ as follows (see Lavallée, 1991; Lavallée et al., 1992):

$$K(q, \eta) = K(q\eta, 1) - qK(\eta, 1) \tag{8.13}$$

$K(q, \eta)$ reduces to the usual $K(q, 1)$ when $\eta = 1$ and the right side of Equation 8.12b to the usual trace moments (Schertzer and Lovejoy, 1987, 1989a). The exponent $K(q,1)$ is the usual scaling exponent defined in Equation 8.10. Up to now it was written $K(q)$ for brevity and because no confusion was possible; for the rest of the paper, $K(q,1)$ will be used. Using Equation 8.10, the expression for $K(q, \eta)$ becomes

$$K(q, \eta) = \eta^\alpha K(q, 1) = \begin{cases} \dfrac{C_1}{\alpha - 1}\eta^a (q^a - q), & \alpha \neq 1 \\ C_1 \eta q \log(q) & \alpha = 1 \end{cases} \tag{8.14}$$

with $0 \leq \alpha \leq 2$, and $q \geq 0$ (for $\alpha \neq 2$).

Keeping q fixed (but different from the special values 0 or 1) and studying the scaling properties of the DTM for various values of η allows the determination of the scaling exponent $K(q,\eta)$ as functions of η and the slope of $|K(q, \eta)|$ as a function of η on a log–log graph, yielding a value of the parameter α. On the same log–log plot, the values of C_1 can be estimated by solving the expression for the intercept. For theoretical development and

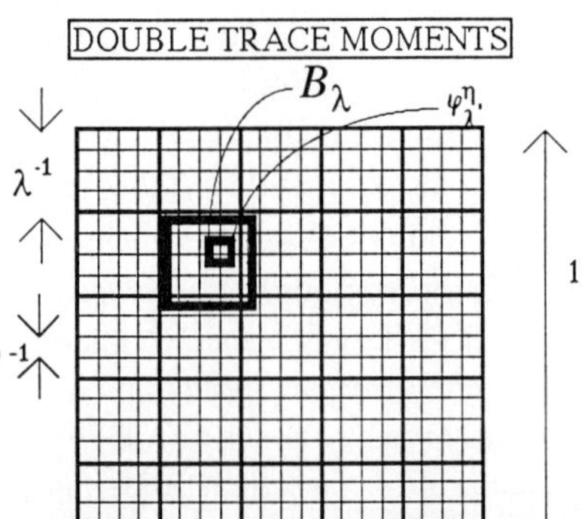

Figure 8.10 A schematic diagram illustrating the different averaging scales used in the double trace moments technique.

extensive numerical testing of the double trace moments technique, see Lavallée, 1991; Lavallée et al., 1992.

Equation 8.13 breaks down when $\max(q\eta, \eta) > \min(q_D, q_{max})$ where q_D and q_{max} characterize, respectively, the divergence of moments and the undersampling (Lavallée et al., 1991a). In these cases, $K(q,1)$ becomes linear for q greater than these critical values, and Equation 8.13 indicates that scaling exponents $K(q, \eta)$ becomes independent of η.

EMPIRICAL RESULTS

The Universal Multifractal Parameters of Deadman's Butte

The DTM technique is now applied to the digital elevation model (DEM) of the Deadman's Butte region of Wyoming in the United States (see figure 8.17). The data were provided by P.J. Muller and D. Rees of University College, London. The height data, in meters, are distributed over a grid of 512 × 512 at 50-m resolution. As shown above, the spectral slope is $\beta \approx 1.93$ (Figure 8.5). Since $\beta > 1$, it is necessary to fractionally differentiate either in Fourier space or by studying the modulus of the gradient of the topography. Denoting the latter by $|\nabla h|$, we have

$$|\nabla h(x, y)| = \sqrt{\left(\frac{\partial h}{\partial x}\right)^2 + \left(\frac{\partial h}{\partial y}\right)^2} \tag{8.15}$$

which can be approximated by the following finite difference:

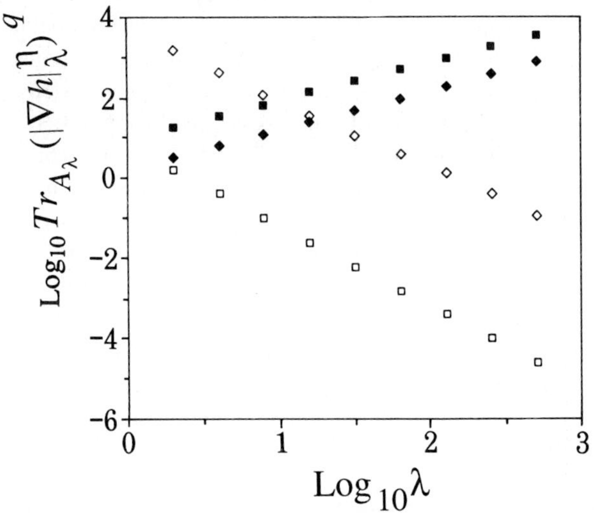

Figure 8.11 The scaling behavior of the double trace moments of the Deadman's Butte data is illustrated here by the straightness of the curves of the log $[Tr_{A\lambda}(|\nabla h|_\lambda{}^\eta)^q]$ as functions of the $\log(\lambda)$. The curves with positive slopes are those for $q = 0.5$, and from top to bottom $\eta = 2, 0.5$. The curves with negatives slopes are those for $q = 2$, and from top to bottom $\eta = 2, 0.5$. The slopes are $K(q, \eta)$ $(q - 1)D$.

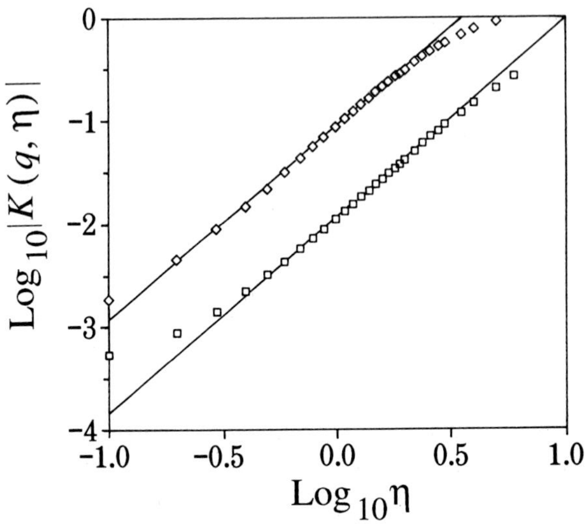

Figure 8.12 Log $|K(q, \eta)|$ versus log η for the Deadman's Butte data. [q is 2 (top) and 0.5 (bottom)]. All the curves are parallel as predicted by Equation 8.14. Their slopes, with $0.5 < \eta < 2.4$ are $\alpha \approx 1.90, 1.89$, and the corresponding C_1 are $\approx 0.044, 0.045$, respectively. The consistency between the estimates of α and C_1 for various values of q is a good indication that they are accurate. For values of η too high or too low, the curve $K(q, \eta)$ becomes fairly constant, as expected, and these values of log $|K(q, \eta)|$ must not be considered to estimate α.

$$|\nabla h(i,j)| = \sqrt{[h(i+1,j) - h(i-1,j)]^2 + [h(i,j+1) - h(i,j-1)]^2} \tag{8.16}$$

where $\Delta x = \Delta y = 1$. The indexes i and j are, respectively, the horizontal and vertical coordinates, in our case going both from 1 to 512. The difference operations are applied without privileging any particular direction; problems related to anisotropy are neglected. The field $|\nabla h|$ is raised to the power η at its smallest scale length, and the qth statistical trace moment of the field is estimated from $\lambda = \lambda' = 2^9 = 512$ to $\lambda = 2$. The rescaled field intensities are obtained by averaging over the $|\nabla h|^\eta$ at the finest resolution.

The scaling behavior of the log $[Tr_{A\lambda}(|\nabla h|_\lambda{}^\eta)^q]$ against log(λ) is given in Figure 8.11 for several values of q and η. The slopes of these curves give the estimated scaling exponents $K(q, \eta)$. Log $[|K(q, \eta)|]$ as a function of log η is shown in Figure 8.12. From this curve, the estimated parameters are $C_1 \approx 0.045 \pm 0.005$ and $\alpha = 1.9 \pm 0.1$. Finally, using our estimate $\beta = 1.93$, we can use Equation 8.11 to estimate the parameter $H = (1.93 - 1)/2 + 0.045(2^{1.9} - 2)/(2 \times 0.9) \approx 0.51 \approx 1/2$.

Universal Multifractal Parameters for French Topography

For comparison with the 50-m resolution Deadman's Butte data, we also applied the DTM technique to a DEM of French topography of 512 × 512 pixel at 1-km resolution. Figure 8.13 shows various DTM as functions of the resolution, indicating the accuracy with which the scaling is respected when λ varies from 2^9 to 2, and Figure 8.14 shows the curves log $K(q, \eta)$ against log η, indicating that over the range $0.01 < \eta < 3$ universality holds very well. As expected at both large and small η, the small sample size (a single realization) leads to serious sampling problems for these extreme moments. For $q = 0.5$, 2, 4, we have very parallel lines with slopes very near $\alpha = 1.7$; analysis of the intercepts with $\log \eta = 0$ indicates that $C_1 = 0.075 \pm 0.005$. These values should be compared with the values for Deadman's Butte ($\alpha = 1.9$, $C_1 = 0.045 \pm 0.005$), recalling that these are statistical exponents and that single realizations of multifractal processes will be extremely variable. To compare the two $K(q, \eta)$ functions, we plot (Figure 8.15) for $q = 2$ and show the intermediate slope $\alpha = 1.8$ for reference. Although we can be quite confident that the values of C_1 of the corresponding multifractal processes really are different, the values of the more fundamental multifractal index α may in both cases be compatible with the common value $\alpha = 1.8$, although many more data sets will need to be analyzed to be sure.

Discussion

The obtained values of H, C_1, and α indicate that the height field has seemingly paradoxical properties that go a long way toward explaining both the successes and the limitations of monofractal analyses and models. First, the multifractal index α is nearly its maximum, indicating that singularities of all orders contribute significantly to the process: we are almost as far as possible from the monofractal case $\alpha = 0$. However, the modest values of C_1 indicate that the mean of φ is not too sparse (a space-filling mean would have $C_1 = 0$). Due to the fractional integration ($H = 1/2$) necessary to obtain the height field from φ, even if $C_1 = 0$, we would still obtain a fractal height field; however, its properties would be independent of the altitude and yield fractal dimensions $2 - H = 1.5$ for lines of constant

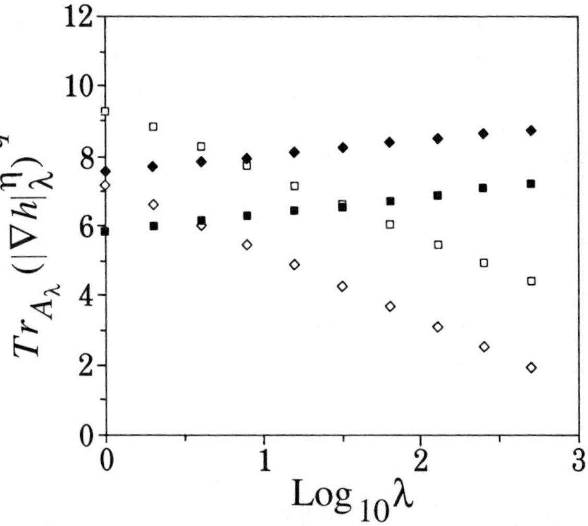

Figure 8.13 Same as Figure 8.11 but for the French topography; when $q = 0.5$ (curves with a positive slope), $\eta = 2.26$, 0.56 from top to bottom; when $q = 2$ (curves with a negative slope), $\eta = 1.6$, 0.56 from top to bottom.

altitude. Hence, our value of C_1 is in perfect accord with the standard empirical results for those statistics which do not depend sensitively on the extremes. However, once we start examining isolines at high altitude or high-order moments, the large value of α will ensure that the multifractal nature of the process will be important.

Actually, the low values estimated for C_1 in both cases may not be very significant because (as mentioned in the section on the DTM technique) the DTM technique cannot distinguish between the conserved process φ and an arbitrary power a of the process. In other words, the fundamental conserved process responsible for topography could have significantly larger $C_1 a^\alpha$. Indeed, this is exactly the case for the velocity and temperature fields analyzed in Schmitt et al. (1992a and b), where the value $a = 3$ is determined by dimensional arguments and low values (≈ 0.1) for the C_1 of the velocity field are associated with significantly larger values (≈ 0.25) for the underlying energy flux ε.

Numerical Simulations of Isotropic, Self-similar Universal Topography

As explained earlier, universal multifractals arise when scaling multiplicative processes are mixed: either by nonlinear interactions between different processes or by nonlinear interactions between structures at intermediate scales (that is, a *densification* of the cascade leading to a continuous cascade). Because of the existence of stable, attractive universal multifractal generators, the results of such processes are independent of most of the details; hence, we may expect to observe universal multifractals in many diverse areas

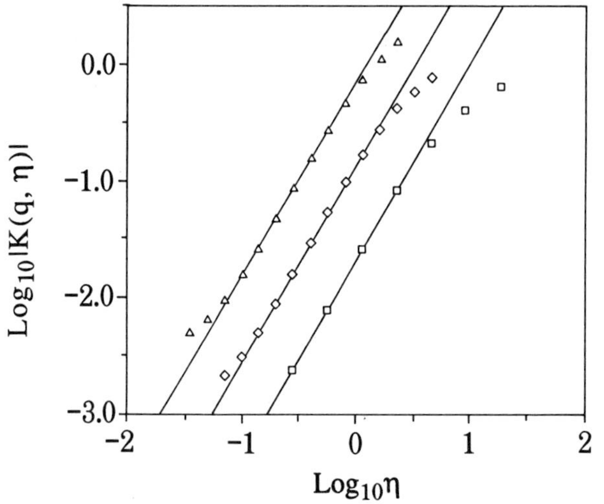

Figure 8.14 Same as Figure 8.12 for the French topography. From top to bottom, q takes the following values: 4, 2, and 0.5. The estimates of the slope for $q = 4$, with η taking values between 0.1 and 0.5, give $\alpha = 1.67$. The same analysis for $q = 2$ and 0.5 (with η taking values between 0.12 and 1.1 in the first case and between 0.3 and 2.3 in the second) yields respectively, $\alpha = 1.69$, 1.7. The values of C_1, obtained by solving the expression for the log $\eta = 0$ intercept given by the log $|K(q, 1)|$ and using Equation 8.10, are, respectively; 0.0078, 0.076, and 0.076. Here also the values of α, C_1 determined are independent of the q values.

where the dynamics are strongly nonlinear and occur over wide ranges of scale. In turbulence, the physical basis of these universal multifractals is the scaling symmetry, conservation of energy flux, and cascading phenomena, which implies that the dynamics at one scale mainly modulate the nearby scales. In topography and most other geophysical and geographical fields, the exact dynamical equations and the corresponding conserved areas of statistics are usually not known. However, just as Gaussian distributions occur in many areas of statistics, so we may expect universal multifractals to arise in many scaling nonlinear dynamical systems. Indeed, the prevalence of nearly log normal phenomena in geography and geophysics suggests the ubiquity of multifractals.

Wilson et al. (1991) describe the numerical simulation of clouds and topography, including how to iteratively zoom in and calculate details to arbitrary resolution in selected regions. Although we will not repeat these details here, enough information has been given in the previous sections to understand how they work. First, recall that a conserved (stationary) multifractal process φ_λ has a generator $\Gamma_\lambda = \log \varphi_\lambda$, which is exactly a $1/f$ noise; that is, its spectrum is $E(k) \approx k^{-1}$ (this is sufficient to ensure the multiple scaling of the moments of φ_λ). Therefore, to produce a generator, we start with a stationary Gaussian or Lévy *subgenerator.* The subgenerator is a noise consisting of independent random variables with either Gaussian ($\alpha = 2$) or extreme Lévy distributions (characterized by the Lévy index α), whose amplitude (for example, variance in the Gaussian case) is determined by C_1. The subgenerator is then fractionally integrated to give a k^{-1} spectrum. This generator is then exponentiated to give the conserved φ_λ, which will thus depend on both C_1 and α. Finally, to obtain a nonconserved process with spectral slope β, the result is fractionally integrated by multiplying the Fourier transform by k^{-H} where H is given in Equation 8.11. The entire process involves two fractional integrations and hence four FFTs. Fields of 512 × 512 can easily be modeled on personal computers (they take about 3

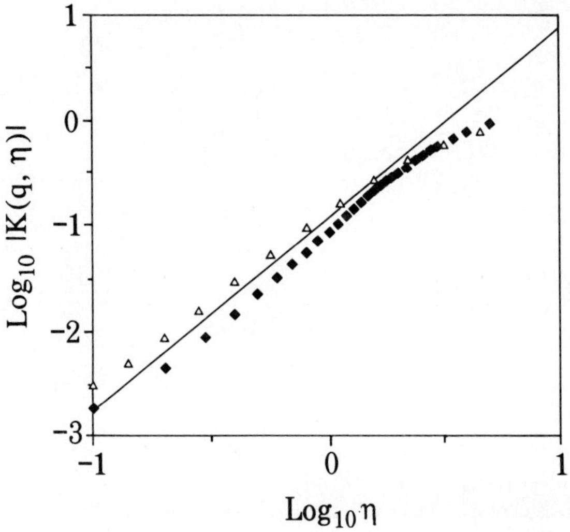

Figure 8.15 Log $|K(q, \eta)|$ for the French topography (top) and Deadman's Butte (bottom) are compared with $q = 2$. The lines have slopes equal to 1.8.

minutes on a Mac II), and 256 × 256 × 256 fields, for example, for the temporal evolution of clouds, have been produced on a Cray 2. Such cloud simulations have been turned into a video called *multifractal dynamics* (Brenier, 1990; Brenier et al., 1990).

Figures 8.7a–c show the results for parameters similar to those found for Deadman's Butte and France. Another example with $\alpha = 1.6$ (Figure 8.16) is shown to indicate the range of behavior achieved by varying the multifractal index α. The commonly used ray-tracing technique has been used to visualize the simulations since this effectively highlights the surface texture. Perhaps one of the most striking differences between the figures is the existence of smooth low "plains" when $\alpha < 2$ (most visible for $\alpha \approx 1.6$), which is a direct consequence of the extremal Lévy character of the random variables used in the corresponding generator. In effect, when $\alpha < 2$, there must be a more and more marked asymmetry between positive and negative fluctuations in the generator: the negative fluctuations becoming more and more extreme with respect to the positive ones as α decreases. When the generator is exponentiated to give the conserved process, the result is strong *holes*, some of which can be large, in which the conserved process will nearly vanish. Finally, the power-law filter (k^{-H}) applied to generate the topography field is a smooth-

Figure 8.16 Simulation of topography with multiplicative cascade processes visualized using ray tracing techniques, with $\alpha = 1.6$, $C_1 = 0.1$, and $H = 1/2$.

ing operation that tends to round out the holes somewhat; nevertheless, their effect is clearly visible in the simulations.

Figure 8.17 shows the Deadman's Butte data with the same format. The most obvious visual difference between this and the preceding simulations is that the Butte data have far more interesting texture due to their anisotropy, which is present at all length scales. Clearly, more realistic topography models will have to go beyond the restrictive isotropic (self-similar) framework described here. Following the formalism of generalized scale invariance (mentioned earlier), this will require introducing a new generator for the scale-changing operations.

Figure 8.17 The topography of Deadman's Butte shown with ray tracing techniques.

CONCLUSION

The problem of adequately conceptualizing and modeling the extremely variable geophysical and geographical fields in our environment has a long history. Of the various ideas that have been put forward to deal with the problem, the notion of scale invariance is undoubtedly the most seductive because it is the simplest hypothesis that seems to be capable of accounting for the ubiquity of complex geographical fields spanning many orders of magnitude in scale.

Early ideas of scaling involved restrictive scale-changing operations: structures were assumed statistically isotropic (self-similar). This means that they were unable to deal with differential rotation, compression, or texture. Furthermore, restrictive types of scaling appropriate for fractal sets, but not for multifractal fields, were used. In the first part of this paper, we reviewed these issues, demonstrating explicitly how many commonly used methods for estimating fractal dimensions (such as spectral or variogram–structure function exponents) can only be used for dimension estimates in monofractal systems.

We continued our review with an outline of multiplicative processes, indicating briefly how they lead to universal multifractals when they are mixed. Stable and attractive generators of multiplicative processes exist, and they are likely to arise from diverse nonlinear mechanisms; the details are unimportant in the limit where interactions between many different processes occur or where interactions occur over a continuum of length scales.

We then outlined the new double trace moments analysis technique, which is the first to be specifically designed to estimate the universal multifractal indexes and tested it on multifractal simulations. We applied this technique to digital elevation models of Deadman's Butte (50-m resolution) and French topography (1-km resolution), obtaining the estimates $\alpha = 1.9 \pm 0.1$ and $\alpha = 1.7 \pm 0.1$, respectively, for the multifractal indexes, and $C_1 = 0.045 \pm 0.005$ and $C_1 = 0.075 \pm 0.005$ for the codimension of the mean of the corresponding conserved process. The multiple scaling of the data sets was accurately confirmed over the entire range of scale (factor 512 in both cases), and the universality of the scaling exponents over the range 0.1 to 3 is identified. For exponents larger than this, we have an undersampling problem: many realizations must be used to get more accurate estimates for these extreme exponents, for exponents smaller than this the results are dominated by noise or measurement errors.

Finally, we briefly described how to simulate such fields and produce simulations with the same α, C_1, and spectral exponents β (≈ 2 here). We argue that the fundamental difference between the simulations and the data is that the former are constructed to be isotropic (self-similar), whereas the latter are highly anisotropic. More realistic models can easily introduce scaling anisotropy (including stochastic anisotropy to deal with locally varying differential compression and rotation of structures) by using anisotropic scale-changing operators in the framework of generalized scale invariance (GSI) (see Lovejoy et al., 1992; Pflug et al., 1993). This is the current subject of empirical, numerical, and theoretical investigation.

ACKNOWLEDGEMENTS

We thank A. Davis, C. Hooge, K. Pflug, F. Schmitt, Y. Tessier, and B. Watson for helpful comments, discussion, and technical assistance. G. Sarma is thanked for the simulation and D. Rees and J. P. Muller for the Deadman's Butte DEM data and for helpful criticism.

REFERENCES

Aviles, C. A., Scholz, C. H., and Boatwright, J. 1987. Fractal analysis applied to characteristic segments of the San Andreas Fault. *Journal of Geophysical Research* 92:331–334.

Bell, T. H. 1975. Statistical features of sea floor topography. *Deep Sea Research* 22:883–891.

Benzi, R., et al., 1984. On the multifractal nature of fully developed turbulence and chaotic systems. *Journal of Physics A* 17:3521–3531.

Bills, B. G., and Kobrick, M. 1985. Venus topography: A harmonic analysis. *Journal of Geophysical Research* 90:827–836.

Brenier, P. 1990. Simulations du dynamique multifractal des nuages. Master's Thesis, Ecole Normale du Science et Technologie Avancée. Paris, France.

———, et al. 1990. Continuous multiplicative cascade models of passive scalar clouds. *Annales Geophysicae* 8, Special Issue: 320.

Clarke, K.C. 1988. Scale-based simulation of topographic relief. *The American Cartographer* 15: 173–181.

De Cola, L. 1989. Multiscale data models for spatial analysis, with applications to multifractal phenomena. In *Proceedings, Ninth International Symposium on Computer-assisted Cartography (Auto-Carto 9)*, Baltimore, pp. 313–323.

———. 1990. Fractal analysis of a classified Landsat scene. *Photogrammetric Engineering and Remote Sensing* 55(5):601–610.

Falconer, K. 1990. *Fractal Geometry: Mathematical Foundations and Applications*. New York: J. Wiley and Sons.

Fan, A. H. 1989. Chaos additif et multiplicatif de Lévy. *Comptes Rendus. Academie des Sciences (Paris) I* 308:151–154.

Fournier, A., Fussell, D., and Carpenter, L. 1982. Computer rendering of stochastic models. *Communications of the Association of Computing Machinery* 6:371–374.

Fox, C. G. 1989. Empirically derived relationships between fractal dimension and power law from frequency spectra. *Pure and Applied Geophysics* 131:211–239.

Frisch, U., Sulem, P. L., and Nelkin, M. 1978. A simple dynamical model of intermittency in fully developed turbulence. *Journal of Fluid Mechanics* 87:719–724.

Gabriel, P., et al., 1988. Multifractal analysis of resolution dependence in satellite imagery. *Geophysical Research Letters* 15:1373–1376.

Gilbert, L. E. 1989. Are topographic data sets fractal? *Pure and Applied Geophysics* 131:241–254.

Goodchild, M. F. 1980. Fractals and the accuracy of geographical measures. *Mathematical Geology* 12:85–98.

Halsey, et al., 1986. Fractal measures and their singularities: the characterization of strange sets. *Physics Review A* 33:1141-1151.

Klinkenberg, B., and Clarke, K.C. 1992. Exploring the fractal mountains. In *Automated Pattern Analysis in Petroleum Exploration*, I. Palaz, and S. Sengupta, eds., pp 261–212. New York: Springer-Verlag.

———, and Goodchild, M. G. 1992. The fractal properties of topography: A comparison of methods. *Earth Surficial Processes and Landforms* 17: 217–234

Kolmogorov, A. N., 1940. Wienersche spiralen and einige andere interessante kurven in Hilbertschen Raum. *C. R. (Doklady) Acad. Sci. URSS (N.S.)* 26: 115–118.

———, and Tihomirov, V. M. 1959. Epsilon entropy and epsilon capacity of sets in functional spaces. *Uspeckhi Matematicheskihk Nauk (N.S.)* 14:3–86. Translated in *American Mathematical Society Translations (Series 2)* 17:277–364.

Korcak, J. 1938. Deux types fondamentaux de distribution statistique. *Bulletin de l'Institut International de Statistique* III: 295–299.

Ladoy, P., Schertzer, D., and Lovejoy, S. 1990. Analyse multifractale du relief Français. Manuscript.

Lamperti, J. 1962. Semi-stable stochastic processes. *Transactions, American Mathematics Association* 204: 62–78.

Lavellée, D. 1991. Multifractal techniques: Analysis and simulations of turbulents fields. Ph.D. Thesis, McGill University, Montréal, Canada.

———, Schertzer, D., and Lovejoy, S. 1991. On the determination of the codimension function. In *Non-linear Variability in Geophysics: Scaling and Fractals*, D. Schertzer, and S. Lovejoy, eds. Dordrecht, the Netherlands: Kluwer.

———, ———, ———, and Schmitt, F. 1992. On the determination of universal multifractal parameters in turbulence. In *Topological aspects of the dynamics of fluids and plasmas*. H. K. Moffal, et al., eds., pp. 463–478. Dordrecht-Boston: Kluwer.

Lovejoy, S., Schertzer. D., and Tsonis, A. A. 1987: Funcional box counting and multiple elliptical dimensions in rain, *Science, 235*, 1036–1038

———, and ———. 1988a. Extreme variability, scaling and fractals in remote sensing: Analysis and simulation. In *Digital Image Processing in Remote Sensing*, J. P. Muller, ed., pp. 177–212. London: Taylor & Francis.

———, and ———. 1988b. Meeting report: Scaling, fractals, and non-linear variability in geophysics. *EOS* 69:143–145.

———, and ———. 1990a. Our multifractal atmosphere: A unique laboratory for nonlinear dynamics. *Physics in Canada* 46:62–71.

———, and ———. 1990b. Multifractals, universality classes and satellite and radar measurements of cloud and rain fields. *Journal of Geophysical Research* 95:2021–2034.

———, and ———. 1991. Multifractal analysis techniques and the rain and cloud fields from 10^{-3} to 10^{-6} m. In *Non-linear Variability in Geophysics: Scaling and Fractals*, D. Schertzer and S. Lovejoy, eds., pp. 111–144. Dordrecht, The Netherlands: Kluwer.

———, ———, and Ladoy, P. 1986a. Fractal characterization of inhomogeneous measuring networks, *Nature* 319:43–44.

———, ———, and ———. 1986b. Brighter outlook for weather forecasts. *Nature* 320:401.

———, ———, and Pflug, K. 1992. Generalized scale invariance and differential rotation if cloud radiances. *Physica A* 185: 121–128.

———, Lavallée, D., Schertzer, D., and Ladoy, Pl 1993. Topography as a universal multifractal process and the $1^{1/2}$ law. Manuscript.

Mandelbrot, B. 1975. Stochastic models for the earth's relief, the shape and the fractal dimension of the coastlines and the number-area rule for islands. *Proceedings of the National Academy of Sciences, USA* 72:3825–3828.

———. 1967. How long is the coastline of Britain? Statistical self-similarity and fractional dimension. *Science* 155:636–638.

———. 1974. Intermittent turbulence in self-similar cascades: Divergence of high moments and dimensions of the carrier. *Journal of Fluid Mechanics* 62:331–358.

———. 1977. *Fractals—Form, Chance and Dimension*. San Francisco: W. H. Freeman & Co.

———. 1982. *The Fractal Geometry of Nature*. San Francisco: W. H. Freeman & Co.

———, and Van Ness, W. 1968. Fractional Brownian motions, fractional noises and applications. *SIAM Review* 10: 422–450.

Mareschal, J. C. 1989. Fractal reconstruction of sea-floor topography. *Pure and Applied Geophysics* 131:197–210.

Mark, D. M., and Aronson, P. B. 1984. Scale dependent fractal dimensions of topographic surfaces: An empirical investigation, with applications in geomorphology and computer mapping. *Mathematical Geology* 16:671–683.

Meneveau, C., and Sreenivasan, K. R. 1987. Simple multifractal cascade model for fully developed turbulence. *Physical Review Letters* 59(13):1424–1427.

Novikov, E. A., and Stewart, R. 1964. Intermittency of turbulence and spectrum of fluctuations in energy-dissipation. *Izvestiya Akademii Nauk SSSR Serija Geofizicheskaya [USSR]* 3:408–412.

Okubo, P. G., and Aki, A. 1987. Fractal geometry in the San Andreas fault system. *Journal of Geophysical Research* 92:345–355.

Parisi, G., and Frisch, U. 1985. A multifractal model of intermittency. In *Turbulence and Predictability in Geophysical Fluid Dynamics and Climate Dynamics*, M. Ghil, R. Benzi, and G. Parisi, eds., pp. 84–88. Amsterdam: North-Holland.

Perrin, J. 1913. *Les Atomes*. Paris: NRF-Gallimard.

Pflug, L. Lovejoy, S. and Schertzer, D. 1993. Differential rotation and cloud testure: analysis using generalized scale invariance. *Journal of Atmospheric Science* 50:538–553.

Rees, D., and Muller, J. P. 1990. Anomalies resulting from the characterization of terrain by fractional Brownian motion. Manuscript.

Richardson, L. F. 1922. (Republished by Dover, New York, 1965). *Weather Prediction by Numerical Process*. London: Cambridge University Press.

———. 1961. The problem of contiguity: An appendix of statistics of deadly quarrels. *General Systems Yearbook* 6:139–187.

Sarma, et al. 1990. Universal multifractal cascade models of rain and clouds. Preprint vol., American Meteorological Society Conference on Cloud Physics, San Francisco, July 24–27, pp. 255–262.

Schertzer, D., and Lovejoy, S. 1983. Elliptical turbulence in the atmosphere. Proceedings of the *4th Symposium on Turbulent Shear Flows* 11.1–11.8, Karlsrhule, Germany.

——— and ———. 1985. Generalized scale invariance in turbulent phenomena. *Physico-Chemical Hydrodynamics Journal* 6:623–635.

——— and ———. 1987. Physically based rain and cloud modeling by anisotropic, multiplicative turbulent cascades. *Journal of Geophysical Research* 92: 9693–9714.

——— and ———, et al., 1988. Universal multifractals in turbulence. In *Fractal Aspects of Materials: Disordered Systems*, D. A. Weitz, L. M. Sander, and B. B. Mandelbrot, eds., pp. 267–269. Pittsburg, PA: Materials Research Society.

——— and ———. 1988. Multifractal simulations and analysis of clouds by multiplicative processes. *Atmospheric Research* 21:337–361.

——— and ———. 1989a. Nonlinear variability in geophysics: multifractal analysis and simulations. In *Fractals: Their Physical Origins and Properties*, L. Pietronero, ed., pp. 49–79. New York: Plenum Press.

——— and ———. 1989b. Generalized scale invariance and multiplicative processes in the atmosphere *Pure and Applied Geophysics* 130:57–81.

——— and ———. 1991a. Preface. In *Non-linear Variability in Geophysics: Scaling and Fractals*, D. Schertzer, and S. Lovejoy, eds., pp. vii–ix. Dordrecht, The Netherlands: Kluwer.

——— and ———. 1991b. Nonlinear geodynamical variability: Multiple singularities, universality and observables. In *Non-linear Variability in Geophysics: Scaling and Fractals*, D. Schertzer, and S. Lovejoy, eds., pp. 41–82. Dordrecht, The Netherlands: Kluwer.

——— and ———. 1992. Hard and soft multifractal processes. *Physica A* 185: 187–194.

——— et al. 1991. Universal hard multifractal turbulence: Theory and observations. In *Nonlinear Dynamics of Structures*, R. Z. Sagdeev, et al., eds., pp. 213–236, Teaneck, New Jersey: World Scientific.

Schmitt, F., et al. 1992a. Estimates of Universal multifractal indices for velocity and temperature fields. *Comptes Rendus. Academie des Sciences (Paris), II-56*: 749–754.

———, ———. 1992a. Empirical determination of universal multifractal exponents in turbulent velocity fields, Physical Review Letters 68: 305–308.

Steinhaus, H. 1954. Length, shape and area. *Colloquium Mathematicum* 3:1–13.

———. 1962. *Mathématiques en instantanés*. Paris: Flammorion.

Tessier, Y., Lovejoy, S. and Schertzer. D. 1993. Universal Multifractals: Theory and observations for rain and clouds. *Journal of Applied Meteorology* 32:223–250.

Turcotte, D. L. 1989. Fractals in geology and geophysics. *Pure and Applied Geophysics* 131:171–196.

Venig-Meinesz, F. A. 1951. A remarkable feature of the Earth's topography. *Proceedings, Koninklijke Nederlandse Akademie Van Wetenschappen. Series B. Physical Sciences* 54:212–228.

Voss, R. F. 1983. Fourier synthesis of Gaussian fields: 1/f noises, landscapes and flakes. Preprints, Siggraph Conference, Detroit, 21 pp.

Wilson, J., Schertzer, D., and Lovejoy, S. 1991. Physically based cloud modeling by multiplicative cascade processes. In *Non-linear Variability in Geophysics: Scaling and Fractals*, D. Schertzer, and S. Lovejoy, eds., Dordrecht, The Netherlands: Kluwer.

Yaglom, A. M. 1966. The influence of the fluctuation in energy dissipation on the shape of turbulent characteristics in the inertial interval, *Soviet Physic. Doklady* 2:26–30.

CHAPTER 9

FRACTAL ANALYSIS OF GEOLOGIC TIME SERIES

Roy E. Plotnick
Department of Geological Sciences, University of Illinois at Chicago

Karen Prestegaard
Department of Geology, University of Maryland, College Park

ABSTRACT

Time series of natural phenomena can exhibit fractal patterns. Methods used for the analysis of spatial fractals are readily applied to temporal fractals. Continuous fractal time series can be modeled using discrete fractional Gaussian noises and fractional Brownian motions, while discontinuous fractal time series can be viewed as Lévy dusts. Both of these types of time series exhibit distinctive statistical properties. One of the most important properties is long-range correlations in a continuous series. We review the use of autocorrelation, Fourier analysis, semivariograms, and rescaled range (*R/S*) analysis in the identification of fractal time series. We then apply these techniques to the analysis of time series data of bedload transport collected in a gravel-bed river. The structure of the bedload transport time series offers insight into the underlying mechanics of bedload movement and provides information that can be used to establish guidelines for bedload sampling.

INTRODUCTION

Geological processes commonly exhibit complex patterns of temporal variability over many orders of magnitude. Fractal mathematics, which provides a simple description of patterns with high variability over a wide range of scales, thus seems ideal for the depiction of such processes.

Fractals have attracted increasing interest within the geologic community. A substantial number of papers have appeared describing fractal applications in geomorphology, such as landform description, and in geophysics, especially earthquake frequency and fracture geometry (Turcotte, 1989). There is also a growing body of literature on flow through porous sediments, which is of interest to petroleum and groundwater geologists (Wong, 1988; Lenormand, 1989). A short theoretical paper by Plotnick (1986; see also 1988) examined the implications of fractals to stratigraphic problems.

In this chapter we will discuss some implications of fractal mathematics for the collection, analysis, and interpretation of geologic time series. We will review the basic properties of both continuous and discontinuous fractal time series and describe techniques for their recognition and study. We will illustrate these methods by analyzing a time series of sediment transport.

FRACTAL TIME SERIES

The ideas developed to describe the fractal properties of objects in space can also be applied to processes in time. Just as spatial fractals show statistical similarity over a wide range of size scales, temporal fractals show statistical similarity over a wide range of time scales. These time series, therefore, have no characteristic temporal scale, and a portion of the time series resembles, in some way, the entire sequence.

Spatial and temporal fractals, however, differ in one important characteristic. For spatial fractals, the vertical and horizontal measurement units are usually the same. For example, landform geometry can be measured in meters for both elevation and distance. In addition, these objects (such as coastlines) usually exhibit self-similarity when the axes are magnified or rescaled by equal amounts.

In contrast, the coordinates used in time series are time and the value of some variable, which have different units. For example, if we measure the change in elevation at a point over time, elevation may be measured in meters and time in years. Consequently, for time series at different levels of resolution to "look" the same, the time and variable axes may have to be rescaled by different amounts. Time series thus show self-affinity, rather than self-similarity, and are *self-affine* (Mandelbrot, 1983; Feder, 1988). Moreover, the usual box-counting dimension applied to self-similar fractals cannot be used for self-affine fractals (Voss, 1988).

Time series can be continuous, where the changing values of a particular variable, such as river flow rates, are measured over time. Alternatively, they can be discontinuous, where the spacing of particular events, such as floods, is determined. Fractal models exist for both types of series and, as we will see, are related via the concept of the zeroset. Both have relevance for the study of geological processes.

FRACTAL NOISES AND MOTIONS

Imagine a variable whose value X fluctuates over time t; this variable we will designate as $X(t)$. Let the initial value at $t = 0$ of the variable equal zero [$X(0) = 0$]. Now assume that the

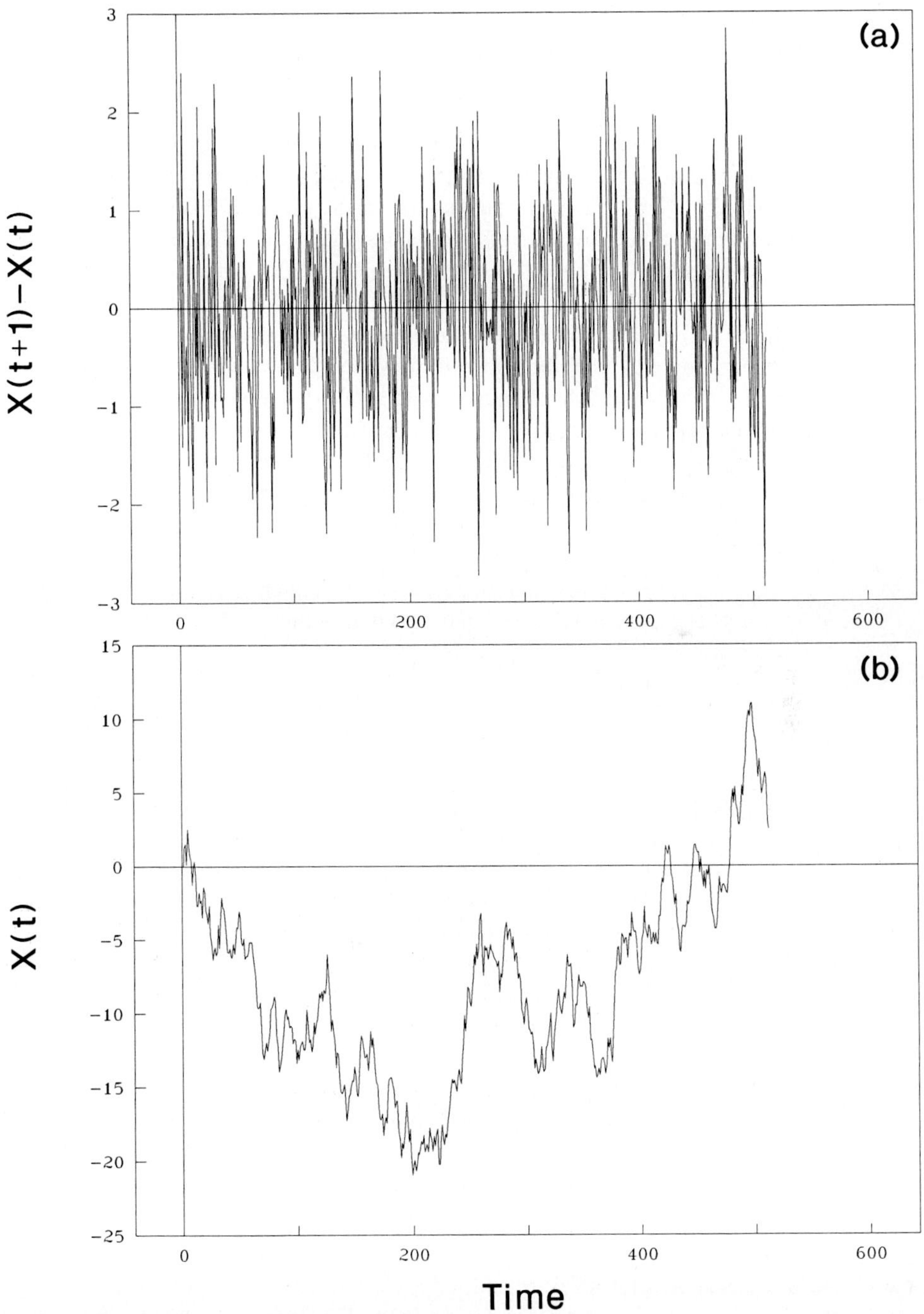

Figure 9.1 (a) A 512-point white noise, representing the increments of the Brownian motion shown in Figure 9.1b. and (b) A 512-point Brownian motion, generated using the method described in Saupe (1988).

value of the variable at the next instant of time has an equal probability of being greater or less than zero and the size of the change is one unit. Therefore, $X(1)$ has a 50% chance of equaling +1 or -1. Repeat the process; $X(2)$ can now equal -2, +2, or 0, depending on the value of $X(1)$ and the direction of change. The process can be repeated indefinitely. Notice that although the direction of the change is random the value of $X(t)$ at any time depends, to a large extent, on the "history" of previous values; for example, if $X(2) = -2$, then $X(3)$ can equal -3 or -1, but cannot equal 0 or any positive value. What we have just described is usually called a random walk, and the changing value of the variable as it moves through time is a *motion*.

In the situation we have just described, the size of the change in value was always 1. This is the step or increment size. Alternatively, the size of the increment could be variable. For example, it could be drawn from what is known as a Gaussian or normal distribution (the familiar bell-shaped curve) with an average value of zero. The probabilities of change in a positive or negative direction are still equal, and the amount and direction of change of one increment does not influence the next increment; they are independent. A series of independent random numbers generated from a Gaussian distribution form a *white noise* (Figure 9.1a). If we use these numbers as the increments of a random walk, we produce a *Brownian motion* (Figure 9.1b).

Although the size and sign of successive increments are independent, the values of $X(t)$ are not. The average relationship between the values of $X(t)$ at time t and a later time $t + d$, where d is known as the *lag*, is given by

$$E[X(t) - X(t + d)] \propto d^{H} \tag{9.1}$$

where $H = 1/2$ (Voss, 1988). In other words, the average difference between two values of a Brownian motion is proportional to the square root of the number of steps between them.

Mandelbrot and Van Ness (1968) and Mandelbrot and Wallis (1968, 1969a) generalized the above relationship to cases where $H \neq 1/2$ $(0 < H < 1)$; that is, the differences between two values of $X(t)$ need no longer be proportional to the square root of the number of steps. The sequence of values taken on by $X(t)$ is a *fractional Brownian motion* (fBm). The increments produced by these relationships, which are still normally distributed, are termed *discrete fractional Gaussian noises* (dfGn). Both white and discrete fractional Gaussian noises are stationary series; their mean and variance are constant over the entire series (Pankratz, 1983). This is not the case for the corresponding motions; they are nonstationary.

Figures 9.2a and 9.2b illustrate a discrete fractional Gaussian noise and corresponding fractional Brownian motion for $H = 0.8$. Notice that the fBm appears to have higher peaks and deeper valleys than the Brownian motion (Figure 9.1b). As we will see, this is due to the long-range correlations of the fractional Gaussian noise that form the increments of the fractional Brownian motion.

White noises, fractional Gaussian noises, Brownian motions, and fractional Brownian motions are all self-affine fractal series (Saupe, 1988). For example, examine the 512-step Brownian motion shown in Figure 9.1b. The range of values is approximately 30 units (-20 to 10). Figure 9.3 shows a 50-step portion of this, with both axes expanded 10 times so that the length is the same as that of the original series. Notice that the range is

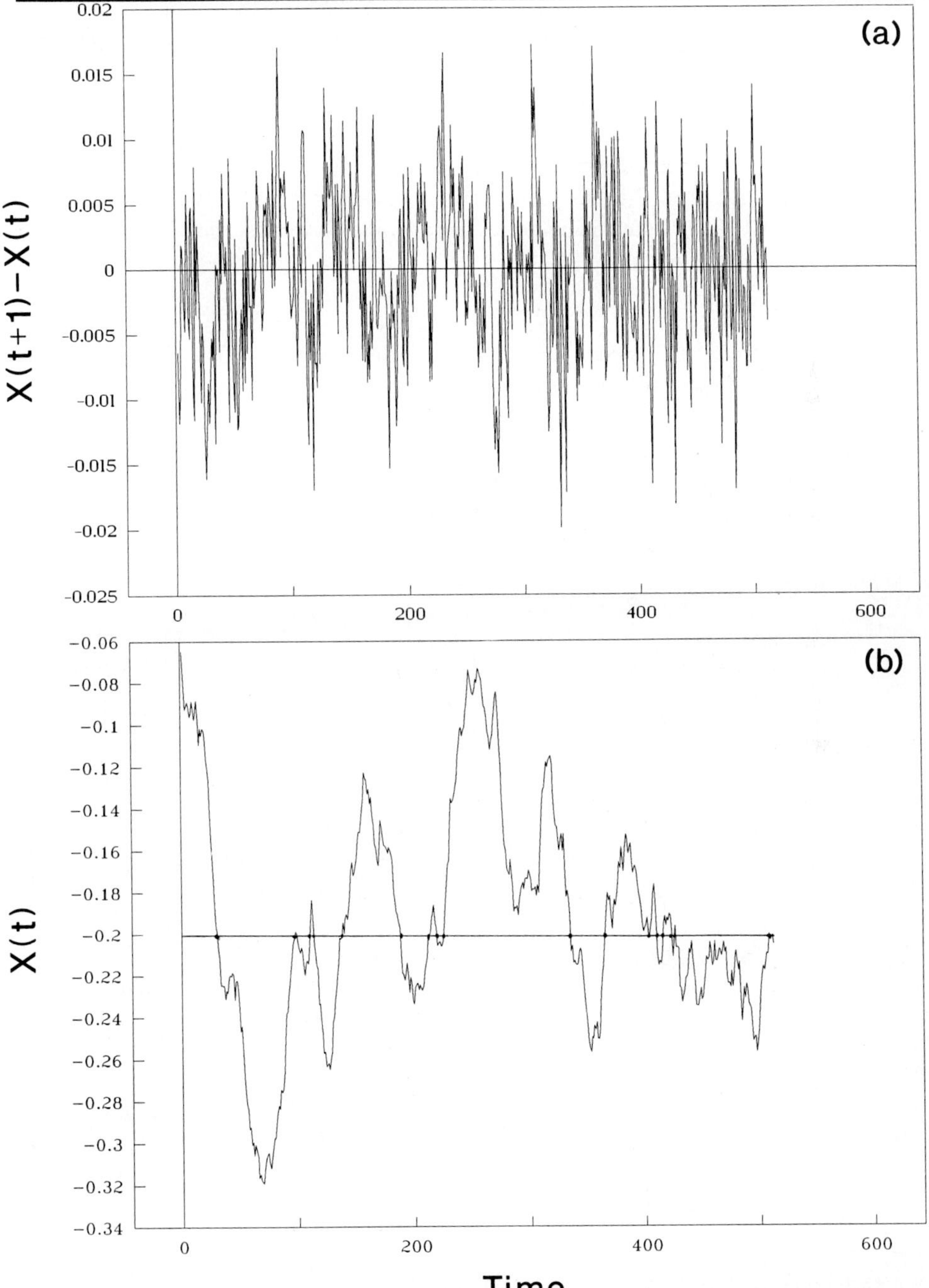

Figure 9.2 (a) A discrete fractional Gaussian noise, with $H = 0.7$, representing the increments of the fractional Brownian motion in Figure 9.2b. and (b) A 512-point fractional Brownian motion, with $H = 0.7$. Generated by the successive random addition method described in Saupe (1988). The points represented by the intersection of the motion and the horizontal line represent a zeroset (more precisely, an isoset) for $X(t) = -0.2$.

now nearly 100 units. Also shown is the same series, with the Y axis expanded by $10^{1/2}$. The range of this series is about 25 units, close to that of the original.

White noises and Brownian motions differ from their fractional counterparts in that fractional increments have very long range correlations. Let $X(0)$, $X(T)$, and $X(-T)$ be the values of a fractional Brownian motion at an arbitrary time 0, at some time T in the future, and an equal time $-T$ in the past. The increments are $X(0) - X(-T)$, the past increment, and $X(T) - X(0)$, the future increment. The average correlation between all past and future increments is given by

$$C(t) = 2^{2H-1} - 1 \tag{9.2}$$

(Mandelbrot, 1983; Feder, 1988). Notice that, when $H = 0.5$, $C(t) = 0$; for Brownian motions, future increments are independent of past increments, and are a white noise.

For fractional Brownian motions, however, all future increments are correlated with past increments and this correlation is independent of time. In cases where $H > 0.5$, $C(t)$ is positive, and positive increments in the past predict positive increments in the future, and negative increments in the past predict negative increments in the future. This feature is known as *persistence*; trends in one direction tend to persist (Feder, 1988). This persistence can be clearly seen in Figure 9.2b. Mandelbrot and Wallis (1968) termed the effects

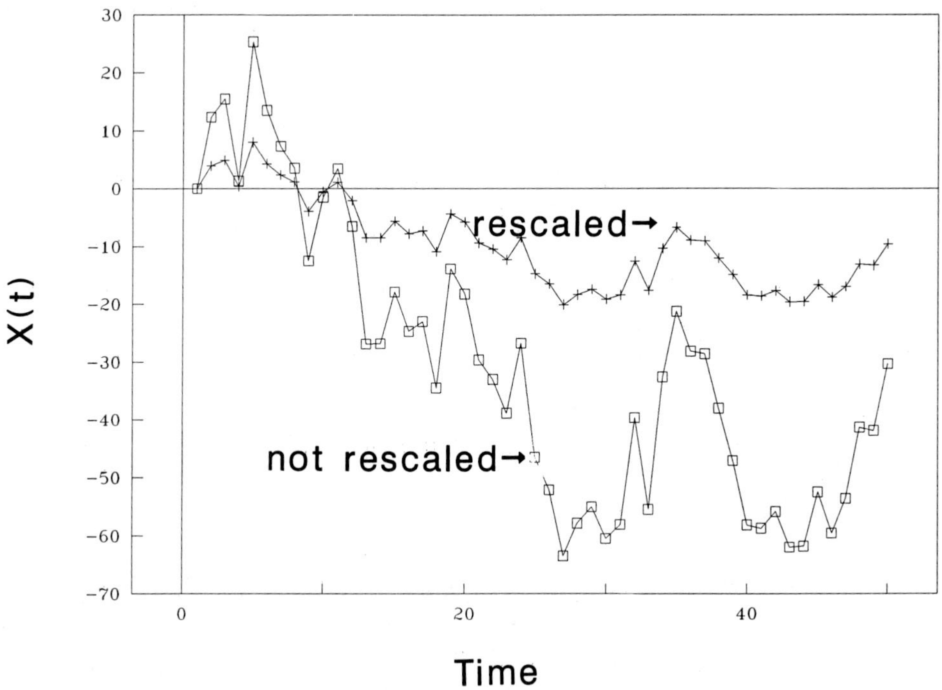

Figure 9.3 The first 50 points of the Brownian motion shown in Figure 9.1b. The line depicted with boxes represents a 10 times expansion of both axes. The line depicted with crosses represents a 10 times expansion of the time axis and a square root of 10 expansion of the $X(t)$ axis.

of persistence in flow and precipitation records the Joseph effect (seven good years, then seven bad years). Conversely, antipersistence holds for $H < 1/2$; $C(t)$ is negative, so that positive increments in the past predict negative increments in the future, and vice versa.

ZEROSETS AND LÉVY DUSTS

One of the very first fractal models developed is the *Lévy dust* (Mandelbrot, 1965, 1983). This model assumes that events occur during a large number of episodes of short duration, which are distinctly clustered in time and are separated by intervals (*gaps*) of various lengths. The distribution of gap lengths U_ε greater or equal to length u is

$$P(U_\varepsilon \geq u) = \left(\frac{u}{\varepsilon}\right)^{-D} \tag{9.3}$$

where ε is the minimum level of resolution and D is the fractal dimension. Lévy dusts are truly self-similar, and this relationship is known as the *uniformly self-similar law of probability* (Mandelbrot, 1965). In addition, the positions of the gaps are random with respect to each other.

Lévy dusts and fractional Brownian motions are related by the concept of the *zeroset*, the set of all points of $X(t)$ at which $X(t) = 0$ (Figure 9.2b). The distances between points in the zeroset follow the uniformly self-similar probability distribution and are thus a Lévy dust. The relationship between D for the Lévy dust and H for the fBm is $D = 1 - H$ (Mandelbrot, 1983; Voss, 1988). Note that constants other than zero can also be used; these sets of points are *isosets* for that particular value. The uniformly self-similar law of probability illustrates a major difference between fractal models and such standard probability models as the Poisson distribution. Fractal probability distributions have much larger and longer tails.

In sum, fractional Gaussian noises and corresponding fractional Brownian motions and Lévy dusts comprise the basic mathematical models for fractal time series. Fractional noises and motions are self-affine and differ from the familiar white noise and Brownian motions in possessing long-range correlations.

TECHNIQUES FOR THE ANALYSIS OF FRACTAL TIME SERIES

Any geologic system that is known to exhibit irregularities over several orders of temporal magnitude can be examined to determine whether this pattern is consistent with a fractal structure. Many of the techniques for time series analysis, such as Fourier analysis and autocorrelation, are by now quite standard (Yevjevich, 1972). In this section, we will examine the results of these methods when applied to fractal time series. In addition, we will review a technique, rescaled range (*R/S*) analysis, which provides a measure of persistence. These methods allow us to determine whether a given real-time series can be mathematically described as being fractal.

Fourier Analysis

Time series can be represented as the sum of sinusoidal curves of different frequencies and amplitudes (Yevjevich, 1972; Davis, 1986). The contribution of each component curve to the total variance of the series is plotted against its frequency to produce a power spectrum or periodogram, where power is the square of the amplitude. The technique that produces a periodogram from a time series is termed spectral or Fourier analysis.

A simple periodic sine wave will be represented as a single peak at a particular frequency in the spectrum. A time series with several embedded periodicities will have several distinct peaks. In contrast, since white noise has no characteristic periodicities, its power spectrum will be essentially a straight horizontal line, representing a broad-band spectrum with no distinct peaks.

The key feature of fractal time series is that they are self-affine and thus show fluctuations over multiple ranges of temporal resolution. As a result, these time series show a distinct power spectrum, the $1/f$ noise. Rather than showing distinct peaks, these spectra are broadband (Goldberger and West, 1987). More importantly, the power of each frequency component is inversely proportional to some power β (between 0 and 3) of the frequency; thus, the term $1/f$. $1/f$ noises appear to be ubiquitous in nature; they are found in Nile stream flows (Mandelbrot and Wallis, 1969a), heart physiology (Goldberger and West, 1987), and even music (Voss, 1988).

For white noise, β equals 0, so that $1/f^{\beta}$ equals 1 and the spectrum is horizontal. In the case of Brownian motion, β equals 2 and the power declines proportionally to the frequency squared, and similarly for fractional Gaussian noises $-1 < \beta < 1$ and for fractional Brownian motions $1 < \beta < 3$ (Voss, 1988; Shtatland, personal communication). Note that the β value for a motion is equal to the value of the corresponding noise +2.

The relationship between H and β for a fractional Brownian motion is given by

$$\beta = 2H + 1 \tag{9.4}$$

(Saupe, 1988; Voss, 1988), while for the corresponding fractional Gaussian noise,

$$\beta = 2H - 1 \tag{9.5}$$

(Mandelbrot and Van Ness, 1968; Mandelbrot and Wallis, 1969a).

Fourier analysis thus allows the researcher to distinguish whether a series is a noise or a motion and should be the first technique applied (Voss, personal communication). The value of β for a given time series is determined by converting both frequency and power to logarithms and determining the slope of the relationship.

Autocorrelation

We next turn to three techniques that examine lags in time series. Autocorrelation is the correlation, for a particular time series, between the values $X(t)$ and $X(t + d)$, where d is the lag. An autocorrelogram shows the values of the autocorrelation over a range of lags. The expected correlation for a standardized (mean = 0, variance = 1) fractional Gaussian noise is a function of d and H:

$$C(d, H) = 2^{-1}|d+1|^{2H} - (2|d|^{2H} + |d-1|^{2H}) \tag{9.6}$$

(Mandelbrot, 1971, 1983). Note that when H equals 0.5 (white noise) $C(d, H)$ equals 0 (as required), and the increments are uncorrelated. For H values greater than 0.5, however, $C(d,H)$ is always positive and never reaches 0; if H is less than 0.5, $C(d, H)$ is always negative and never reaches 0. This again illustrates the long-range correlation of fractional Gaussian noises.

For Brownian and fractional Brownian motions, the value at a particular time is highly dependent on values at previous times. As a result, these nonstationary series show a very slow decline of the autocorrelation with the lag (Pankratz, 1983). They are best analyzed by converting them to their corresponding noises and examining the autocorrelation of the noises. If the series can be represented as a fractional Brownian motion, then its increments will be a fractional Gaussian noise with long-range correlations.

Semivariograms

The second technique to measure time lags use semivariograms. Semivariograms are a statistical technique developed to study the distribution of ore reserves (Clark, 1979; Davis, 1986). They measure the dependence among points in a spatial or temporal series that are separated by a lag d. For example, consider 10 elevations (1 to 10) an equal distance apart along a transect. The pairs of points a lag of 2 apart are 1–3, 2–4, 3–5, 4–6, 5–7, 6–8, 7–9, and 8–10. The semivariance for that lag is equal to the variance of the differences in elevations between the pairs of points. A semivariogram is a plot of the semivariances against different values of the lag.

Because semivariograms measure the degree of dependence between points an arbitrary distance apart, they are readily applied to the study of fractal series. For a stationary series, the semivariance at a particular lag can be directly calculated from the autocorrelation by

$$\text{semivariance} = (1 - \text{autocorrelation}) \bullet (\text{variance of entire series}) \tag{9.7}$$

(Davis, 1986). Since the autocorrelation is zero for a white noise, the semivariance is equal to the variance of the entire series. For discrete fractional Gaussian noises, the semivariance asymptotically approaches this value.

As is the case for autocorrelation analyses, fractional Brownian motions should be transformed to their Gaussian noise equivalents. An excellent review of the use of semivariograms in fractal analysis is given by Hewett (1986).

Rescaled Range

A third technique that examines the relationship between values a given lag apart is rescaled range (R/S) analysis, which was developed by Hurst (1951) for describing the empirical distribution of river flows. He hoped to develop a method for determining the size of a reservoir that, given a particular stream discharge record, would be necessary to guarantee a fixed outflow. Hurst (1951) and Mandelbrot and Wallis (1969b and c) applied this method of analysis to numerous long-term records (see also Feder, 1988).

Consider a time series of length T, comprising a series of independent observations $x(t)$. The running sum of the series, from its beginning to some time u, where $u \leq T$, is

$$X^{*}(u) = \sum_{t=0}^{u} x(t) \tag{9.8}$$

and $X^{*}(0) = 0$. The values of $x(t)$ can thus be thought of a noise and $X^{*}(u)$ as the corresponding motion.

The average value of $x(t)$ over the entire interval T equals $X^{*}(T)/T$. We now determine the accumulated departure $D(u)$ of each value of $X^{*}(u)$ from this average by:

$$D(u) = X^{*}(u) - t\frac{X^{*}(T)}{T} \tag{9.9}$$

and let $R = \max D(u) - \min D(u)$ be the range of these accumulated departures. To compare different time series of different scales, we divide R by S, the standard deviation of the original series, and

$$S = \left(\frac{X^{2*}(T)}{T} - \frac{X^{*2}(T)}{T^{2}}\right)^{1/2} \tag{9.10}$$

For short time series, it can be expected that the range of deviations from the mean should be small, whereas longer series should have larger deviations. Hurst found that the ratio R/S generally follows the empirical relationship

$$\frac{R}{S} \propto T^{J} \tag{9.11}$$

where J is known as the *Hurst exponent*. J measures the degree to which the range increases with increasing time span.

Simulations have shown that for a series of Gaussian random variables (white noise) the value of J should be 0.5 (Mandelbrot and Wallis, 1969b, 1969c; Feder, 1988). Hurst (1951) found, however, that most natural processes showed J values in excess of 0.5 and were typically in the range of 0.7 to 0.8. For example, river discharges show an average J value of 0.72. This implies that the range of values increases much more rapidly with series length than would be expected from a Brownian random process.

Mandelbrot and Wallis, in a series of papers (1968; 1969a, b, c), showed that discrete fractional Gaussian noises have J values >0.5, with J equal to H. Therefore, J values greater than 0.5 for a time series indicate that it can be represented as a fractional Gaussian noise, with accompanying long-range correlations or persistence. These correlations are responsible for the large J values, since persistence produces longer periods of deviation from the mean in one direction and therefore wider ranges.

In summary, a continuous time series consistent with a fractional Gaussian noise model exhibits long-range autocorrelations or semivariances, a $1/f$ power spectrum, and J values in rescaled range analysis greater or less than 0.5. Discontinuous events can be examined for fractal structures by determining whether their distribution follows the uniformly self-similar law of probability discussed above.

FRACTALS AND THE REAL WORLD

Fractals, like planes or spheres, are mathematical constructions used to model nature; that is, real-time series are not fractals but instead may have a fractal structure. Are fractal models of time series geologically realistic? We would argue that a fractal distribution of events is the expected product in many systems. A number of possible mechanisms for generating fractal geologic patterns can be suggested.

In many systems, controlling processes act at different time scales. Imagine a light bulb hooked up to a series of switches. The first switch is on approximately 1/3 of the time, the second switch about 1/9 of the time, the third switch 1/27 of the time, and so on. The light is on only when all switches are on at the same time. The periods of time when the light flashes will form a Lévy dust. Such a hierarchical system of controls will thus produce a fractal distribution of hiatuses (Plotnick, 1988). This process is equivalent to the fractal-generating "overlapping trema generators" discussed by Mandelbrot (1983).

Numerous geologic systems may have such a hierarchical structure. For example, Mississippi river subdeltas result from avulsions of the main channel. These breaches occur every few decades. At longer time scales, the entire delta complex switches. Other processes also turn deposition on and off at different time scales. All processes must be "on" at the same time for deposition to occur.

Similarly, it is possible that a fractal pattern in depositional environments could be caused by a hierarchical set of climatic controls, all operating at different frequencies and influenced by plate tectonics, Milankovich cycles, solar variability, and others. For a particular climatically controlled depositional environment to exist, all cycles may need to be in phase.

Imagine a geologic process, such as initiation of sediment transport, that only occurs when a critical value of some driving variable is exceeded. If the value of the driving variable fluctuates around the critical level as a fractional Gaussian noise, then the time intervals between occurrences of the process will follow a Lévy dust distribution.

SELF-ORGANIZED CRITICALITY

The fractal structure of many time series may be created by nonlinear dynamics of the physical system. One example of how this might work has been outlined by Bak and coworkers who have argued that "fractals in nature originate from self-organized critical dynamic processes" (Bak and Chen, 1989; see also Bak et al., 1988). Many physical systems have critical states, which are conditions where the properties of the system change suddenly. An example is the sudden loss of magnetism when a bar magnet is heated above a certain temperature. This temperature defines a critical state. Bak has suggested that natural systems that are open (energy or matter can enter and leave), have spatial and temporal degrees of freedom, and are dissipative (energy is lost within the system) naturally evolve into a critical state, termed *self-organized criticality.* At this critical state, the systems have fractal spatial structure and $1/f$ temporal behavior.

These properties may be characteristic of many systems of geologic interest, such as mass movements and earthquakes (Bak and Chen, 1989), so fractal patterns should be common in geologic phenomena. For example, Bak et al. (1988) used sand piles, such as those produced by children at the beach, as a heuristic model for self-organized criticality. Based on a computer simulation, they suggested that the sizes of avalanches in sand piles should be fractally distributed and that the frequency of avalanches should obey a $1/f$ distribution. Another possibility is that fractal structure is a product of the fractal structure of some controlling variable. For example, turbulence is a nonlinear phenomenon with a fractal structure that directly influences sediment transport and deposition. There is also evidence for a fractal structure for climatic variability, which may produce a fractal occurrence of events such as floods (Nicolis, 1987).

AN EXAMPLE: SEDIMENT TRANSPORT IN GRAVEL-BED STREAMS

We are exploring patterns of sediment transport and deposition in streams, particularly gravel-bed streams, and determining whether these processes can be modeled using techniques of nonlinear dynamics and fractal mathematics. Gravel-bed streams are rivers

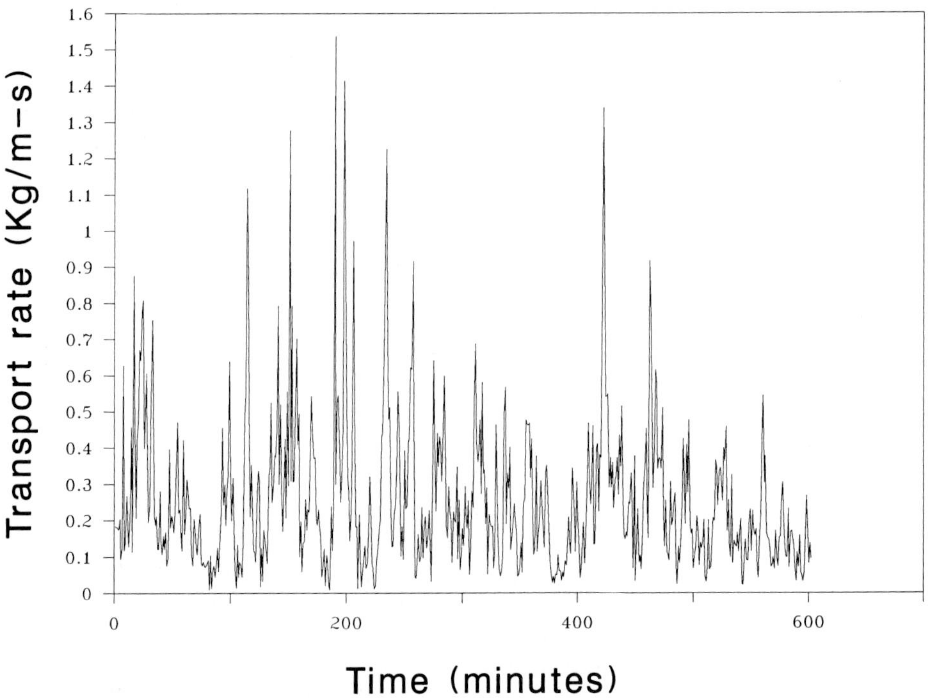

Figure 9.4 Sediment transport rate time series measured every minute for a 10-hour period in the East Rosebud River, Montana. The stream bed is heterogeneous gravel, and flow conditions were constant during the period of measurement.

whose beds are predominantly gravel, although significant amounts of finer material are often hidden in the subsurface. The bed material and the material in transport are thus heterogeneous.

Our technique is to collect field data on hydraulic processes and associated sediment movement in these streams. By analyzing the structure of these time series, we gain insights into the underlying processes. We will now describe one such time series.

Data Collection

Prestegaard and others (1988) collected time series data on bedload transport rates (mass of sediment passing a particular point per unit time) in the East Rosebud River, Montana, in July 1988. One data set was collected on a mid-channel bar with a flat gravel bed. The flow and shear stress were just above the levels required for initiation of motion, and no dunes or other features were observed on the bed. Hydraulic characteristics (water surface gradients, vertical velocity profiles, bed shear stress, mean velocity, and depth) were measured at the site.

Bedload transport rate was measured every minute for 10 hours in a small chute channel 3 m wide and 0.42 m deep. The depth of flow varied less than 5% over the 12 hours spent at the site. Bedload was collected using two identical Helley–Smith samplers. After one sampler had been on the bed for 60 seconds, it was raised and exchanged for the other sampler. It took 5 seconds to change samplers. The bed was kept under observation at all times and care was taken not to disturb the bed when samplers were changed. The samples were weighed, bagged, and retained for particle size analysis.

Figure 9.4 shows the measured bedload transport rate in kilogram/meter-second (kg/m-s) for the 10-hour period. The bedload transport rate varies over several orders of magnitude, but it does not show clearly periodic behavior that might be caused by sediment dunes or waves. We wish to determine whether these variations are simple independent and random variations around a mean (a white noise) or whether there is an underlying fractal structure to the time series.

Data Analysis

A log–log power spectrum derived from the data is shown in Figure 9.5. Also plotted are the theoretical slopes for $H = 0.7$, 0.8, and 0.9. Although somewhat noisy, the transport-rate data appear to show $1/f$ type frequency dependence, with H greater than 0.7 (a linear regression yields an H value of approximately 0.9). These values are consistent with a discrete fractional Gaussian noise.

The result of an autocorrelation analysis of the transport-rate time series is shown in Figure 9.6. Also shown are the predicted autocorrelations of a fractional Gaussian noise with an H value of 0.7, based on Equation 9.6, and an autocorrelogram for a 512-point simulated fractional Gaussian noise. The transport-rate time series indicates good evidence of long-range autocorrelation, again consistent with a discrete fractional Gaussian noise of H value greater than 0.7.

Figure 9.7 shows the results of an R/S analysis for the gravel transport time series in the form of a *pox diagram* (Mandelbrot and Wallis, 1969b). To construct this diagram, the original time series was standardized and then divided into nonoverlapping subseries of varying length T. The values of R/S are then determined for each subseries. A regression line fitted to the plot of R/S versus T finds the value of J by

$$\log\left(\frac{R}{S}\right) \propto J \log(T) \tag{9.12}$$

The J values obtained from the sediment transport time series is 0.67. Also shown in Figure 9.7 is the theoretical slope for $H = J = 0.8$.

In sum, our time series has properties consistent with a fractional Gaussian noise, including long-range autocorrelations, an $1/f$ power spectrum, and a rescaled range analysis J value greater than 0.5. The apparent value of H is in the range of 0.7 to 0.9 and thus indicates persistence.

The representation of the sediment transport data as a fractional Gaussian noise has several implications. First, the presence of persistence indicates that sediment transport

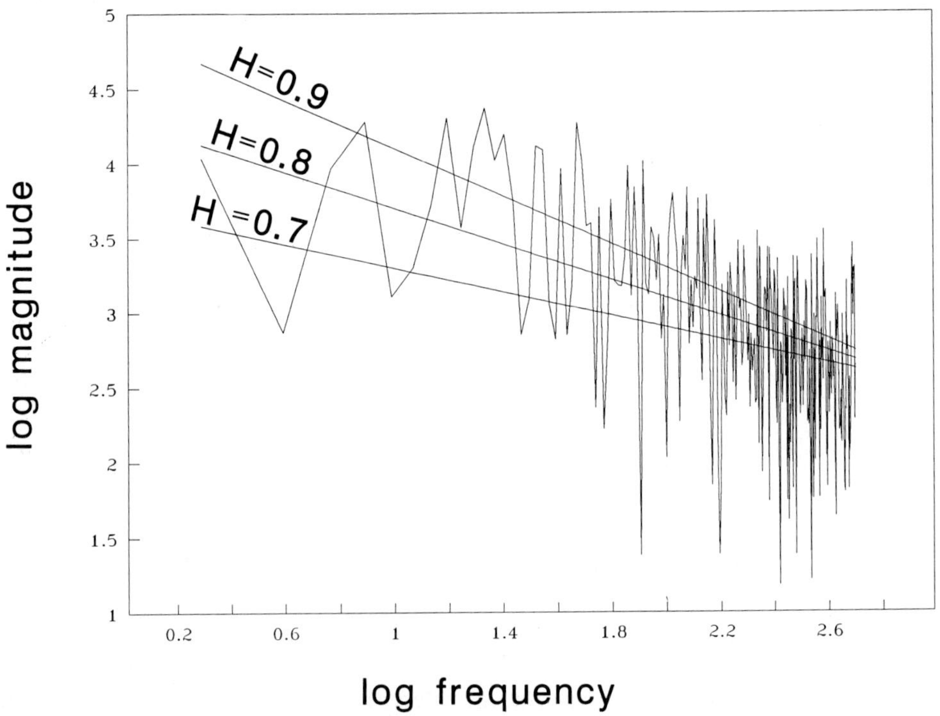

Figure 9.5 Spectral (Fourier) analysis of the sediment transport rate. Log power (magnitude) versus log frequency. Notice $1/f$ type behavior. Also shown are theoretical lines for $H = 0.7$, 0.8, and 0.9.

rates at different times are not independent. This persistence may be generated by the entrainment mechanisms of heterogeneous material. The entrainment of large particles enables hidden smaller particles to be released from the bed, producing large transport events. The resulting release of fine sediment may lead to smoothing of the bed and further entrainment of large particles. Second, because of persistence, short-term averages are likely to be very different from long-term averages. Long observation times are thus necessary to ensure an adequate description of the time series.

Identifications of fractal patterns in the time series data provide us with information about the sediment transport processes. Examination of other bedload transport time series for different bed and flow conditions could provide considerable insight into sediment transport mechanisms in heterogeneous beds. Time series statistics can also be used to develop appropriate bedload sampling schemes.

SUMMARY

Fractal time series models are based on generalizations of white noises and Brownian motions. These discrete fractional Gaussian noises and fractional Brownian motions are characterized by self-affinity and long-range correlations or persistence. Fractal distribu-

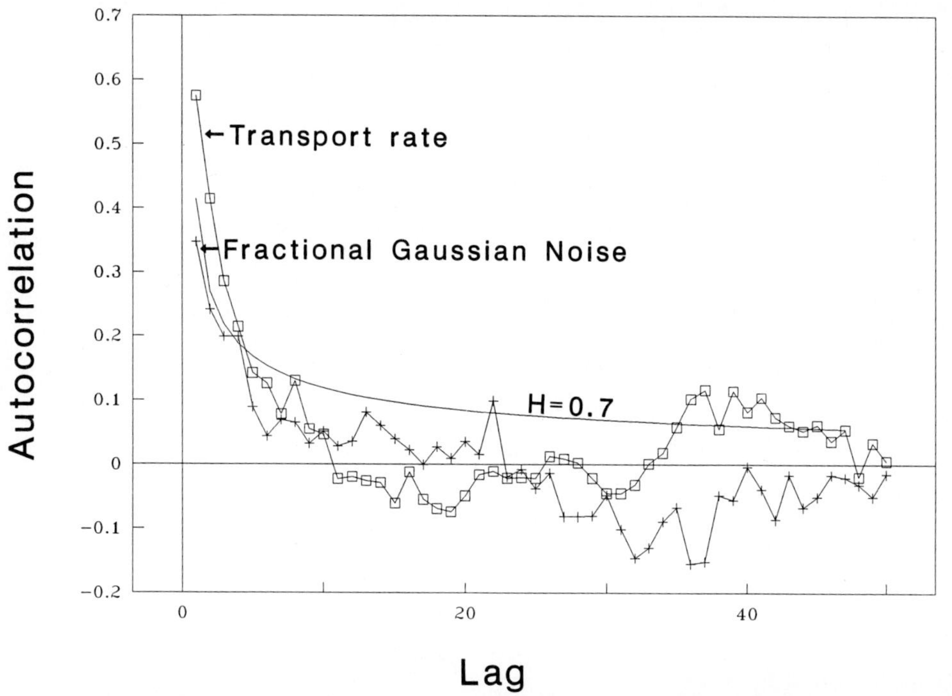

Figure 9.6 Autocorrelograms of the transport rate and of the discrete fractional Gaussian noise illustrated in Figure 9.2a. Also shown is the predicted average autocorrelation for a series with $H = 0.7$.

tion of events can be modeled as Lévy dusts, which are related to fractional Brownian motions by the concept of the zeroset.

Whether a given time series has a fractal structure can be determined using Fourier, autocorrelation, semivariance, and rescaled range analyses. These methods can also be used to determine the magnitude of the persistence or antipersistence exhibited by a time series.

We applied these methods to the analysis of bedload transport data that we collected in a gravel-bed stream. We found that bedload transport fluctuations were not independent. Instead, large values tended to follow large values, and small values tended to follow small values. This pattern of persistence may be generated by the processes controlling the entrainment of heterogeneous material. These results have implications for our understanding of bedload entrainment processes and for sampling bedload materials. These results indicate that, even though a statistical average bedload transport rate exists, this average may be rarely attained over short time scales.

Many natural processes should exhibit time series that have a fractal structure. We believe that future research should be directed not only to uncovering these structures but to understanding their implications and determining the underlying processes responsible for them.

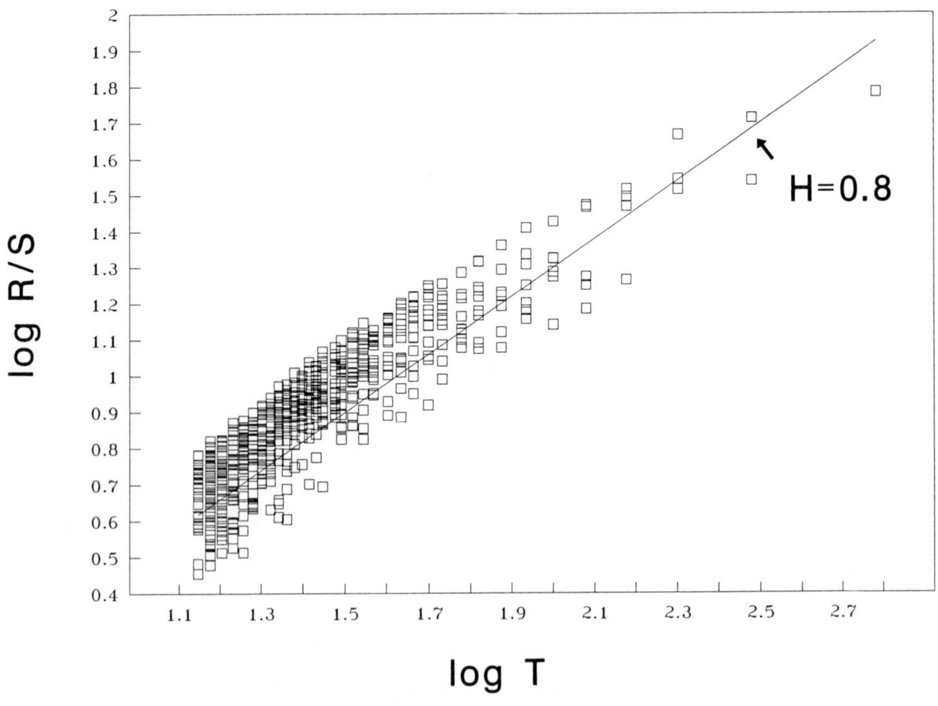

Figure 9.7 Rescaled range (*R/S*) analysis of the sediment transport rate. Also shown is the theoretical line for $H = 0.8$.

ACKNOWLEDGMENTS

Field work for this research was conducted with the able assistance of Alan Bartha, Jason Hicks, Patty Woicek, and Hong-Wei Yang. Useful discussions were held with Richard Voss, Ernest S. Shtatland, and Benoit Mandelbrot. Anne Matherne kindly commented on the manuscript. The editorial comments of Nina Lam and Lee De Cola greatly improved the paper. The research was partially funded by NSF grant EAR-890484.

REFERENCES

Bak, P., and Chen, K. 1989. The physics of fractals. *Physica D* 38:5–12.

———, Tang, C., and Wiesenfeld, K. 1988. Self-organized criticality. *Physical Review A* 38:364–374.

Clark, I. 1979. *Practical Geostatistics*. London: Applied Science Publishers.

Davis, J. C. 1986. *Statistics and Data Analysis in Geology*. 2nd ed. New York: John Wiley & Sons.

Feder, J. 1988. *Fractals*. New York: Plenum Press.

Goldberger, A., and West, B. 1987. Fractals in physiology and medicine. *Yale Journal of Biology and Medicine* 60:421–435.

Hewett, T. A. 1986. Fractal distribution of reservoir heterogeneity and their influence on fluid transport. *Society of Petroleum Engineers, SPE* 15386:1–16.

Hurst, J. E. 1951. Long-term storage capacity of reservoirs. *Transactions of the American Society of Civil Engineers* 116:770–808.

Lenormand, R. 1989. Applications of fractal concepts in petroleum engineering. *Physica D* 38: 230–234.

Mandelbrot, B. 1965. Self-similar error clusters in communications systems and the concept of conditional stationarity. *IEEE Transactions on Communications Technology* 13:71–90.

———. 1971. A fast fractional Gaussian noise generator. *Water Resources Research* 7:543–553.

———. 1983. *The Fractal Geometry of Nature*. San Francisco: W. H. Freeman & Co.

———, and Van Ness, J. 1968. Fractional Brownian motions, fractional noises, and applications. *SIAM (Society for Industrial and Applied Mathematics) Review* 10:422–437.

———, and ———. 1968. Noah, Joseph, and operational hydrology. *Water Resources Research* 4:909–918.

———, and ———. 1969a. Computer experiments with fractional Gaussian noises. *Water Resources Research* 5:228–267.

———, and ———. 1969b. Some long-run properties of geophysical records. *Water Resources Research* 5:321–340.

———, and ———. 1969c. Robustness of the rescaled range R/S in the measurement of long-range statistical dependency. *Water Resources Research* 5:967–988.

Nicolis, C. 1987. Climatic predictability and dynamical systems. In *Irreversible Phenomena and Dynamical Systems Analysis in the Geosciences*, C. Nicolis and G. Nicolis, eds., pp. 321–354. Dordrecht, The Netherlands: D. Reidel Publishing Co.

Pankratz, A. 1983. *Forecasting with Univariate Box–Jenkins Models*. New York: John Wiley & Sons.

Plotnick, R. E. 1986. A fractal model for the distribution of stratigraphic hiatuses. *Journal of Geology* 94:885–890.

———. 1988. A fractal model for the distribution of stratigraphic hiatuses: reply. *Journal of Geology* 96:102–103.

Prestegaard, K. L., Bartha, A., and Yang, H. W. 1988. Temporal variations in bedload transport rates and particle sizes. *Eos* 69:1218.

Saupe, D. 1988. Algorithms for random fractals. In *The Science of Fractal Images*, H. Peitgen and D. Saupe, eds., pp. 71–136. New York: Springer-Verlag.

Shtatland, E. S. 1990. Personal communication.

Turcotte, D. L. 1989. Fractals in geology and geophysics. *Pure and Applied Geophysics* 131:171–196.

Voss, R. 1988. Fractals in nature: From characterization to simulation. In *The Science of Fractal Images*, H. Peitgen and D. Saupe, eds., pp. 22–70. New York: Springer-Verlag.

Wong, P. 1988. The statistical physics of sedimentary rock. *Physics Today* 41:24–32.

Yevjevich, V. 1972. *Stochastic Processes in Hydrology.* Fort Collins, CO: Water Resources Publications.

Part III

RESEARCH AND APPLICATIONS IN HUMAN GEOGRAPHY

CHAPTER 10

CENTRAL PLACE FRACTALS: THEORETICAL GEOGRAPHY IN AN URBAN SETTING

Sandra Lach Arlinghaus
Institute of Mathematical Geography, Ann Arbor, MI

ABSTRACT

The geometry of close packings has been used to model the structure of a hierarchy of trade areas resulting from interurban competition in complex, abstract systems of cities. When fractal geometry is used instead of the traditional approach, unsolved problems are resolved, new results are exposed, and new research directions are suggested by this deep connection between geometry and geography.

INTRODUCTION

The problem of how to partition space "evenly" in the Euclidean plane is solved abstractly by considering geometric packings of the plane (Coxeter, 1961); thus, one close packing of the plane sees circles, centered on points of an equilateral triangular lattice, expand outward until all interstitial space is filled with colliding circles compressing into regular hexagons. The efficiently arranged hive of the bee exhibits this phenomenon (von Frisch, 1950), as do soap bubbles blown between parallel plates of glass (Boys, 1962; Courant and Robbins, 1958).

The question of how human settlements might share space, even as a purely theoretical question embedded in an abstract geographical space that might resemble the Euclidean plane, has numerous complexities. There are economic issues facing commercial establishments concerning volume of sales and cost of moving goods to market areas, as well as concerns from consumers involving competitive pricing and distance to market. There are topographic issues, which might involve the ease of travel to market. There are political issues, which might involve gerrymandering of, or within, political units. There are social issues, which might involve religious or food preference or any of a number of elements based on cultural tradition. In short, the various components already present in a society, coupled with different perceived needs, might influence decisions as to how settlements share space and how trade areas might take shape around such settlements. These and related issues are topics on which much geographic research has been based. The reader is referred to references by Parr and Jones (1983) and Mulligan (1984) as entries to this literature and to works by Hanjoul et al. (1988) and Griffith (1989), among others, for mathematical interpretations.

CLASSICAL CENTRAL PLACE THEORY: AN ABSTRACT OVERVIEW

The geometry of the classical central place theory, originated by Walter Christaller in the 1930s and corroborated, from a perspective focusing on local rather than on global economic needs, by August Lösch in the 1940s, presents a theoretical model for examining the geographical issue of how cities and towns might share space (Christaller, 1933, 1966; Lösch, 1940, 1954). Michael Dacey's mid-1960s explanation of this geometry provided a clear and compact explanation of its theoretical foundations (Dacey, 1965).

The focus of this paper is on the geometry of classical central place theory and its linkage with self-similar, geometric fractals (Arlinghaus, 1985; Arlinghaus and Arlinghaus, 1989). Because central place theory rests on the notion of hierarchy and because self-similarity does as well, this is a logical place to look for alignment of theoretical concepts from perhaps otherwise disparate disciplines. An effort is made to show how this alignment happens by concentrating on basic issues of central place theory and fractal geometry, thus referring the reader interested in greater mechanical detail to other sources.

GEOMETRY OF THE CLASSICAL CENTRAL PLACE THEORY

The classical central place theory solution to creating a hierarchy of settlements requiring trade areas of different sizes has been to use the close-packing idea (Christaller, 1966; Lösch, 1954). Loosely speaking, large settlements provide a wider variety of goods and services and therefore have a broader support and service area (that is, trade area) than do smaller places. Thus, a net composed of relatively large close-packed hexagonal trade areas, with each hexagon centered on a large settlement as a node, might represent the abstract pattern of how large settlements partition space. Smaller settlements have smaller trade areas, and therefore an associated hexagonal net of smaller cell size, uniformly distributed throughout the net. Any number of nets, each of uniform cell size determined by city or town size (or its rank on some list), may be formed.

To see how the entire configuration of cities and towns of varying sizes might share space, imagine a stack of superimposed layers of close-packed hexagonal nets with cells of fixed diameter within a layer, but with diameter varying from layer to layer (Figure 10.1). Evidently, there are an infinite number of orientations of one layer relative to another within this stack. Because the large settlements also offer the goods and services of the small settlements, a hierarchy, based on nesting of services, becomes evident under orientations that align boundaries. Historically, different styles of orientation have been interpreted in different ways. The examples below display some of the detail of the three most commonly considered cases.

In an abstract geographical space in which movement is equally easy in all directions from every point (an isotropic plain), consider a stack of three layers of hexagonal nets (Figure 10.1). The first layer (layer 1) is composed of small towns, each competing with its nearest neighbors for the space between towns as trade areas. The distance between each pair of rival towns in layer 1 is one unit. Rival towns split evenly the space between them, creating a net of small unit hexagons (dotted lines in Figure 10.1) centered on the rival towns (layer 1 towns will not necessarily be evident in all figures to follow; on occasion, boundaries will obscure them).

The next layer (layer 2) is composed of larger rivals (second largest dots in Figure 10.1) that are next nearest neighbors at a distance of $\sqrt{3}$, when measured relative to the layer 1 rivals (Figure 10.1). In an identical manner, a third layer ("layer 3") of even larger hexagonal trade areas may be created by centering hexagons on the next nearest rivals (largest dots in Figure 10.1). The spacing between rivals in layer 3 is also $\sqrt{3}$ (when that

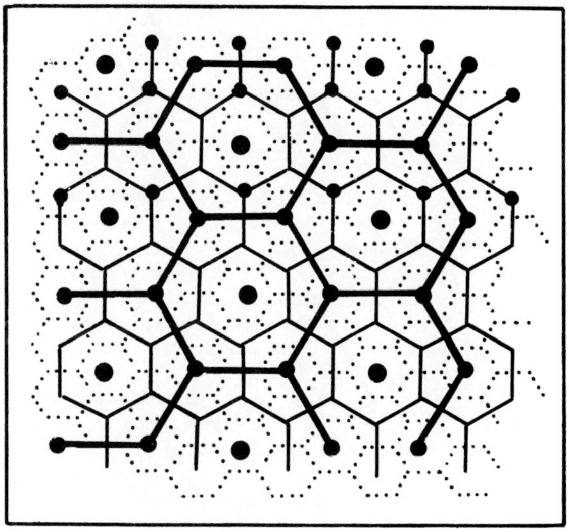

Figure 10.1 A classical central place hierarchy; the orientation is such that each trade area contains the equivalent of three of the next smaller sized trade areas ($K = 3$). Solid lines bound the trade areas of large competing central places (largest dots); dashed lines bound the trade areas of mid-sized competing central places (medium-sized dots, evident near the top of this figure and inferred elsewhere); dotted lines bound the trade areas of smallest competing central places (not shown, but inferred).

spacing is measured relative to layer 2), which is a scaling factor that remains constant from layer to layer, independent of actual trade area size.

When the three layers are superimposed (Figure 10.1), each larger trade area contains the equivalent of exactly three trade areas of the next smallest size. A hexagonal cell in layer 2 contains the equivalent of 1 + (1/3) + (1/3) + (1/3) + (1/3) + (1/3) + (1/3) = 3 (unit) trade areas from layer 1. A hexagonal cell in layer 3 contains the equivalent of 1 + (1/3) + (1/3) + (1/3) + (1/3) + (1/3) + (1/3) = 3 hexagonal trade areas from layer 2, and $3^{(3-1)} = 9$ unit trade areas from layer 1. This stack of hexagonal nets is referred to as a $K = 3$ central place hierarchy because, continuing this process, the trade area of a central place in layer n contains the equivalent of three hexagonal trade areas of layer $(n - 1)$. More generally, the trade area of layer n contains the equivalent of $3^{(n-k)}$ trade areas of layer k. This constant of the hierarchy is simply the square of the distance between rival central places. In this case, the distance between rivals is $\sqrt{3}$; therefore, the constant of the hierarchy is 3.

The orientation of this particular stack of layers was determined by the choice of nearest neighbors as rivals in each layer of the stack. In this case, "nearest" meant closest within the Euclidean metric applied to two-dimensional Euclidean space: trade areas were determined by splitting the distance between rivals. One interpretation of this orientation of nets is a marketing one; consumers are drawn to the nearest market (in the plane), and

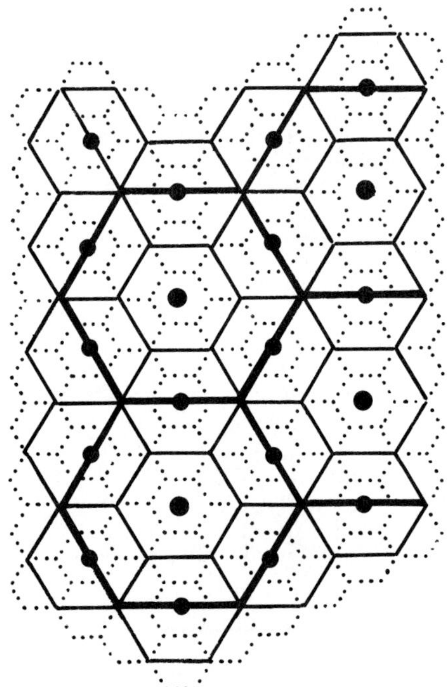

Figure 10.2 A classical central place hierarchy; the orientation is such that each trade area contains the equivalent of four of the next smaller sized trade areas ($K = 4$). Solid, dashed, and dotted lines bounding trade areas are as described in the caption for Figure 10.1.

commercial enterprises target customers nearest them (as cited in numerous texts; for example, Haggett et al., 1977).

When rivals are spaced as closely as possible along lines of the triangular lattice points, a different orientation appears in the stack of three layers (Figure 10.2). The unit hexagons of layer 1 are as before (dotted lines in Figure 10.2); each trade area in layer 2 is centered on a town whose nearest rival is two units away; these trade areas (surrounded by solid lines) contain the equivalent of $1 + (1/2) + (1/2) + (1/2) + (1/2) + (1/2) + (1/2) = 4$ (unit) trade areas. Each trade area of layer 3 (heavy line) is 2 units from its nearest rival and contains the equivalent of $1 + (1/2) + (1/2) + (1/2) + (1/2) + (1/2) + (1/2) = 4$ layer 2 trade areas, and $4^{(3-1)} = 16$ unit trade areas. This stack of hexagonal nets is thus a $K = 4$ central place hierarchy; 4, and its square root, are scaling constants of the hierarchy.

In this configuration, a linear linking of cities is evident. Thus, transportation needs, where linear penetration of settlements is desirable, are traditionally regarded as those that are emphasized with the $K = 4$ orientation. We might suppose that this configuration provided Christaller a backdrop against which to consider why there are successions of towns along routes, progressing from small to large and back to small again (Berry and Harris, 1970; Christaller, 1966).

When the rivals chosen to compete with each other are the third-closest neighbors outside the underlying net (and so the next choice after those that led to $K = 3$ and $K = 4$ configurations), yet another orientation appears in the stack of layers (Figure 10.3). The unit areas are as before (dotted lines in Figure 10.3); each trade area in layer 2 (solid line) contains the equivalent of 7 (unit) trade areas and the spacing between layer 2 rivals is $\sqrt{7}$. Each trade area of layer 3 (heavy line) contains the equivalent of 7 layer 2 trade areas, and $7^{(3-1)} = 49$ unit trade areas. The spacing between layer 3 rivals is also $\sqrt{7}$. This stack of hexagonal nets is thus a $K = 7$ central place hierarchy; 7, and its square root, are scaling constants of the hierarchy.

Because trade area boundaries, in this case, do not pass through relatively large central places, the lines of the network penetrate few of the larger towns, and administrative unity is preserved. Indeed, we might imagine Hitler's reported use of Christaller's central place theory (Berry and Harris, 1970) in planning locations for Eastern European settlements to have employed this orientation designed to focus on tight, central, administrative control. In today's world, we might reinterpret this notion to suggest locations for unwanted land uses such as toxic waste sites.

Abstractly, the construction of central place hierarchies can continue forever, simply by choosing different spacings for rivals in the underlying, infinite, triangular lattice. The infinite set of corresponding K values is characterized by the Diophantine equation $x^2 + xy + y^2 = K$, where x and y are integers (Dacey, 1965). When ordered pairs of integers (x and y), serving as coordinates of points on a triangular lattice, are used as inputs in this Diophantine equation, the resultant K is the central place K value. Classically, there is no particular geometric transformation that will completely characterize each (entire) hierarchy individually, although each has its unique numerical designation in terms of a scaling (constant) K value.

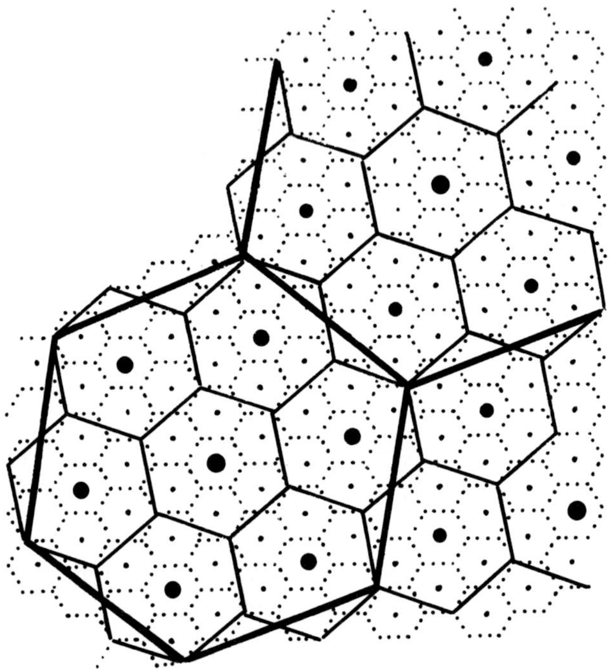

Figure 10.3 A classical central place hierarchy; the orientation is such that each trade area contains the equivalent of seven of the next smaller sized trade areas ($K = 7$). Solid, dashed, and dotted lines bounding trade areas are as described in the caption for Figure 10.1.

FRACTAL GENERATION OF THREE CENTRAL PLACE HIERARCHIES

The following transformation shows how to generate entire central place hierarchies from fractal geometry; a single generator, applied to a hexagonal initiator (as shown below) will produce the hierarchies noted in the section above. Different generators produce different hierarchies; however, the transformation remains the same, independent of the shape of the generator or the size of the initiator. The fractal dimension (Mandelbrot, 1983), determined from geometric entities invariant under this transformation, will serve as a single value to characterize each hierarchy (Arlinghaus, 1985).

Fractal Generation of the K = 3 Hierarchy

Consider scaling a two-sided wedge with vertex angle of 120° to span the six sides of a regular hexagon (Figure 10.4). This wedge is the generator, and the hexagon (Figure 10.4a) is the initiator. Apply the generator alternately to the inside and the outside of the initiator. Remove the edges of the initiator, leaving a shape more highly crenulated than the initiator: this is called the first teragon (Figure 10.4b). The highlighting surrounding the first teragon is designed to make the first teragon stand out, and to suggest the inside–

outside manner of applying the generator to the initiator. The outer edges of the darkened areas are regions of the previous stage "underfit" by the transformation, thereby suggesting the outline of the shape in the previous frame. When the lines suggested by the symmetry of the first teragon are also included (as dashed lines in the middle frame, Figure 10.4b), this figure comprises three hexagons: three scaled down images of the preceding stage. Next, scale the same wedge-shaped generator to fit the side of the first teragon. Apply it, alternately, inside and outside, and remove the edges of the first teragon. The result is the second teragon, shown in the last frame (Figure 10.4c). The highlighting suggests the boundary of the first teragon, and it illustrates the manner in which the second teragon was derived from the first. When interior lines suggested by the symmetry are added, it becomes clear that the second teragon contains three scaled-down copies of the first teragon (and, therefore, $3 \times 3 = 9$ scaled-down copies of the initiator).

The cell sizes and numerical properties within successive frames of this transformation (of applying a wedge-shaped generator to an initiator) meet the classical central place, $K = 3$, criteria outlined above. Iteration of this transformation permits the rapid generation of the layers of the $K = 3$ hierarchy to any level of detail.

Each fractally generated layer contains hexagonal cells of the correct size so that, when resultant nets are superimposed on the initial hexagon on the point distinguished in the interior of each stage in Figure 10.4, the orientation of the nets is also correct for the $K = 3$ configuration (Figure 10.5). This small stack might be regarded as a $K = 3$ tile (Grünbaum and Shepard, 1987). When this tile and identical copies of it are placed side by side in the plane, they fit perfectly with no overlap and no gap. This is equivalent to using a lattice of hexagonal initiators and applying the inside–outside transformation of Figure 10.4 to the entire set. Figure 10.6 displays the stages of how this transformation might proceed, displaying in increasing detail, from left to right, a mosaic of $K = 3$ tiles. The highlighting is intended to suggest how elements of the tiles fit together (rather than to correspond with previous highlighting used to display different relationships). This fractally generated central place landscape (Figure 10.6, right-hand side) is identical to the classical $K = 3$ net as shown in Figure 10.1.

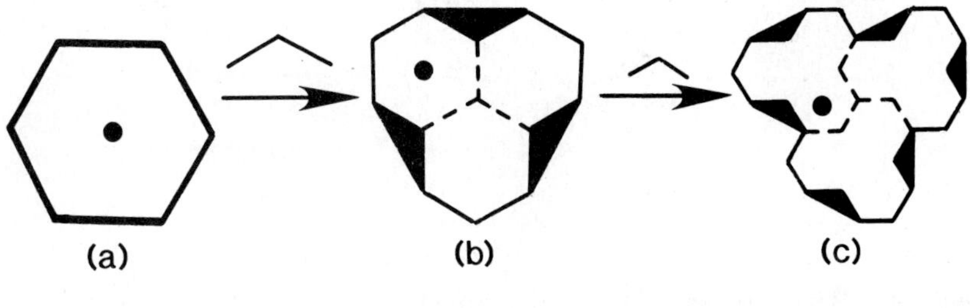

Figure 10.4 Fractal transformation leading to the $K = 3$ central place hierarchy: (a) hexagonal initiator; (b) result of applying the generator (above the left arrow), successively inside and outside, to the initiator; the three adjacent white cells are the first teragon; (c) second teragon, derived from the first teragon, using the scaled-down generator above the right arrow. The dot is distinguished as a point on which the shapes are to be stacked.

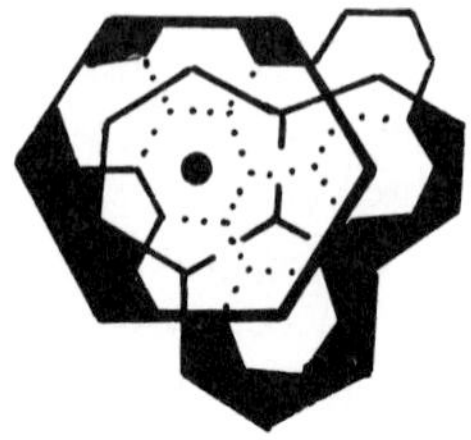

Figure 10.5 Initiator, first teragon, and second teragon from Figure 10.4 are stacked on the distinguished point, forming a tile of overlain nets with the correct $K = 3$ relationship.

Fractal Generation of the K = 4 Hierarchy

In the same vein, when a three-sided trapezoidal generator, with included angles of 120°, is scaled to span the sides of a hexagonal initiator (Figure 10.7a) and applied successively inside and outside the initiator, the resulting first teragon contains four self-similar copies of the initiator (Figure 10.7b). Rescaling the generator to fit the sides of the first teragon and applying it in the inside–outside scheme produces a second teragon (Figure 10.7c) containing four scaled-down copies of the first teragon (and, therefore, $4 \times 4 = 16$ scaled-down copies of the initiator). (This pattern is not unique to central place fractals; see Holroyd, 1983, for example). When the teragons are stacked on the initiator, on the point distinguished in Figure 10.7, a tile showing three layers of the $K = 4$ central place hierarchy emerges with cells of the correct size and in the proper relative orientation (Figure 10.8). When these tiles are placed side by side (that is, when the transformation of Figure 10.7 is applied to a net of hexagonal initiators), a fractally generated $K = 4$ central place landscape emerges (Figure 10.9) that is identical to the $K = 4$ net (Figure 10.2).

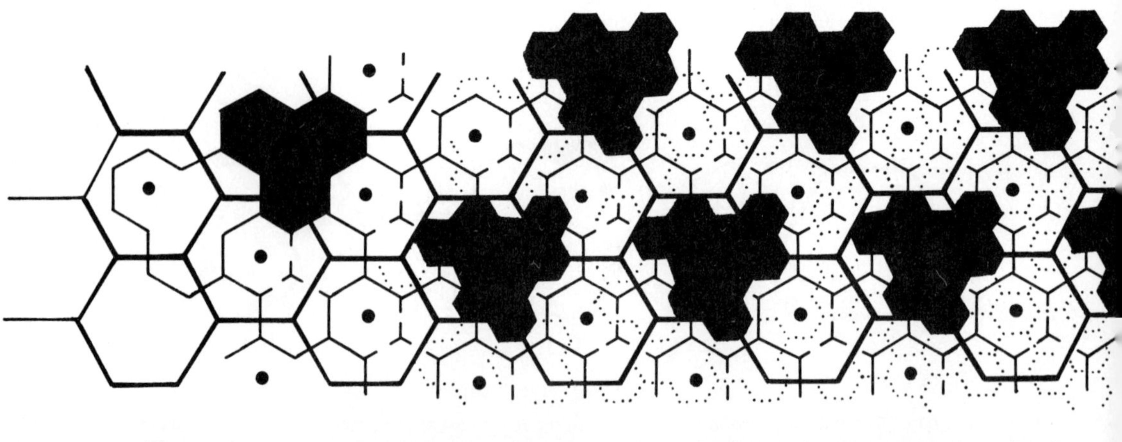

Figure 10.6 A fractally generated central place landscape. The blackened regions suggest how tiles fit together to cover the plane. The level of detail, shown as a continuum, increases from left to right. The white portion of the right side displays detail identical to that of the $K = 3$ central place hierarchy in Figure 10.1.

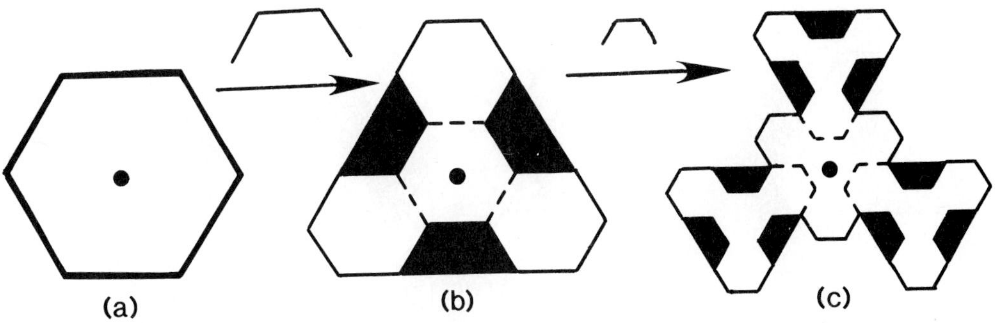

Figure 10.7 Fractal transformation leading to the $K = 4$ central place hierarchy. Details of explanation in text and in caption for Figure 10.4.

Fractal Generation of the K = 7 Hierarchy

When a three-sided, zigzag generator with included angles of 120° is scaled to fit the sides of a hexagonal initiator (Figure 10.10a) and applied in the inside–outside pattern, a first teragon containing seven smaller hexagons is generated (Figure 10.10b). Repeating the procedure produces a second teragon containing seven scaled-down copies of the first teragon (and hence $7 \times 7 = 49$ self-similar copies of the initiator) (Figure 10.10c). When these configurations of Figure 10.10, which exhibit the $K = 7$ characteristics, are superimposed on the point distinguished in Figure 10.10, a portion of a $K = 7$ three-layer stack is formed (Figure 10.11). When this and copies of it are used to tile a plane, the $K = 7$ central place landscape emerges (Figure 10.12) in the same manner it would have had the generator been applied to an entire net of hexagonal initiators (Figures 10.12 and 10.3 are identical).

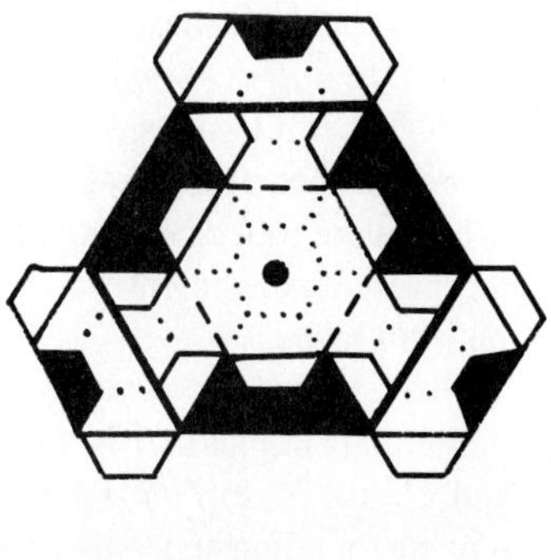

Figure 10.8 Initiator, first teragon, and second teragon from Figure 10.7 are stacked on the distinguished point, forming a tile of overlain nets with the correct $K = 4$ relationship.

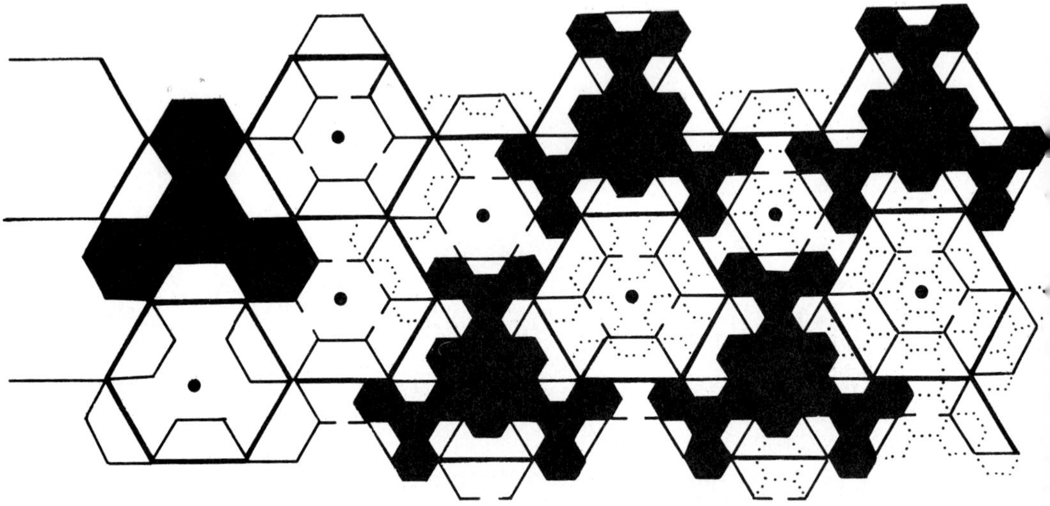

Figure 10.9 A fractally generated central place landscape. The blackened regions suggest how tiles fit together to cover the plane. The level of detail, shown as a continuum, increases from left to right. The white portion of the right side displays detail identical to that of the $K = 4$ central place hierarchy in Figure 10.2.

THE DIMENSION OF CENTRAL PLACE HIERARCHIES

Mandelbrot's (1983) formula for calculating the dimension of fractals

$$D = \frac{\ln(N)}{\ln(\frac{1}{r})} \tag{10.1}$$

where N is generally a count of the number of units of size $1/r$, may be used to calculate a single value to characterize each central place hierarchy (Arlinghaus, 1985; Arlinghaus and Arlinghaus, 1989). Two constants that are independent of size of trade area within the central place fractal construction for a single K value (but which change for different K values) are the number of generator sides and the spacing between rival centers. Thus, when N is taken to be the number of generator sides, and $1/r$ is taken to be the spacing between rival centers (the square root of K), Mandelbrot's D will yield a value to associate with each K value that represents the extent to which space would "fill-in" if the transformation were carried out infinitely.

Thus, for the K = 3 central place fractal, $N = 2$ and 1/r is $\sqrt{3}$, so $D = 1.2618595$. For the $K = 4$ central place fractal, $N = 3$ and $1/r$ is $\sqrt{4}$, so $D = 1.5849625$. For the $K = 7$ central place fractal, $N = 3$ and $1/r$ is $\sqrt{7}$, so $D = 1.1291501$. The higher the value of D is the greater the penetration of surrounding space by network lines. Thus, fractional dimensions systematically support our subjective impressions that the boundaries of the $K = 4$ system are more complicated that those of the $K = 7$ system.

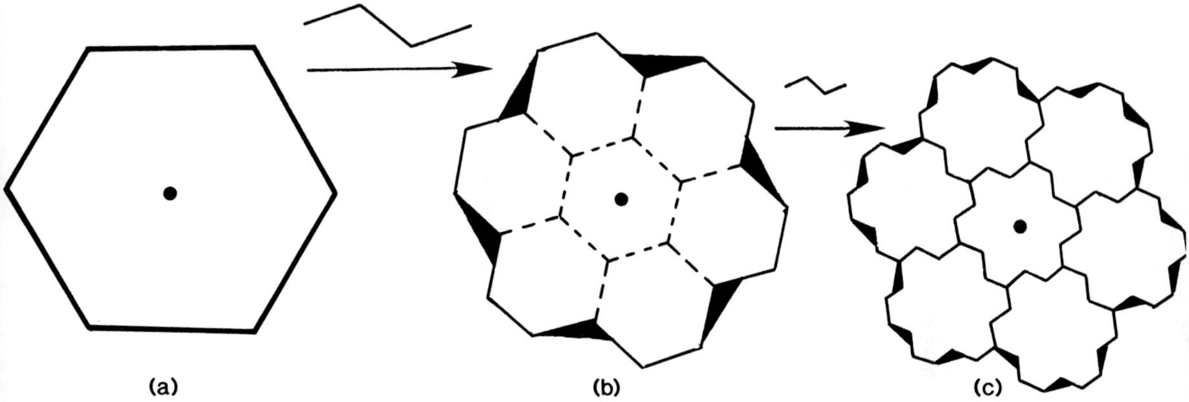

Figure 10.10 Fractal transformation leading to the $K = 7$ central place hierarchy. Details of explanation in text and in caption for Figure 10.4.

GENERALIZATIONS TO HIGHER *K* VALUES

It is possible to construct a central place fractal and to calculate its fractal dimension for all central place *K* values. What is difficult is to choose a generator from the set of infinite possibilities that will yield the desired result. How to do this was conjectured in Arlinghaus (1985); number theoretic properties of the *K* values provide all the information required to produce a generator that will guarantee the correct cell size, shape, and orientation for corresponding nets once a particular *K* has been selected. John U. Marshall proved a theorem in 1975 showing that integers with prime divisors with specified number-theo-

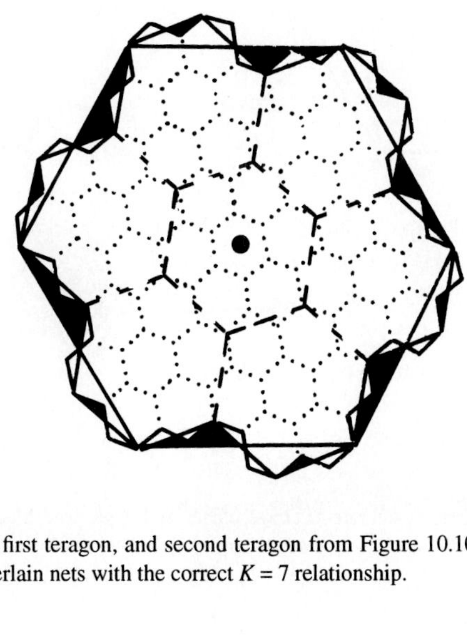

Figure 10.11 Initiator, first teragon, and second teragon from Figure 10.10 are stacked on the distinguished point, forming a tile of overlain nets with the correct $K = 7$ relationship.

retic properties could not serve as central place K values (Marshall, 1975). William Arlinghaus (1989) cast Marshall's theorem in a different number-theoretic light so that he might also prove its converse (Arlinghaus and Arlinghaus, 1989). This latter theorem permitted the author to demonstrate a systematic basis for 1985 conjectures concerning the size and shape of central place fractal generators (and fractal dimensions), and to show how to select "correct" generators to yield central place fractals for arbitrary K (Arlinghaus and Arlinghaus, 1989). Figure 10.13 shows, for example, a generator and first teragon for $K = 76$.

As a side benefit, this theorem also permitted the author to solve Dacey's (1965) open questions concerning which K values have nonunique solutions to the Diophantine generating equation. The reader interested in the mechanics of the solutions to these open questions and of generator choice for arbitrary K is referred to the aforementioned references.

DIRECTIONS FOR FURTHER RESEARCH

A natural, but mechanical, next step would be to automate this procedure to yield clear graphic, computerized displays of central place fractals. The reasons for doing so are twofold. First, it is clear that hexagonal trade areas do not exist as real-world entities anymore than do contour lines. The contour line is an approximation to a level line of a surface.

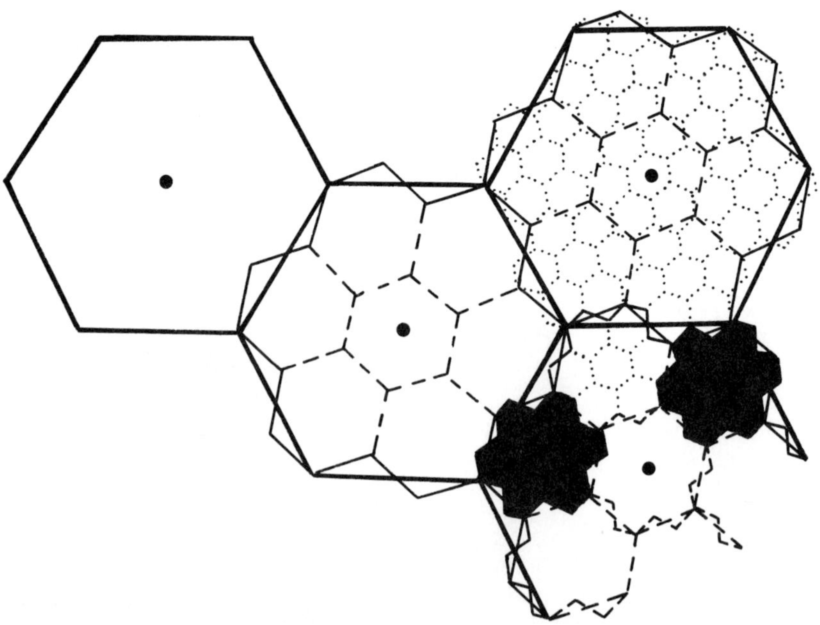

Figure 10.12 A fractally generated central place landscape. The blackened regions suggest how tiles fit together to cover the plane. The level of detail shown increases from left to right. The white portion of the right side displays detail identical to that of the $K = 7$ central place hierarchy in Figure 10.3.

Because the earth is not a regular mathematical surface, equations that describe actual mountains do not exist. This fact does not, however, deny the need to study techniques of integration in order to have a sound grasp of what contouring is all about. Similarly, a sound grasp of central place fractals, from the abstract to the visual, might lead in a number of directions: (1) If we had a computerized system with hexagonal pixels, central place fractals would address directly issues associated with the filling in of shapes formed from these pixels. The theory presented here might be immediately applicable to software based on a pixel arrangement of this sort. (2) Furthermore, it might be productive to modify central place fractals by deriving them from initiators of other shapes: squares to correspond to square pixels, or any number of other shapes of regular polygons or polyhedra. These extensions provide innovative ways of recursively subdividing space for GIS data management and multiscale spatial analysis.

Second, in an effort to remove the condition of polygonal regularity for the initiator, a geometry similar to this one might be based on centrally symmetric hexagons (ones with opposite sides parallel and of equal length, in pairs, or Dirichlet regions) (Kasner and Newman, 1956). Because centrally symmetric hexagons can be derived from triangulations of the plane as well of more general mathematical spaces (Spanier, 1966), it might also follow that the corresponding geometry exhibits properties parallel to those in the central place fractal geometry. Such extensions may suggest new ways of generating useful nested irregular networks from point data for interpolation and visualization. Research on all these topics is currently underway and some has been a continuing effort over the past 16 years.

The above outlines some theoretical directions for future research. We might also wonder what other real-world issues are likely to prompt responses of this sort: (1) The notion of estuary action describes a "crossing" of the land–water boundary, much as the discriminant of a quadratic form in mathematics describes a "crossing" of the x axis; this

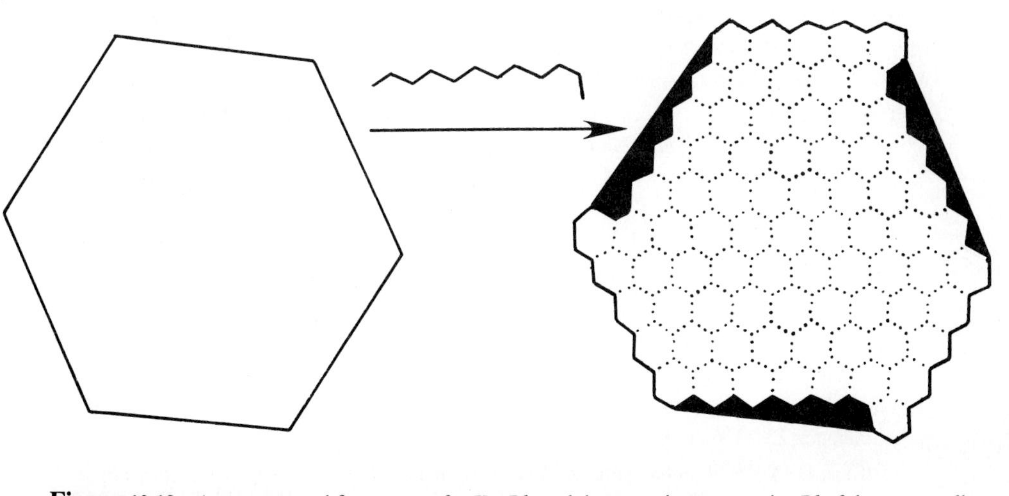

Figure 10.13 A generator and first teragon for $K = 76$; each larger trade area contains 76 of the next smaller trade areas.

might suggest the appropriateness of looking for a linkage between algebra and geometry, such as (but not necessarily limited to) the fractal model, in order to gain insight into problems of acceptable levels of salinity in fresh water. For example, the shape of the Suisun Bay area is similar to the shape of the larger San Francisco Bay area containing it (Arlinghaus and Nystuen, 1986, unpublished paper). Thus, we might use this noted self-similarity to look for corresponding self-similarity of estuary action in these bodies of water that might reflect levels of salinity. (2) The notion of self-similarity and corresponding ideas from fractal geometry might be applied to the idea of measuring rates of energy transference across a soil profile through the spaces between self-similar particles within that profile (Outcalt, 1972; Gleason et al., 1986; Arlinghaus and Nystuen, 1986, unpublished paper; Wheatcraft and Tyler, 1988).

With theory, methodology, application, and possible extension of application in hand, the discovery process will have come full circle. As is characteristic of scientific theory in general, and of spatial theory in particular (dealt with by Christaller, 1966; Lösch, 1954; Bunge, 1966; and others), this discovery process relies on a set of mathematical theorems to capture elements critical to that (spatial) theory. The promise of the wide range of pure mathematical theory (of which fractals are a small part) untapped by geographers offers exciting vistas in spatial theory.

REFERENCES

Arlinghaus, S. 1985. Fractals take a central place. *Geografiska Annaler* 67B(2): 83–88.

———, and Arlinghaus, W. 1989. The fractal theory of central place geometry: A Diophantine analysis of fractal generators for arbitrary Löschian numbers. *Geographical Analysis* 21(2):103–121.

———, and Nystuen, J. 1986. Fractal Constructions. Unpublished.

Berry, B. J. L., and Harris, C. 1970. Walter Christaller: Obituary. *Geographical Review* 60:116–118.

Boys, C. V. 1962. *Soap Bubbles: Their Colours and the Forces Which Mould Them.* New York: T. Y. Crowell. (Based on lectures delivered originally before the London Institution, 1889–90.)

Bunge, W. W. 1966. *Theoretical Geography.* Lund, Sweden: Lund Studies in Geography C, no. 1.

Christaller, W. 1966. *Central Places in Southern Germany,* translated from *Die zen-tralen Orte in Süddeutschland: Eine ökonomisch-geographische Untersuchung über die Gesetzmässigkeit der Verbreitung und Entwicklung der Siedlungen mit städtischen Funktionen* (Jena 1933), by C. W. Baskin. Englewood Cliffs, NJ: Prentice Hall.

Courant, R., and Robbins, H. 1958. *What Is Mathematics*? New York: Oxford University Press.

Coxeter, H. S. M. 1961. *Introduction to Geometry.* New York: Macmillan, Inc.

Dacey, M. F. 1965. The geometry of central place theory. *Geografiska Annaler* 47:111–124.

von Frisch, K. 1950. *Bees: Their Vision, Chemical Senses, and Language.* Ithaca, NY: Cornell University Press (revised edition, 1971).

Gleason, K. J., et al. 1986. Geometrical aspects of sorted patterned ground in recurrently frozen soil. *Science* 232:216–220.

Griffith, D. A. 1989. *Simulating Christaller Central Place Structures: An Algorithm Using a Constant Elasticity of Substitution Consumption Function.* Ann Arbor, MI: Institute of Mathematical Geography.

Grünbaum, B., and Shephard, G. C. 1987. *Tilings and Patterns.* San Francisco: W. H. Freeman & Co.

Hanjoul, P., Beguin, H., and Thill, J.-C. 1988. *Theoretical Market Areas Under Euclidean Distance.* Ann Arbor, MI: Institute of Mathematical Geography.

Haggett, P., Cliff, A. D., and Frey, A. 1977. *Locational Analysis in Human Geography.* London: Edward Arnold.

Holroyd, F. 1983. The geometry of tiling hierarchies. *Ars Combinatoria* 16B:211–244.

Kasner, E., and Newman, J. R. 1956. New names for old. In *The World of Mathematics*, J. R. Newman, ed. New York: Simon & Schuster.

Lösch, A. 1954. *The Economics of Location*, translated from Die räumliche Ordnung der Wirtschaft (Jena 1940), by W. H. Woglom and W. R. Stolper. New Haven, CT: Yale University Press.

Mandelbrot, B. R. 1983. *The Fractal Geometry of Nature.* San Francisco: W. H. Freeman & Co.

Marshall, J. U. 1975. The Löschian numbers as a problem in number theory. *Geographical Analysis* 7:421–426.

Mulligan, G. F. 1984. Central place populations: Some implications of consumer shopping behavior. *Annals of the Association of American Geographers* 74:44–56.

Outcalt, S. I. 1972. The development and application of a simple digital surface-climate simulator. *Journal of Applied Meteorology* 11(4):629–636.

Parr, J. B., and Jones, C. 1983. City size distributions and urban density functions: Some interrelationships. *Journal of Regional Science* 23:283–307.

Spanier, E. H. 1966. *Algebraic Topology.* New York: McGraw-Hill Book Co.

Wheatcraft, S. W., and Tyler, S. W. 1988. An explanation of scale-dependent dispersivity in heterogeneous aquifers using concepts of fractal geometry. *Water Resources Research* 24(4):566–578.

CHAPTER 11

FRACTAL GEOMETRY AND URBAN MORPHOLOGY

Michael Batty

A. Stewart Fotheringham

National Center for Geographic Information and Analysis, Department of Geography, University at Buffalo, NY

Paul Longley

Department of Geography, University of Bristol, U.K.

ABSTRACT

An urban system is viewed here as a fractal structure resulting from the outcome of innumerable and essentially unpredictable decisions. These decisions are captured in terms of urban development by a variety of cluster growth models. These models lead to realistic looking urban forms and provide very different ways of investigating urban growth from traditional models. The urban forms that result from these clustering processes have fractal properties, and the fractal dimension of each structure is shown to be related to a density gradient. This paper thus demonstrates a modeling approach that sheds new light on the urban growth process and links process to form through fractal analysis. The paper highlights attention on the value of simulation and diffusion in the growth of urban systems in much the same vein as Hagerstrand's seminal work. It also recognizes and utilizes the increasing role of visualization in developing urban models.

INTRODUCTION: NEW APPROACHES TO UNDERSTANDING URBAN STRUCTURE

In the last decade, progress in building applicable models of urban spatial structure has been slow. Traditional approaches in land use–transportation modeling based on simulating the location of demographic and economic activities have been little different from their prototypes developed in the 1960s. These models still allocate activities to large tracts often based on census zones using a combination of spatial interaction and macro economic relations. Progress, such as it has been, revolves around model structures that are consistent with micro economic theory (Anas, 1987) and capable of being estimated more rigorously using state of the art econometric methods (Putman, 1990). Developments in urban economic theory, too, have not departed from the paradigm that was first sketched out in the 1960s by Alonso (1964), Muth (1969), and Beckmann (1969), among others, and there has been little progress in generalizing these models to spatial systems that are more realistic than those based on the monocentric isotropic plane. Although there have been some exciting developments in designing dynamic urban models using catastrophe and bifurcation theory (Wilson, 1981; Dendrinos with Mullally, 1985; and Allen, 1982), these developments have been mainly focused on demonstrating the value of thinking about the evolution of urban systems using various types of nonlinear dynamics from which surprising and novel discontinuities in development might emerge.

There are, however, several important issues in the design of more applicable urban theory and models that have not been broached in any comprehensive way. One of these involves the physical form of the spatial system in which urban development takes place. Existing approaches simply assume that urban development can be modeled at a scale where the physical picture of development is perceived as being of little importance, as for example in the case of spatial interaction models. In similar fashion, theoretical models of urban economic systems are based on systems in which the space has to be simplified to the abstract geometry of the line and the plane, where the basic assumption is that such representation is sufficient to demonstrate the quintessential form of spatial structure. Moreover, in none of these traditions is there concern for the ways in which the process of development structures the space in which it occurs.

These difficulties have persisted despite the fact that much of the descriptive literature of urban geography is concerned with the physical form of cities, and the predominant instruments of urban planning continue to be couched in terms of the manipulation of physical space. Rapid progress in computer cartography, geographical information systems, and scientific visualization, generally following new ideas in computer graphics, have not yet been widely employed in making the outputs of computer-based urban models more tractable in visual terms. This is due largely to the fact that the many scales at which such models are operated make their predictions difficult to visualize. There is thus a clear need to rethink and extend some of the physical bases on which such theory and models are postulated if urban models are to be made more relevant and if the new developments in computer-based simulations and graphics are to be properly exploited.

In this chapter we begin to explore some possibilities for modeling urban structure at the physical level of development and for applying and building on a variety of new approaches to physical form that are being pioneered in the physical sciences, particularly in recent research into the growth of particle clusters (Stanley and Ostrowsky, 1986). During the last 10 years there has been a remarkable growth in methods and models for describing and explaining physical structures, particularly those that manifest a common degree of irregularity and self-similarity across a range of scales. These ideas are attractive in that the models are able to generate rich and complex forms from relatively simple processes, thus providing a robust and easily understandable base line from which to proceed (Jullien and Botet, 1987).

In these models, process and form are inextricably linked, thus providing the essential connection between dynamics and spatial structure. The physical systems that are structured by these processes are scaled and measured using fractal geometry (Mandelbrot, 1982; Feder, 1988; Goodchild and Mark, 1987). But these approaches are of much wider significance in that they tie together ideas involving diffusion, far-from-equilibrium structures, irreversibility, allometry, and density, ideas that have been important to urban geography and regional science for nearly half a century. Thus, this line of research has far-reaching implications for theories and models of the central place, rank–size relations between cities, population density gradients, the functioning of urban economic markets, and the development of transportation infrastructure.

MODELS OF URBAN GROWTH AND CLUSTER GROWTH

Analogies between Forms

Urban areas exhibit several characteristics that are not captured by our present models of urban growth and development but that can be understood through cluster growth models. The quickest and most dramatic way to illustrate how recent developments in the simulation of particle clusters provide useful analogies with urban growth patterns is to directly compare urban form with the forms that are generated by the mechanisms of cluster growth. In Figures 11.1 and 11.2, we present four examples of urban forms and contrast these with three typical structures that have been "grown" using the new methods of cluster aggregation and that might be used to visualize urban growth and change.

Leaving aside for the moment the mechanisms that underlie these forms, it is immediately clear from Figures 11.1 and 11.2 that both cities and particle clusters have three spatial characteristics in common. First, cities and particle clusters have the same sorts of irregularity, although the characteristic dendritic patterns in both types of form indicate that there is strong self–similarity across a range of scales. These dendrites are characteristic of transport infrastructure in cities, and they are related to the nodal structure representing the location of commercial–retail centers from the neighborhood to the CBD (central business district). The dendritic pattern is common to every urban area we have examined and is captured in various cluster growth models. Clearly, our models do not replicate the shape of a particular urban area exactly (although we can be very close given certain constraints on our growth process), but they replicate fairly accurately the essence of urban form.

A second feature of urban areas that is captured by cluster growth models is that they exhibit increasing detail with increasing resolution. Urban areas viewed at low resolution appear as bloblike structures with little internal differentiation. With increasing resolution, it becomes clear that what we call urban areas are actually composed of a mixture of urban and nonurban land and that some cities have quite large areas of nonurban land within their boundaries. Cluster growth structures also have this fractal property.

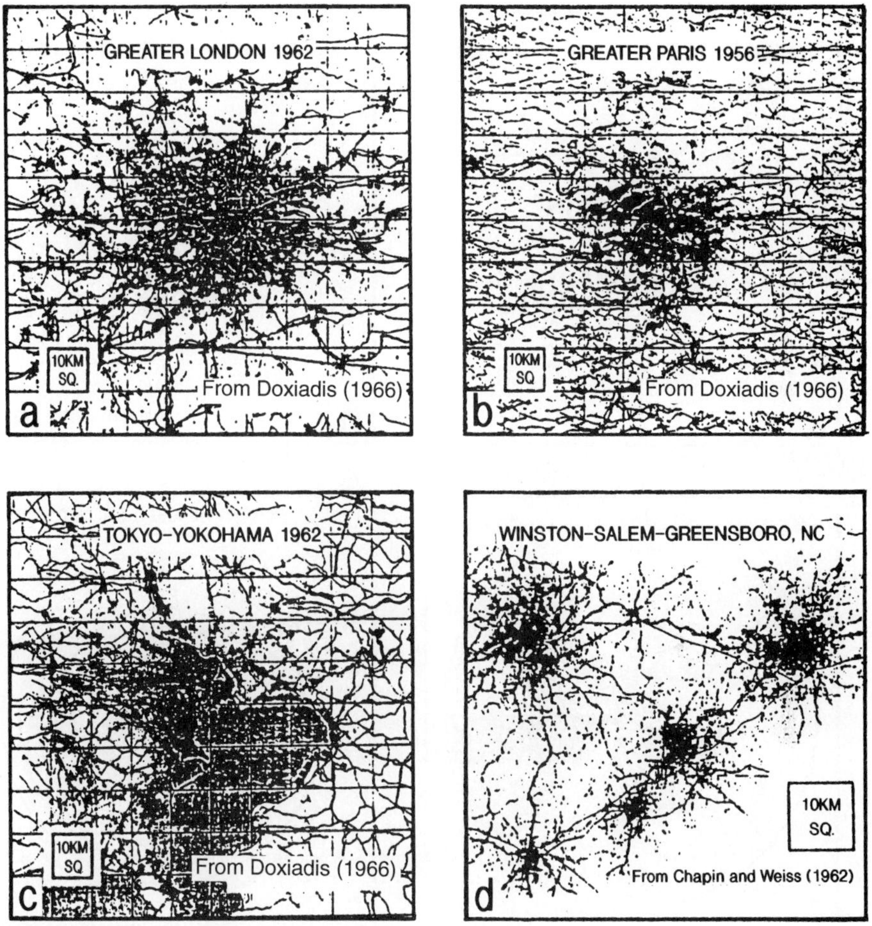

Figure 11.1 Four examples of real urban forms.

A third feature of urban areas is that, despite their obvious uniqueness, many urban areas share such common properties as the presence of density gradients, allometric growth, and the extent to which urban development fills the available space. In terms of this latter property, it is clear that development does not fill the two-dimensional plane, but it clearly fills more space than the one-dimensional line. (Urban development also has a vertical dimension, not analyzed here.) It is in this sense then that the dimension of such growth patterns is said to be fractal, the dimension lying between 1 and 2 in value. In previous work (Fotheringham et al., 1989), we explored the relationships between the fractal dimension of urban structures, their density gradients, and the theory of relative growth or allometry that is manifest in these scaling relations. As such, we have begun to develop a framework that enables cities and clusters of different size to be compared (Mills, 1970; Woldenberg, 1973).

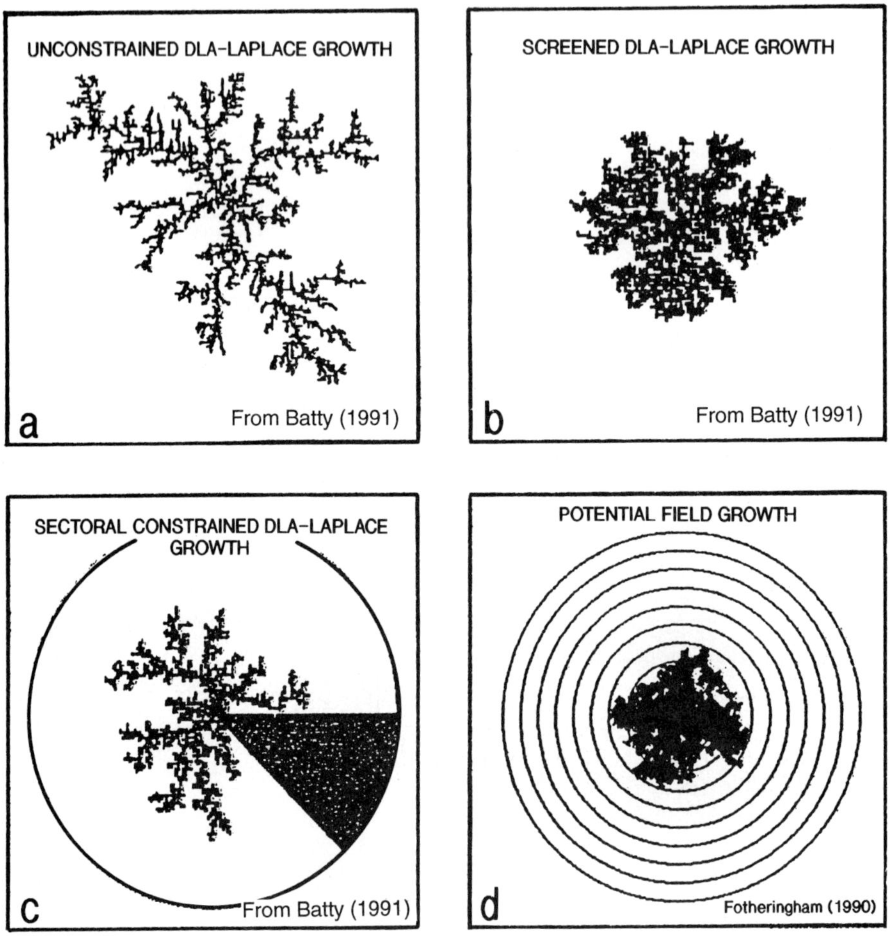

Figure 11.2 Four DLA simulations of urban forms.

Mechanisms of Growth: Process and Form

For the analogy between urban growth and cluster growth to be both useful and powerful, the sorts of mechanisms that give rise to each form of growth must be similar, and it is here that the most difficult comparisons emerge. The processes used to grow particle clusters are irreversible in that, once a particle is fixed in space, it does not move again. This is clearly not the case in cities, where development is reversible (although in the development of the European industrial city since the late eighteenth century, the urban pattern has generally not been renewed). While existing research (Fotheringham et al., 1989; Batty et al., 1989) has only considered irreversible growth processes, there is no reason in principle why these models cannot be extended to deal with reversible development.

The processes that underlie the development of particle clusters are essentially based on diffusion, which is limited by both the extent of the two-dimensional space in which the cluster is grown and the development of the cluster so far. In fact, the processes used to simulate such diffusion are not necessarily those that are at work in real physical systems for, in one sense, they are too simple. This is borne out by the fact that there are several different ways of growing similar clusters that have been exploited in their simulation, each demonstrating the profound and deeply stimulating notion that complexity of form can be generated from the repetitive application of extremely simple processes. The processes in question can be divided into three distinct types: those based on random walks, those based on the breakdown of various fields that the particles both influence and are influenced by, and those based on ad hoc approximations to the interaction of both global and local effects. We shall deal with these by discussing four examples.

First, it is possible to grow dendritic structures such as that shown in Figure 11.2a by initiating a random walk on a lattice in which a focal site has been seeded. This process has been called *diffusion-limited aggregation* (DLA) by Witten and Sander (1981), who first outlined the model. In essence, the process is one in which a particle is released at a far distance from the seed site, beginning a random walk on the lattice moving from one particular site to an adjacent one at each time step. If the particle wanders off the lattice, it is killed and another one begins its walk. Once a particle becomes adjacent to one that already forms part of the growing cluster, it "sticks" (it is fixed) and another one is released. This process continues until a cluster of a certain size has been grown.

The resultant form is not the approximately circular structure or even amorphous mass that might be expected a priori because, once the beginning of branches appear, the chances of parcels of space between the branches being developed are reduced and this further accelerates the branching process. In effect, it becomes practically impossible for the interstices to be filled, and thus gaps appear in the simulated urban fabric. A plausible urban analogy here is that development often takes place along transport corridors, while parcels of land between these corridors remain undeveloped; however, it should be stressed that at this stage the process we are describing is independent of any preexisting transportation network. The relative importance of these gaps increases with distance from the center of the structure, which is analogous to a common finding that urban densities decline with increasing distance from the center of cities.

The types of structure that are grown are dendrites with a fractal dimension of about 1.71 (Witten and Sander, 1981). Although the structure depicted in Figure 11.2a is not particularly citylike, it does exhibit a characteristic of urban structure in that density declines with increasing distance from the center, and it could be argued that it is representative of ribbon development. However, the generating mechanism, that of geographic individuals moving in a completely random manner across space, is unrealistic, and in the sequel we begin to relax this assumption and begin to develop a taxonomy of urban structures.

The second type of mechanism that can be used to grow seemingly identical clusters is based on a potential field around the cluster in which the potential for further growth in the existing cluster is zero, while the potential for growth at a far distance from the cluster is at a maximum. Growth occurs at the interface of the cluster where the probability of growth is given by the gradient of the potential. As in the DLA model, the development of a particular perimeter site on the cluster is randomly chosen. This noise introduced into the field leads the sites of highest potential on the perimeter of the cluster to be those that are at the tips of the growing dendrite, due to the fact that sites nearer the center of the cluster, in the fissures between the tentacles, are effectively screened from growth. This process is similar to the breakdown of an electric field and as such it has been called the *dielectric breakdown model* (DBM) (Niemeyer, et al.,Weismann 1984). Figures 11.2a–c show typical variations in structures from DLA and DBM simulations.

A third set of processes for simulating cluster growth, which will be referred to as *direct screening models* (DSM), are based on evaluating the probability of sites for growth on the cluster's perimeter by assuming a spatial function relating both the local and global properties of the cluster. Such functions might approximate the characteristics of a potential field in which the cluster is growing. For example, Meakin (1983) has introduced a family of functions that simulate the screening properties of every particle in the cluster on every perimeter site; these bear some relation to well-established ideas of population potential and migration fields that have been used fairly extensively to both describe and model urban form (Dorigo and Tobler, 1983). There are a variety of such ad hoc models in the literature.

A fourth mechanism for cluster growth, the *potential field growth model* (PFGM), is a modification of the third and involves the calculation of a potential development field around the growing structure based on the time at which adjacent land parcels were developed. That is, the probability of a land parcel being developed on the periphery of a growing structure is a function of the time when the adjacent parcel was developed: the more recent the development of the adjacent parcel, the greater the probability of the peripheral parcel being developed in the current time period. The process is updated every time a parcel is developed. In this way, a process is simulated whereby development on one edge of the city increases the probability of neighboring land parcels being developed. Figure 11.2d represents a structure developed by this potential field growth model.

It is a relatively simple matter to adapt each of the above processes to simulate a continuum of forms from the linear with a dimension near 1 to a nearly filled aggregate with a dimension near 2. This is done for the DLA model by changing the probabilities that a particle will "stick" once it reaches the growing cluster; for the DBM, by systematically altering the magnitude of the potential gradient that determines the growth probabilities of the perimeter sites; for the DSM, by altering the parameters governing the power of the screening; and for the PFGM, by altering the time dependency of the process. Thus, a continuum of forms from the line to the plane can be simulated, enabling a "pattern book" or "look-up table" of spatial forms to be constructed. In the following section, we begin to consider how these processes can be used to analyze the nature of urban forms.

These processes of cluster growth tell us much about the development of urban form. A key notion is that growth is determined as a balance between the need for space on the edge of a cluster and the need for that growth to be part of the cluster in physical terms. Growth is clearly a centrifugal process that is limited by the constraints of physical space and the economies and externalities of the extant city. The provision of transport infrastructure, too, as mirrored in the effects that rail, tram, and bus lines have had on development in the early industrial city and in the effects of freeways on the contemporary city, seems consistent with the tendencies in cluster aggregation to emphasize dendritic growth. Some aggregation processes are likely to be of little relevance to the way in which cities are formed: cities do, after all, grow from the inner core outward, rather than by random peripheral accretion of migrants, and thus DLA is apparently at variance with conventional wisdom concerning filtering through the urban fabric. However, the DBM, DSM, and PFGM models, which rely on the notion of continually changing fields, appear more directly relevant, promising further insights into migration and other forms of potential fields (Shepherd, 1979; Tobler, 1981).

Fractal Nature of Urban Forms

There exists a clear need to develop our understanding of the analogies between urban growth and form and cluster aggregation. Previous work (Batty and Longley, 1988; Longley and Batty, 1989) has considered spatial irregularity and the statistical self-similarity of urban boundaries and edges. Such measurements have also been used to simulate the partitioning of land-use parcels within cities as a means of generating spatial forecasts based on conventional urban models (Batty and Longley, 1986). However, in this simulation work, the geometry of urban form has remained no more than a means of articulating model forecasts, and no attempt has been made to relate the partitioning of urban space to the dynamics of urban growth. As such, space regrettably remains a mere model appendage, and existing research does not explicitly address the evolution of urban morphology over time.

More recently, some preliminary attempts have been made to remedy this deficiency and to integrate fractal measurements of urban form to established thinking about urban density gradients and allometric growth (Batty et al., 1989; Fotheringham et al., 1989). We can begin by imagining a city at one point in time that has grown outward from an initial seed site and where parcels of land have been developed through a contiguous process. Let $N(r)$ be the number of parcels of land that are now developed within radius r of the center such that

$$N(r) \sim r^D \tag{11.1}$$

and

$$\frac{\partial N(r)}{\partial r} \sim r^{D-1} \tag{11.2}$$

where D is a parameter that scales developed area to distance from the center of the city. If all the land were developed, then $N(r)$ in Equation 11.1 could be replaced by $A(r)$, where

$$A(r) \sim r^2 \tag{11.3}$$

the total amount of land within radius r; then clearly D would be equal to 2. However, if there are parcels of land that are undeveloped and the density of such undeveloped parcels increases with distance from the center, D will be less than 2. We call D in this case the fractal dimension of the structure.

It is a simple matter to relate D to urban density gradient and allometric growth. If we define the density of developed land within radius r as

$$\rho(r) = \frac{N(r)}{A(r)} \sim r^{D-2} = r^{\beta} \tag{11.4}$$

then the urban density gradient ß will always be negative when D is less than 2.

In terms of allometry, the ratio of total land area to developed land area is given by $A(r) : N(r)$ (Equation 11.4). Since total area has dimension 2 and developed area has dimension D,

$$N(r)^{\gamma} \sim r^2 \tag{11.5}$$

where $\gamma = 2/D$. The allometric growth parameter, γ, is always greater than 1 when D is less than 2. That is, total area grows at a faster rate than the area of developing land, implying that the margins of the structure become increasingly sparse as the structure grows. This may question conventional thinking in allometric analysis (for example, Dutton,1973) in which development as measured by population is taken to have dimension 3 (a volume) so that isometry is indicated by $\gamma = 2/3$. Thus, in DLA structures the urban density gradient and allometric growth parameters are directly related to the fractal dimension. Full details are available in Fotheringham et al. (1989).

Batty et al. (1989) have drawn comparisons between the classic fractal dendrite form generated by DLA and the form of the small English town of Taunton (the population in 1981 being around 50,000 resident individuals), so chosen because its development appeared to be relatively undistorted by strong physical constraints. In this study, comparisons were drawn between the DLA form and Taunton with respect to various scaling relationships based on development and distance; these computations all relate to fractal dimension. A variety of estimates of dimension based on Equation 11.1 were computed, although the general conclusion was that the dimensional analysis was confounded by problems concerning the definition of development and the scale at which it was recorded. This was, in itself, an important conclusion since it highlighted the fact that our present methods of measuring development and its density are far from accurate and require much clearer definition. What is clear is that measurement prescribes analysis in this work, and this makes the precise and unambiguous definition of "urbanity" and "population" of overriding priority in inductive modeling work of this nature.

Other work has focused on the generation of a typology of forms that differ from the base-line fractal case. Fotheringham et al. (1989) have explored ways in which the assumption of completely random walks in the DLA model can be relaxed. Figure 11.3 illustrates how, for an individual seeded due west of the center of the circle, the movement of individuals across the grid placed over the region can be controlled by two parameters. (Appropriate adjustments are made for individuals seeded elsewhere on the bounding circle, and the two parameters each range between 0 and 1.) The parameter x controls the strength of directional drift, with backward and forward movement equally probable at $x = 0$ and only forward movement possible when $x = 1$. The parameter y controls the degree of focal attraction, with $y = 0$ indicating no predominance of movement directly toward the center and $y = 1$ indicating unidirectional movement toward the focal point. Figure 11.4

illustrates a range of simulations based on different x values (with y fixed at 0.0) and shows that high x values are associated with high fractal dimensions because the structure fills more space within a given area. Figure 11.4 also records the association between high fractal dimensions, low allometric growth parameters (α) and smaller negative density gradients (β). The effects of changing the y parameter (with x fixed at 1.0) are illustrated in Figure 11.5. The structures become increasingly compact as y increases and would reach the limit of an approximately circular dense mass in the case of unidirectional movement.

Batty (1990) has also explored the ways in which the DBM models can consistently generate a range of urban forms from the linear to the concentric, as well as demonstrating how the model can be applied to a real urban simulation using data from the town of

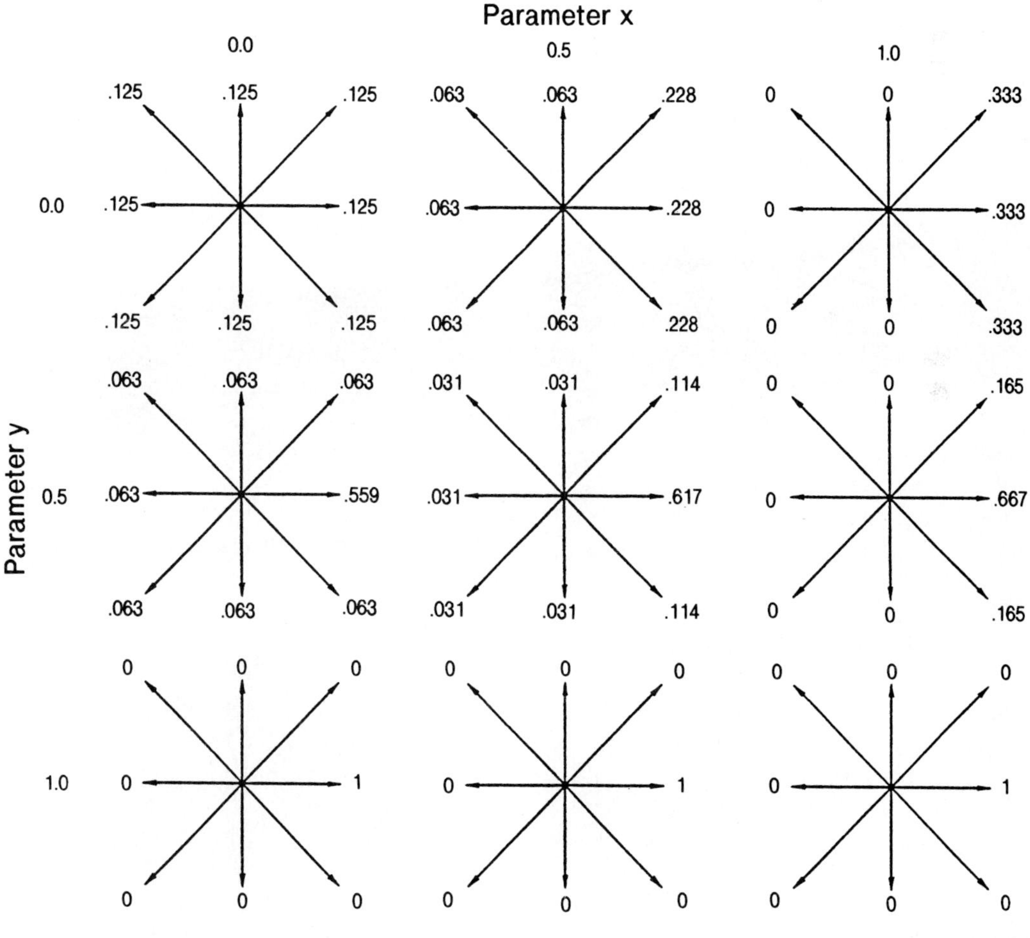

Figure 11.3 Random walk generators. Values indicate probabilities of movement in each direction from a starting position due west of the fixed point.

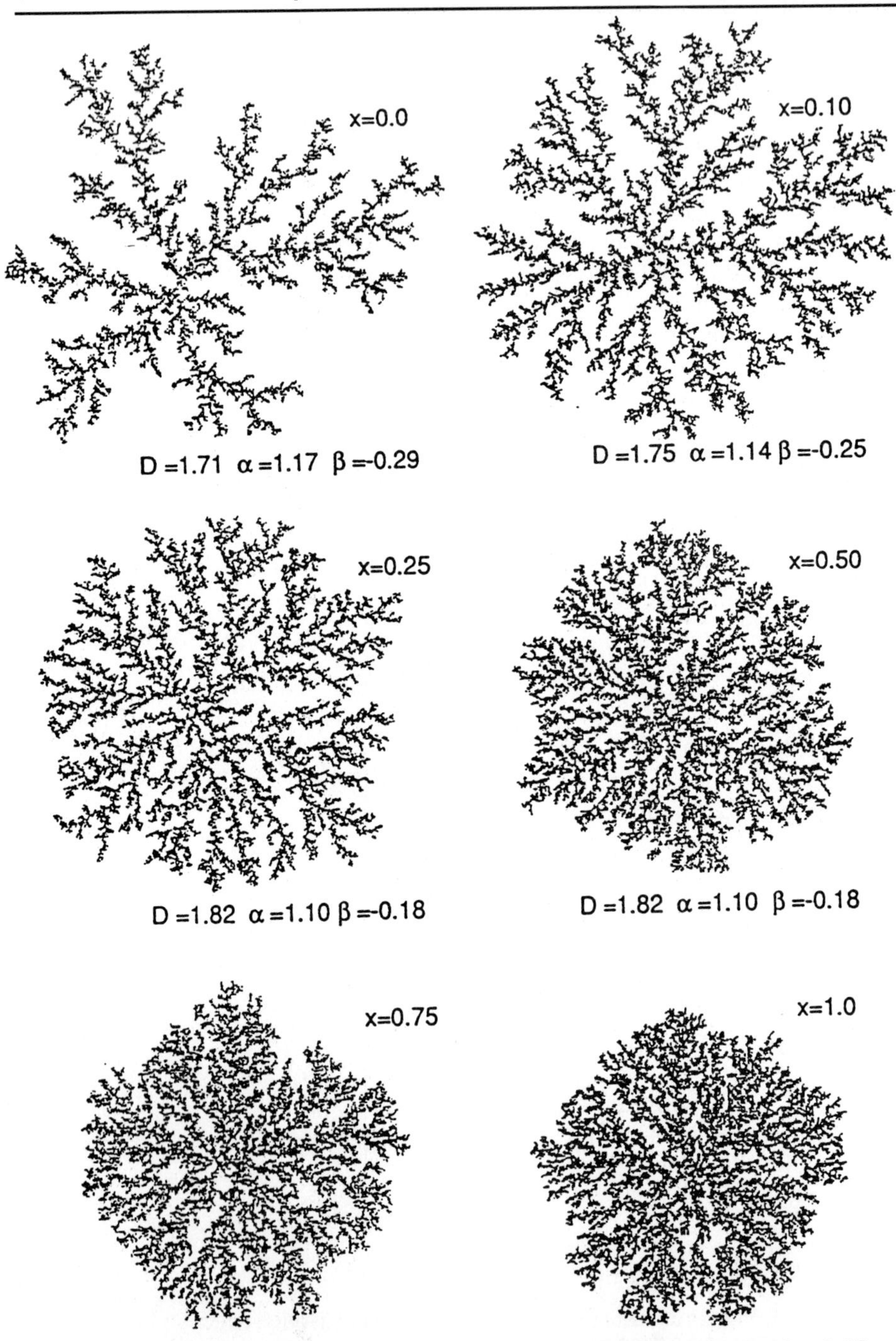

Figure 11.4 Variations in DLA structures with different degrees of forward motion (i.e., x varies and y fixed at 0.0). α and β are allometric growth and density gradient parameters, respectively.

Cardiff. At present the authors are working with variants of the DLA, DBM, DSM, and PGFM models in an effort to explore the extent to which these models might provide insights into the macro forces that motivate urban growth. Figures 11.2a–d illustrate the types of structures that have been simulated so far. The statistical properties of these structures have also been researched (Fotheringham et al., 1989; Batty, 1990).

Such urban research is in the spirit of other work within the broad domain of fractals and chaos theory. Those working with chaotic phenomena in urban and regional systems have mainly based their research on the application of nonlinear dynamics to the evolution and growth of urban systems. Among many researchers, significant contributions have been made by A.G. Wilson at Leeds, P.M. Allen at Cranfield, D.S. Dendrinos at Kansas, G. Haag at Stuttgart, and O. Batten and T. Puu at Umea. The central thrusts of these various projects, however, are concerned less with spatial and physical form and more with the dynamics of change, particularly novel change. Nevertheless, the common thread to these studies is a shared interest in systems and models that manifest self-similarity across a range of scales. Other research has considered the physical morphologies of two-dimensional space more explicitly. For example, Woldenberg (1988) has studied extensively analogies among river, city, lung, and other treelike systems; Tobler (1978) has been concerned with the definition of space and geographic fields at the urban and regional levels; and Hawkes et al. (1987) have explicitly addressed changes in urban form from the standpoint of energy efficiency and transportation requirements. There is also a strong relation between the DLA models and models of cellular automata (Toffoli and Margolis, 1987), and spatial systems have been conceived of in these terms by Tobler (1979) and Couclelis (1985), among others.

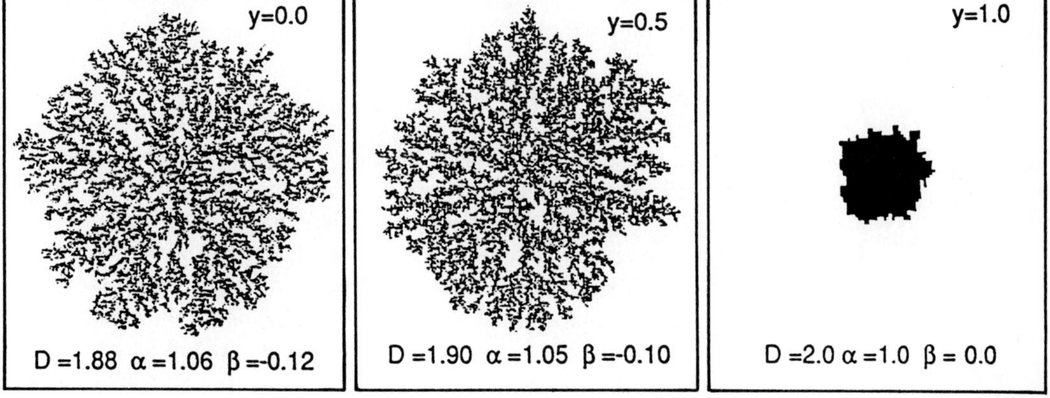

Figure 11.5 Variations in DLA structures with different degrees of directly forward motion (i.e., y varies and x fixed at 1.0). α and β are allometric growth and density gradient parameters, respectively.

ISSUES IN ANALYTICAL URBAN MORPHOLOGY RESEARCH

Although some progress has been made in formulating a coherent approach toward the study of urban morphology, there is clearly some way to go. In thc previous sections we discussed how fractal geometry holds the key both to a description of urban form and the development of analogies between normative (yet nonetheless visually plausible) structures and real-world urban forms. In this section, we examine how this research might be invigorated by addressing a number of general themes, including (1) the investigation of model processes, (2) the development of new model structures, (3) the classification of urban forms, (4) the development and refinement of statistical measures and comparisons of forms, (5) the use of analogies with urban development and population density models, and (6) further consideration of urban dynamics and model dynamics. We now consider each of these in turn.

Investigation of Model Processes

The four varieties of model based on their processes of cluster aggregation, the DLA, DBM, DSM, and PFGM models, require thorough investigation with respect to the way different forms with different fractal and other dimensions can be generated (Nittman et al., 1985; Sherwood and Nittman, 1986). What are required are analogous models that can capture the local and global laws of development and are closer to what we know about real urban development. We require more insight into how the time paths of both real and simulated growth can be compared and contrasted, as well as the ways in which external physical constraints affect development. There is also an urgent need to explore the extent to which the underlying spatial system contributes to the form of development. Already there is evidence that very large clusters with greater than 10^5 particles (grown on equally large lattices) are sensitive to the shape and configuration of the lattice grid. Such anisotropy is highly undesirable and must be thoroughly understood before effects to counter it can be introduced (Meakin, 1986).

Development of New Model Structures

The urban models that have been developed in analogy to cluster growth in physics are all predicated on the basis of a single seed site around which the cluster grows. Although this is a useful starting point, there is a need to explore the possibilities of multiple sites that clearly characterize the spatial structure of real cities. This will involve very selective seeding, as well as a careful exploration of the effects of different seeds on one another, thus demonstrating how complex it will be to initiate simulations of the real urban growth with such models. These spatial effects are also paralleled by several issues affecting the way these types of model deal with the temporal dimension. As we noted earlier, these types of model simulate systems that are far-from-equilibrium in that once particles are located they remain fixed. They do not attempt to equilibrate with respect to the whole cluster, and although much of the real urban system is similarly irreversible over long time periods, it is characteristic of cities that sites do change in terms of their use and their physical development; processes involving redevelopment and renewal that combine physical obsolescence with the economics of the marketplace must be incorporated if such models are to progress toward real applications.

Accordingly, it is necessary to explore how such models might incorporate reversibility. As soon as this is broached, it would appear that another process of diffusion is laid across the original one that simulates growth. The connections between the two processes are likely to be based on strong coupling with respect to their competition for the same units of development, but, nevertheless, the complexity of the model will be greatly increased when reversibility is added. There is another issue involving the way time is treated in these models. Because only one site is occupied at a time, the growth profile of the system being modeled is uniform in that development never builds up faster than one unit at a time. This is clearly unrealistic; it is necessary to provide these types of models with mechanisms in which the growth profile can be controlled and in which the system is able to handle growth to certain limits. At present, the clusters are grown up to arbitrary levels, whereas in real systems the pace of growth and the size of the cluster or city at any point in time are important. There are mechanisms for structuring the time frame of the model to deal with varying amounts of development, but these, once again, complicate the model.

Classification of Urban Forms

The four examples of real urban forms shown in Figure 11.1 are each sufficiently different from one another to suggest that, although there may be similar growth processes at work, these processes may be operating across a spectrum of extremes. This spectrum is already a feature of the various cluster models being explored, that is, the DLA, DBM, DSM, and PGFM models, and thus there is a clear possibility that these processes might be used to classify the actual forms observed. Of course, Figure 11.1 shows forms at a point in time and does not reveal the actual growth pattern through time, which, in turn, might represent another dimension for classification. In essence, then, a goal of future research must be to classify a range of actual urban forms that can be already classified in terms of their size, location, level of economic activity, and so on; in this context, the classification would be in terms of spatial indexes such as fractal dimension. It is also possible to classify cities in terms of the spatial growth processes at work. This is especially important if fractal structures are perceived as resulting from growth processes in which there are a large number of random events (or at least events that are unpredictable; Fotheringham, 1990). Thus, features such as canals have a fractal dimension not much greater than 1, far lower than that of rivers. Similarly, planned urban areas probably do not have fractal characteristics, but those that have grown as the result of innumerable decisions probably do. Thus, it would appear that cities might be classified according to an index of their form such as fractal dimension.

Statistical Measures and Comparisons of Forms

The limited existing research that has compared real-world settlements with diffusive forms has, at best, simply examined the coincidence or otherwise between gross physical characteristics of real and simulated city-cluster patterns. The only measure of spatial coincidence linking process to form has been through the fractal dimension and other related scaling parameters (such as that on density). There is need for many more measures of form and, in this context, research into cluster growth in physics has already indicated the way forward. Recent work with the cluster growth model has shown that a suitable basis for comparison between real and simulated patterns is through the use of the growth probability profiles from the perimeter site data (Meakin et al., 1986; Stanley and Meakin, 1988). The probability distribution governing the growth potential of the perimeter sites for a fractal cluster has a characteristic profile that can be fully described by the moments of the distribution. These moments give rise to a hierarchy of scaling parameters, which in fact are also referred to as fractal dimensions. These can be associated with different locations on the perimeter sites around the cluster, indicating that the tips of the cluster's dendrites have different scaling than those sites in the cluster's interior, deep inside its fissures.

There are now demonstrations of the coincidence between real clusters, such as those formed from the interpenetration of liquids of different viscosity, and their simulated forms. The perimeter site distributions of the simulated forms provide a rich source of comparative tests between the real and the simulated. So far, there is no published work on the changing structure of these perimeter site distributions as the cluster grows, and this represents a line of research that is worthy of further study as an example of the increasing importance of the broader area of visualization. For example, how do we assess when we have a reasonable "picture" of reality, and under what conditions is local matching more important than global matching?

Analogies with Urban Development and Population Density Models

Thus far in our discussion, we have alluded to development and density quite freely without providing any precise meaning. In one sense, our use of the term urban development is clear in that we are defining such development as any area of land within the bounding envelope of a town or city that is not specifically agricultural or resource based. In fact, this is not a particularly good definition for these purposes because development here constitutes the built-up structures comprising the town, not the open space of various sorts that permeates all urban areas. In our previous work, where we compared DLA and DBM models to Taunton and Cardiff, we used this latter definition. Density then is simply the area of land on which the built-up structure exists as a ratio of the total land area associated with the unit in question.

However, this definition is still problematic for it depends on what the model is allocating, that is, what constitutes the growth. Units of development might be satisfactory for phenomenological analysis, but in relating these models to the research tradition dealing with population density, each unit would have to be treated as a unit of population, a person or a household. There is a need to examine this question very carefully since this will have an effect on the scale at which researchers record basic density data. To enable appropriate levels of density to be simulated, the system being simulated must have the potential for aggregations of units of development to be made that will mirror existing densities.

In the spirit of this approach, there exists the additional possibility of modifying cluster growth models to replicate the three-dimensional aspect of development that is present in most urban areas.

A related feature of current research involves comparisons between traditional forms of population density models and the types of density profile that emerge from the cluster aggregation approach. Because most of the work in population density modeling as well as spatial interaction modeling has been concentrated on the use of exponential rather than power functions relating activity to distance, it is difficult to make exact comparisons through estimated parameter values. A fruitful avenue of research might nevertheless be to explore this question thoroughly and attempt a comparison with traditional density models (Mills, 1970; Parr, 1985).

Urban Dynamics and Model Dynamics

One of the most problematic characteristics of any macro model is reconciling its mechanism with the kinds of processes that occur at the micro level. In this context, how can allocation processes based on the definition of overall potentials be reconciled with the sorts of ways developers, public agencies, house buyers, and home owners engage in the market for land and buildings? A traditional perspective on such questions suggests that it is sufficient to demonstrate that a careful definition of potential appears consistent with the pattern of urban growth. This represents a sensible point of view, since there is no practical way in which we can add up all the micro decisions that give rise to urban development, nor should there be. Thus, although it is right to be interested in the behavioral foundations of these models and although future research will doubtless search for such consistencies between processes of growth and change involving the whole spectrum of migration behaviors, we do not believe that any single line of reasoning linking micro to macro in this context will ever be found.

This is not to say, however, that it will not be possible to develop a better understanding of urban growth dynamics. This will nevertheless require a rich data base, something like that used by Chapin and Weiss (1962) for Greensboro, North Carolina, in their urban simulation research, which they initiated about 30 years ago. What is required is a detailed inventory of urban change, ideally at the level of the site, so that the growth of a city through time can be traced.

CONCLUSIONS

In this chapter, we have discussed the ways in which fractal measurement is central to the development of new approaches to an understanding of urban structure. Such understanding is important for several reasons. Primarily, the concepts of cluster dynamics incorporate new ways of thinking about the process of urban growth and lead to models with richer and more realistic structures than the classical frameworks that have formed the basis of our understanding of how urban areas develop. In investigating the application of cluster growth techniques to the development of urban areas, we can refocus attention on the important linkages between urban form and the processes operating within the urban environment. In so doing, it should be possible to provide a means of classifying cities based on a geographical property of their form, fractal dimension, that is related to a city's development history.

Current research has begun to highlight some of the problems in measuring urban characteristics, such as what constitutes urban land as opposed to nonurban land, and the appropriate unit of area over which a density gradient should be measured. While others have claimed variations in density gradients to be "among the most pervasive and best documented trends in the developed world" (Mills and Tan, 1980, p. 316), we would argue that severe measurement problems must be overcome before such extravagant claims can be made. Given the volume of research in this area, we feel that our contributions to a better understanding of how density gradients can occur and how they should be measured will be extremely valuable.

More generally, study of fractal urban morphologies will highlight some important aspects of urban modeling that have implications for all geographical modeling. The bases for clustered growth that we are developing fall into a gray area between order and uniqueness. While it is a truism to state that all cities are unique, there are several underlying points of similarity between them, and our task lies in uncovering these commonalities and in understanding the processes that produce them. For instance, our own experiments so far in this area have produced a series of DLA simulations that are all characterized by similar fractal dimension and density gradient yet each remains a unique structure. It should therefore be possible to model the outcomes of innumerable decisions that create urban form in such a way that we can capture the properties of both uniqueness and order; in so doing, we perhaps provide a common ground for those who have argued for one of these points of view to the exclusion of the other.

In summary, fractal morphological research holds the enticing prospect of providing simple, yet potentially powerful, theories of urban growth that can lead to rich structures that are difficult to anticipate. It highlights attention on the value of simulation and diffusion in the growth of geographic systems in much the same vein as Hagerstrand's seminal work, and it recognizes the increasing role of the visualization of results and the use of computer graphics for developing urban theory. In this spirit, we might anticipate geographers generating further important insights surrounding the evolution of spatial systems.

REFERENCES

Allen, P. M. 1982. Evolution, modeling and design in a complex world. *Environment and Planning B* 9:95–111.

Alonso, W. 1964. *Location and Land Use*. Cambridge, MA: Harvard University Press.

Anas, A. 1987. *Modeling in Urban and Regional Economics*. Chur, Switzerland: Harwood Academic Publishers.

Batty, M. 1991. Generating urban forms from diffusive growth. *Environment and Planning A* 23:511–544.

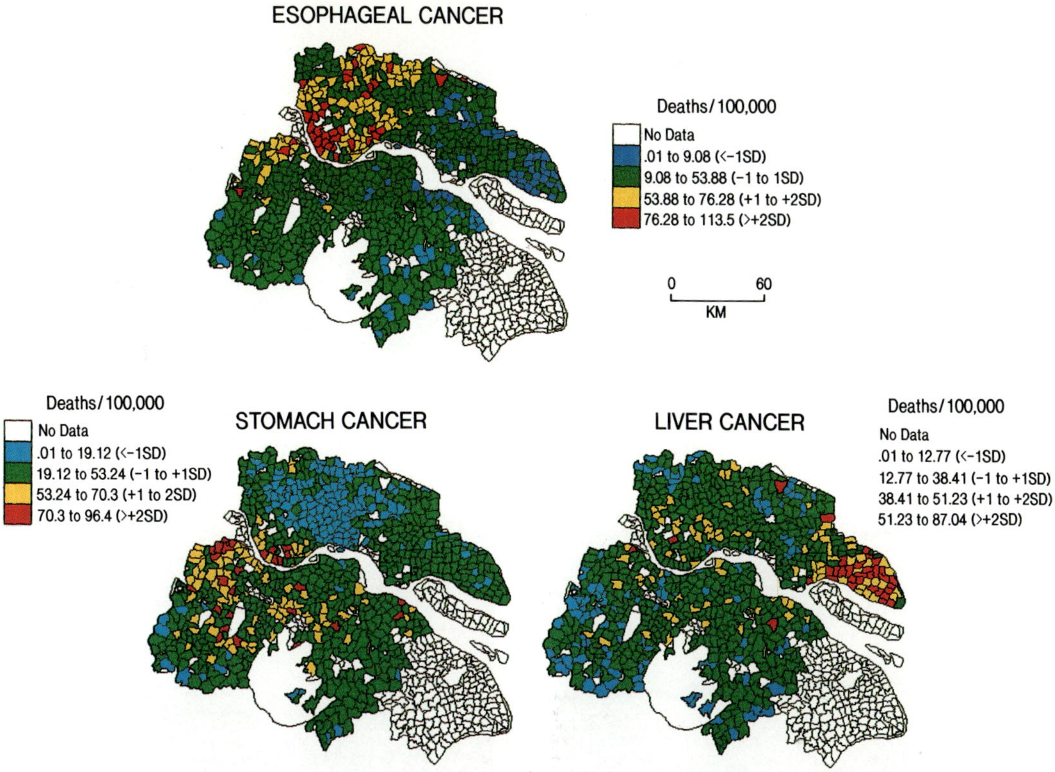

Figure 12.3 Choropleth maps of the three major cancers (esophageal, stomach, liver in Taihu Region, China.

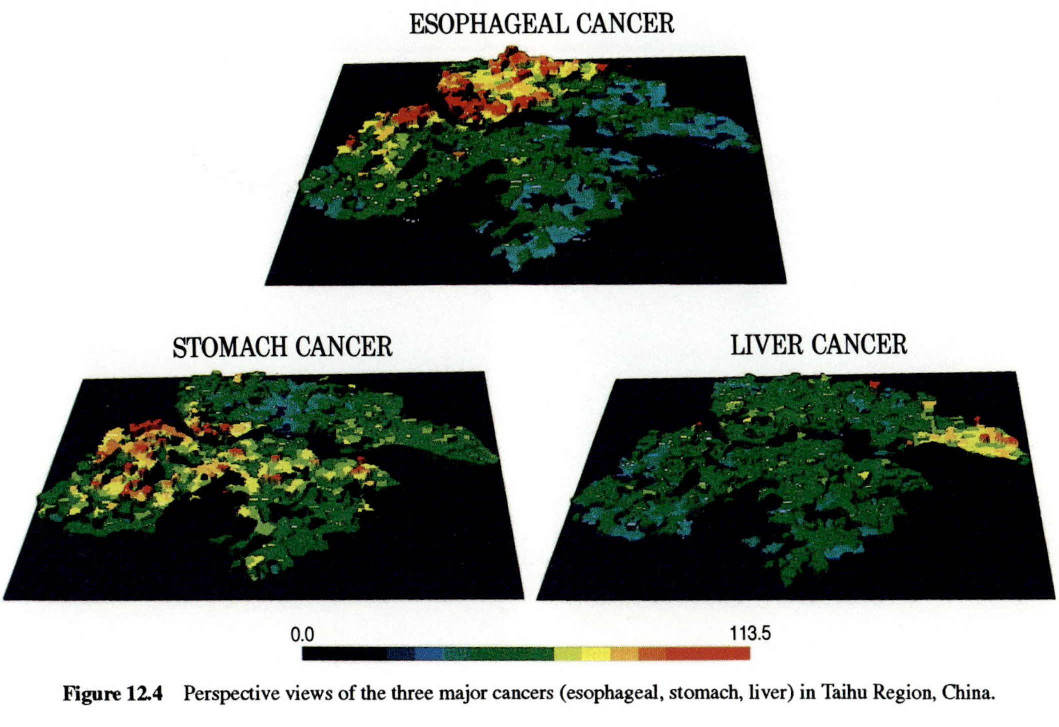

Figure 12.4 Perspective views of the three major cancers (esophageal, stomach, liver) in Taihu Region, China.

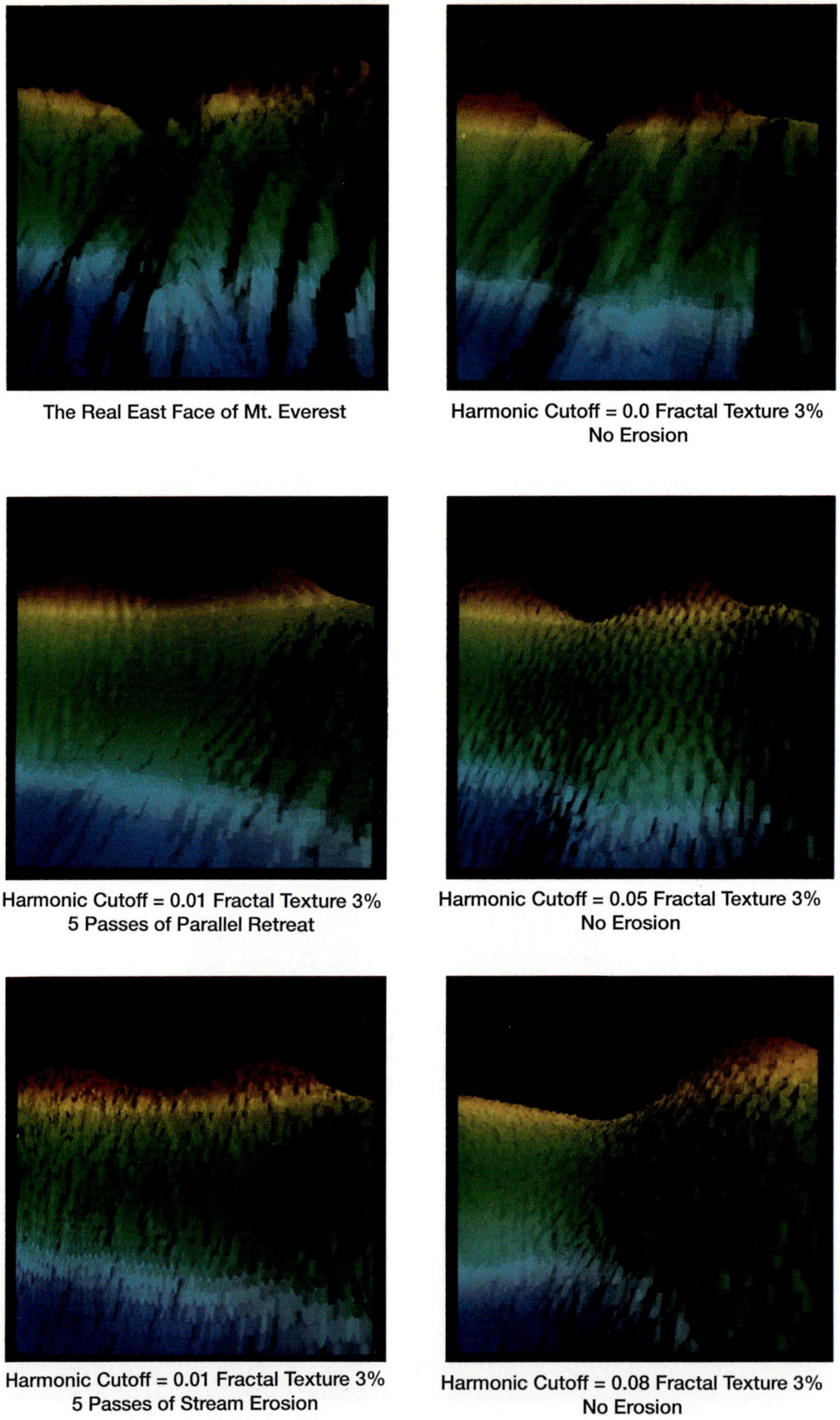

Figure 13.6 Perspective views of simulated terrain used in the test.

———, and Longley, P. A. 1986. The fractal simulation of urban structure. *Environment and Planning A* 18:1143–1179.

———, and ———. 1988. The morphology of urban land use. *Environment and Planning B* 15:461–488.

———, and ———, and Fotheringham, A. S. 1989. Urban growth and form: Scaling, fractal geometry, and diffusion-limited aggregation. *Environment and Planning A* 21:1447–1472.

Beckmann, M. J. 1969. On the distribution of urban rent and residential density. *Journal of Economic Theory* 1:60–67.

Chapin, F. S., and Weiss, S. F., eds. 1962. *Urban Growth Dynamics: In a Regional Cluster of Cities*. New York: John Wiley & Sons.

Couclelis, H. 1985. Cellular worlds: A framework for modeling micro-macro dynamics. *Environment and Planning A* 17:585–596.

Dendrinos, D. S., with Mullally, H. 1985. *Urban Evolution: Studies in the Mathematical Ecology of Cities*. New York: Oxford University Press.

Dorigo, G., and Tobler, W. 1983. Push–pull migration laws. *Annals of the Association of American Geographers* 73:1–17.

Doxiadis, C. A. 1966. *Ekistics: An Introduction to the Science of Human Settlements*. London: Hutchinson

Dutton, G. 1973. Criteria of growth in urban systems. *Ekistics* 215:298–306.

Feder, J. 1988. *Fractals*. New York: Plenum Press.

Fotheringham, A. S. 1990. What's the fuss about fractals. Guest editorial. *Environment and Planning A* 22:716–718.

———, Batty, M., and Longley, P. A. 1989. Diffusion-limited aggregation and the fractal nature of urban growth. *Papers of the Regional Science Association* 67:55–69.

Goodchild, M. F., and Mark, D. M. 1987. The fractal nature of geographic phenomena. *Annals of the Association of American Geographers* 77:265–278.

Hawkes, et al., eds. 1987. *Energy and Urban Built Form*. London: Butterworth.

Jullien, R., and Botet, R. 1987. *Aggregation and Fractal Aggregates*. Singapore: World Scientific Publishing Company.

Longley, P. A., and Batty, M. 1989. Fractal measurement and cartographic line generalization. *Computers and Geosciences* 15:167–183.

Mandelbrot, B. B. 1982. *The Fractal Geometry of Nature*. San Francisco: W. H. Freeman & Co.

Meakin, P. 1983. Cluster-growth processes on a two-dimensional lattice. *Physical Review B* 28:6718–6732.

———. 1986. Computer simulation of growth and aggregation processes. In *On Growth and Form: Fractal and Non-fractal Patterns in Physics*, H. E. Stanley and N. Ostrowsky, eds., pp. 111–135. Boston: Martinus Nijhoff.

———., et al. 1986. Scaling properties for the surfaces of fractal and non-fractal objects: An infinite hierarchy of critical exponents. *Physical Review A* 34:3325–3340.

Mills, E. S. 1970. Urban density functions. *Urban Studies* 7:5–20.

———, and Tan, J. P. 1980. A comparison of urban population density functions in developed and developing countries. *Urban Studies* 17:313–321.

Muth, R. F. 1969. *Cities and Housing*. Chicago: University of Chicago Press.

Niemeyer, L., Pietronero, L., and Weismann, H. J. 1984. Fractal dimension of dielectric breakdown. *Physical Review Letters* 52:1033–1036.

Nittman, J., Daccord, G., and Stanley, H. E. 1985. Fractal growth of viscous fingers: Quantitative characterization of a fluid instability phenomena. *Nature* 314:141–144.

Parr, J. B. 1985. A population density approach to regional spatial structure. *Urban Studies* 22:289–303.

Putman, S. 1990. *Application of Optimization and Dynamics to Integrated Transportation and Location Modeling*. London: Pion Press.

Shepherd, E. S. 1979. Geographic potentials. *Annals of the Association of American Geographers* 69:438–447.

Sherwood, J., and Nittman, J. 1986. Gradient governed growth: The effect of viscosity ratio on stochastic simulations of the Saffman–Taylor instability. *Journal de Physique* 47:15–22.

Stanley, H. E., and Meakin, P. 1988. Multifractal phenomena in physics and chemistry. *Nature* 335:405–409.

———, and Ostrowsky, N., eds. 1986. *On Growth and Form: Fractal and Non-fractal Patterns in Physics*. Boston: Martinus Nijhoff.

Tobler, W. 1978. *A Geographical Migration Probability Density Function*. Santa Barbara: Department of Geography, University of California.

———. 1979. Cellular geography. In *Philosophy in Geography*, S. Gale and G. Olsson, eds., pp. 279–386. Dordrecht, The Netherlands: D. Reidel Publishing Co.

———. 1981. A model of geographical movement. *Geographical Analysis* 13:1–20.

Toffoli, T., and Margolis, N. 1987. *Cellular Automata Machines: A New Environment for Modeling*. Cambridge: The MIT Press.

Wilson, A. G. 1981. *Catastrophe Theory and Bifurcation: Applications to Urban and Regional Systems*. London: Croom Helm.

Witten, T. A., and Sander, L. M. 1981. Diffusion-limited aggregation: A kinetic critical phenomenon. *Physical Review Letters* 47:1400–1403.

Woldenberg, M. J. 1973. An allometric analysis of urban land use in the United States. *Ekistics* 215:282–290.

———. 1988. Microgeographic determination of optimality in mammalian trees. In *Geographical Systems and Systems of Geography: Essays in Honour of William Warntz*, W. J. Coffey, ed., pp. 121–140. London, Canada: Department of Geography, University of Western Ontario.

CHAPTER 12

A FRACTAL ANALYSIS OF CANCER MORTALITY PATTERNS IN CHINA

Nina Siu-Ngan Lam
Department of Geography and Anthropology, Louisiana State University, Baton Rouge

Hong-Lie Qiu
Remote Sensing and Image Processing Laboratory, Louisiana State University, Baton Rouge

Rui Zhao

Nan Jiang
Nanjing Institute of Geography, Academia Sinica, Nanjing, People's Republic of China

ABSTRACT

This paper explores the applicability of fractal models in analyzing socioeconomic data collected and defined in polygonal form. Through the use of a set of Chinese cancer mortality data at the commune scale, we demonstrated the methods and problems in estimating the fractal dimensions of three major cancer patterns in China, including stomach, esophageal, and liver cancers. Both the isarithm and the variogram methods were used. All three cancer surfaces yielded very high fractal dimensions (2.65 to 2.86), and the dimensions derived from the two methods are very similar. While the fractal dimension can serve as an index of complexity for the entire cancer surfaces within certain scale ranges, its use in differentiating the locational variations among these patterns remains unknown.

More studies on different socioeconomic patterns are urged so that certain bench values can be determined and against which patterns can be tested. The utility of fractals will not be fully recognized until these bench values are established.

INTRODUCTION

The other chapters in this book and previous literature indicate that fractal models can be applied in a variety of situations. The use of fractal models in characterizing and analyzing socioeconomic patterns, however, remains largely unexplored. These socioeconomic data are often collected and defined in irregular polygonal form. The nature and the format of these data are quite different from that of the regularly spaced data to which fractal models have been applied, such as digital elevation data and digital remote-sensing data. The primary purpose of this paper is to evaluate the applicability of fractals in analyzing geographical patterns based on areal (polygonal) data. The geographical patterns of the three leading cancer types in China, stomach, esophagus, and liver cancers, for the Taihu (Tai Lake) region are examined in this paper. The data set contains mortality rates (number of deaths per 100,000 people) by commune (now called village) for 1973–1975. In this study, we compared two different methods, the variogram method (Mark and Aronson, 1984) and the isarithm method (Shelberg et al., 1983) in estimating the fractal dimensions of these three cancer patterns.

The cancer data set possesses several characteristics that make it ideal for illustrating the issues and potential problems of applying fractal models to real-world socioeconomic phenomena. All three geographical patterns are highly clustered, but the locations of high mortality vary distinctly from one to another. Also, cancer mortality patterns in particular and medical geography in general are highly scale dependent (that is, the patterns look different at different scales and resolutions). The use of this data set demonstrates the utility as well as the problems of using fractals in analyzing scale-dependent phenomena. Furthermore, the study area has irregular boundaries and holes due to lakes, islands, rivers, and missing data. Existing algorithms will have to be modified to accommodate these irregularities, and their effects on the estimation of fractal dimensions and subsequent interpretation should be considered.

More specifically, the objectives of this study are threefold: (1) We wish to determine the fractal dimensions of the three cancer mortality patterns and the scale ranges (self-similarity ranges) of these patterns. In applying the existing fractal measurement methods (with modifications) to the cancer patterns, we already regard these patterns as surfaces, although we recognize that they are not continuous and are very different from topographic surfaces. Since polygonal data are discrete and have more abrupt changes across polygonal boundaries, it is expected that the fractal dimension values derived from the cancer surfaces for specific ranges of scales would be higher than those derived from most terrain surfaces on the earth. (2) We explore whether different cancer surfaces exhibit distinctly different dimension values and whether the fractal dimension can be used to reflect the locational variations among the three cancer mortality patterns. (3) Since we are still faced with the problem that measurement results vary with methods, we apply two of the existing methods to our data, the isarithm and variogram methods. The results from applying these two methods are compared with those of other surfaces computed in previous literature.

It is hoped that our findings will shed some light on the potential of using the fractal dimension as an effective parameter for describing and quantifying spatial patterns and illustrate that fractal analysis might provide a useful, yet different, perspective on the analysis of cancer mortality patterns and the issue of scale. Although not examined in this paper, the dimension values could be compared to other spatial statistics estimated for the same patterns, such as spatial autocorrelation statistics, information statistics, and other measures.

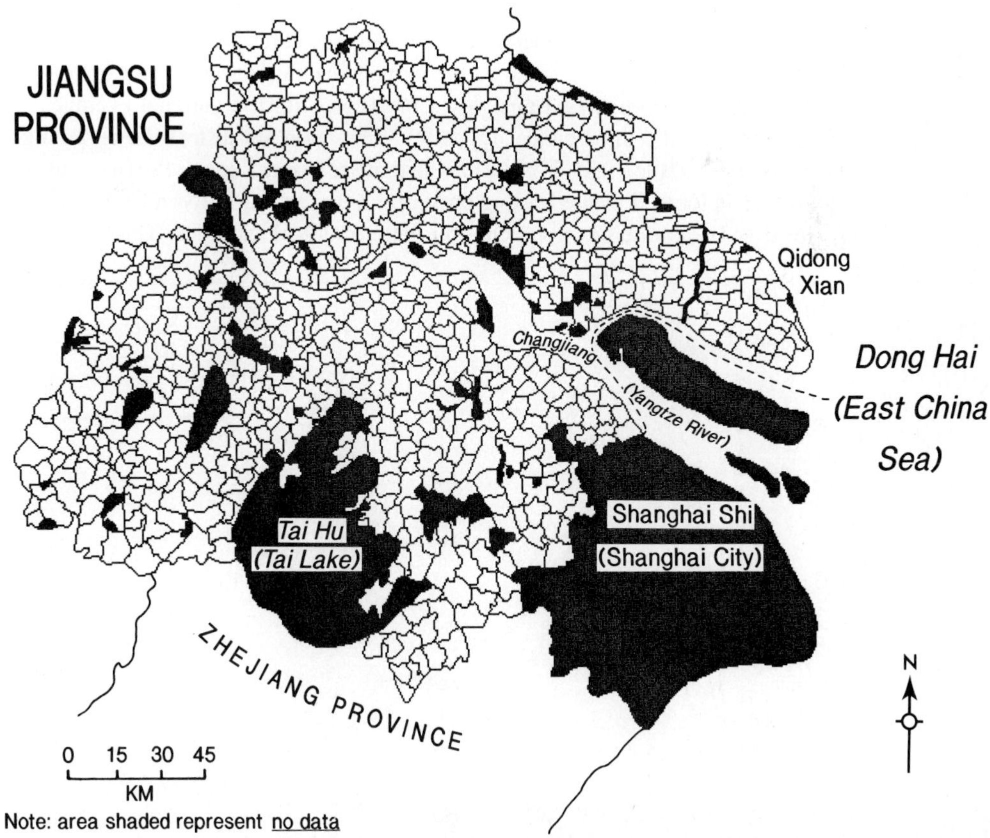

Figure 12.1 Locational map showing the communes in the Taihu Region, China.

STUDY AREA AND DATA

The present data set consists of mortality rates by commune for the three leading cancer types in China, stomach, esophagus, and liver. The mortality data were collected by the Chinese government for 1973–1975 and were further manipulated, aggregated, and mapped by county for the entire nation (Editorial Committee, 1979). The present study area is confined to the Taihu region of Jiangsu Province, which is adjacent to Shanghai and consists of 25 counties encompassing 774 communes and cities (Figure 12.1). On average, a county has about 30 communes. The resultant number of polygons being processed in this study is 777. (There will be 1106 polygons if the Shanghai region is included.) Nationally, there are over 70,000 communes in China, aggregated into 2392 counties and further into 29 provinces and municipalities, excluding Hong Kong, Macau, and Taiwan. Table 12.1 lists the summary statistics for the present data set. The frequency distribution curves of the three cancers for the study area are shown in Figure 12.2. The geographical distributions of the three cancer types are mapped in Figure 12.3 (see color insert) using standard deviation units as the basis for class intervals. Their corresponding three-dimensional representations are included in Figure 12.4 (see color insert).

A brief description of the cancer patterns is deemed necessary for better interpretation of the results from fractal analysis. Compared with the national average, the mean mortality rates of the three cancer types for this region are at least twice as high as the national figures (Table 12.1). For reasons that are unclear, Jiangsu Province, in which the Taihu region is located, is the leading province in cancer mortality in China, in spite of its high level of economic development and better medical facilities. This was the case as

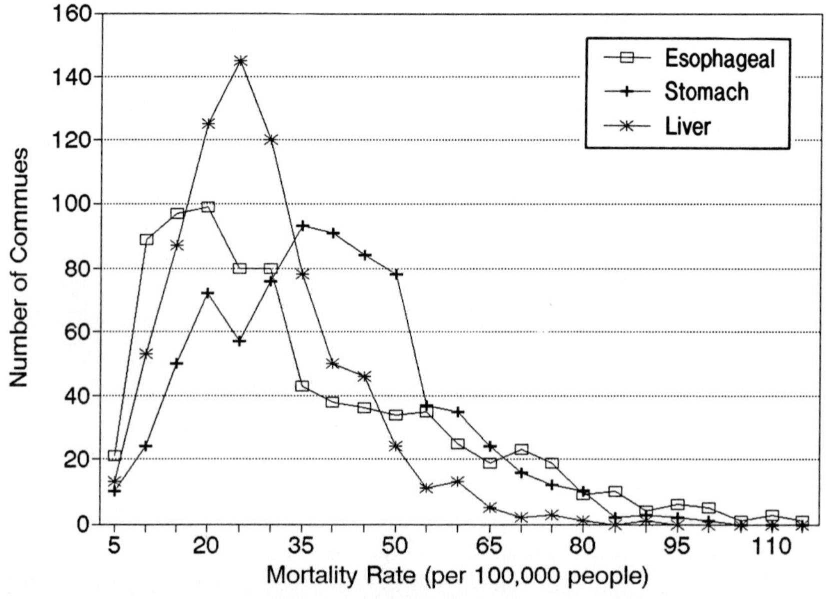

Figure 12.2 Frequency distributions of esophageal, stomach, and liver cancer mortality rates in the Taihu Region, China.

reflected by the 1973–1975 data, and the situation has not improved according to recent statistics (*Sing Tao Daily*, Hong Kong, April 13, 1990). These statistics show that cancer is the number one cause of deaths and approximately 11 people die of cancer every hour in this province. Cancer mortality in Jiangsu Province is about one-ninth of the total cancer mortality of the whole country, and about 70% of cancer mortality was caused by the three leading cancer types examined in this paper. Analysis of the cancer mortality patterns in the Taihu region therefore is not merely an academic exercise; it could potentially be of use to future research on the causes of these cancers.

TABLE 12.1 Summary Statistics of the Mortality Rates (Number of Deaths per 100,000 People in 1973–1975) for the Three Leading Cancers, Derived from 774 Communes and Cities in the Taihu Region, China

Cancer	Mean	National Average[a]	Maximum	Standard Deviation
Esophageal	31.48	15.55	113.50	22.40
Stomach	36.18	14.77	96.40	17.06
Liver	25.59	10.07	87.04	12.82

Source: *Atlas of Cancer Mortality in the People's Republic of China*, 1979.

[a] Derived from averaging the national male and female age-adjusted rates (age adjusted to the 1964 census population of China).

Stomach and Esophageal Cancers

All three cancer patterns are highly clustered, but the locations of high mortality vary from one pattern to another (Figures 12.3 and 12.4). A striking observation on the geographical patterns of stomach and esophageal cancers is that areas of high stomach cancer mortality generally coincide with low esophageal cancer mortality, and vice versa. This dichotomy also seems to exist when the cancer data were aggregated and presented at the provincial and county scales. Previous factor analysis of provincial-level data has demonstrated that stomach and esophageal cancers are associated with two different axes, indicating geographic nonconcurrence between the two patterns (Lam, 1986). Nationally, both stomach and esophageal cancers are more common in the northern part of China than in the south. Changes in mortality rates across space are more gradual for stomach cancer, whereas esophageal cancer is usually found in a few sharply separated areas with considerably higher rates than those recorded for stomach cancer (Foster, 1989). In the Taihu region, esophagal cancer mortality slightly leads, with high rates occurring in areas north of the Yangtse River (that is, Changjiang). Areas south of the Yangtse River are, on the other hand, dominated by high rates of stomach cancer mortality.

Despite the numerous epidemiological studies that have been carried out, the etiologic causes of the three cancer types have not yet been firmly determined (Howe, 1986). Most of the correlations and associations found between cancers and environments (including both physical, such as climate and soil types, and social environments, such as dietary habits and life styles) can only be suggestive, but not conclusive. In the case of esophageal cancer, the risk factors and their importance vary according to the geographical regions. Smoking and drinking, for example, attribute significantly to esophageal cancers in the United States and Europe, but their effects are negligible in China and central Asia (Foster, 1989). Instead, esophageal cancer has been found to correlate closely with calcium and selenium deficiencies in the latter region. High stomach cancer mortality is closely associated with high dietary intake of potassium and sodium, especially in areas where the water is lacking calcium and magnesium. These associations are further complicated and compounded by factors derived from the physical environment, such as drinking water, plants, and crops (see Lam, 1986, for more references). Based on the map patterns here, it is strongly suspected that the risk factors between stomach and esophageal cancers in China are very different.

Liver Cancer

Liver cancer is common in this region and along the east coast of China. At the eastern tip of Yangtse Delta lies Qidong County (Qidong Xian), the county of highest liver cancer mortality in China and in the world (see Figures 12.1, 12.3, and 12.4). Nationally, liver cancer mortality ranks third. Since liver cancer affects a relatively young and productive age group (between 40 and 50) with significant social and economic impacts, the prevention of this cancer has become a major focus of the Chinese government. The Qidong Liver Cancer Institute, established in early 1980s, is part of a major effort in researching and curbing the disease. Various risk factors of liver cancer in this region have been suggested, such as chronic infection with hepatitis B virus, drinking water contamination from shallow wells and dikes, and aflatoxin food contamination, found mostly on crops in storage, such as corn. The humid climate that favors the condition of aflatoxin growth in this region is also considered an important factor. Other studies have shown that liver cancer appears common in areas of sedimentary rock, where susceptible individuals are exposed to high levels of calcium, magnesium, and potassium (Foster, 1989). As a result of intense research, several preventive measures against liver cancer were taken in this region during the early 1980s, such as digging more deep wells to minimize drinking water contamination. But after more than 10 years, the effectiveness of these preventive measures seems minimal, and mortality from liver cancer remains high in this region.

Liver cancer mortality is also common in Southwest China, that is, Guangxi Province. This region is characterized by limestone karst topography and has very different geographical environments (both physical and sociocultural) from that of Qidong County. Therefore, their risk factors would likely be different. It is clear that the etiology of liver cancer, as well as of stomach and esophageal cancers, is multivariate in nature, with dominant risk factors varying from place to place.

Moreover, the geographical patterns of liver cancer mortality vary considerably with scale. When the same set of data was aggregated and reported at the county and provincial levels, the geographical patterns appear random. Previous spatial autocorrelation analysis using data by province has confirmed the randomness of the liver cancer mortality pattern at that scale (Lam et al., 1990). The role of geographical scale in analyzing the cancer mortality patterns is no doubt essential, and the implications of their scaling properties are many. In this case, one is likely to overlook possible geographical relationships if only provincial or county scale data were used. This scale problem is common in studies involving data defined and reported in polygon form and has been described as the modifiable areal unit problem (MAUP) (Openshaw, 1974). It is likely that fractal models could be utilized to illustrate and identify the scaling properties of these cancer patterns so that more reliable interpretation about the patterns and the processes or factors leading to these patterns can be achieved.

MEASUREMENT OF CANCER PATTERNS

Two different methods were applied to estimate the fractal dimensions of the three cancer surfaces. The isarithm method originally developed by Goodchild (1980) and later modified by Shelberg et al. (1983) measures the fractal dimensions of isarithms depicting the surface, such as contours for elevation. The surface's dimension is the resultant line dimension plus 1. The variogram method used by Mark and Aronson (1984) utilizes variograms, either of the entire surface or of certain profiles extracted from the surface, for fractal calculation. These methods were originally developed and used for regular grid lattices and are described in detail in Chapter 2. Applying these two methods to irregular polygonal data, such as the present study, requires some modifications.

Isarithm Method

To apply the isarithm method, a polygon-to-grid conversion was required to transform from an irregular to a regular space. In this study, we used a grid size of 1 km by 1 km, and this resulted in a matrix of 222 rows by 264 columns. On average, each polygon is represented by about 30 to 40 grids. This grid size allows for some variations along the commune boundaries and corresponds to the pixel size of some remote-sensing and GIS data [for example, National Oceanic and Atmospheric Administration (NOAA) satellite data]. The size of the grid will affect the resultant dimension values. The grid size should not be too small to avoid making the surface arbitrarily smooth, leading to an underestimation of the complexity of the surface and a lower fractal dimension. Overrepresentation of the polygonal surface by grids, however, can be detected easily by examining the regression plot, and adjustment of dimension values can be accomplished by selecting only ranges of points on the regression plot (see below and Figure 12.5).

In this study, the algorithm used is slightly different from the one discussed in Chapter 2, but follows closely the one presented in Shelberg et al. (1983). In brief, the present isarithm algorithm requires an input of a matrix of z values (the cancer mortality rates), numbers of rows and columns defining the matrix (222×264), an isarithmic interval for determining the number of isarithms to be examined, and the maximum walk size to be used. A maximum of 50 grids was chosen as the largest walk size, and an isarithmic interval of 5.0 was used for all three cancers. For each isarithm, the algorithm finds its length (the number of boundary cells) at each walk size from 1 to 50 grids and then computes the linear regression between the two variables in a double logarithmic form. The regression fit for each isarithm, the *R* square, is also computed. The fractal dimension (*D*) of each isarithm is derived from the slope (*B*) of the regression, where $D = 1 - B$. The fractal dimension of the entire surface is then determined by averaging the *D*'s for all eligible isarithms and adding 1. For an isarithm to be included in averaging, the number of boundary cells must be greater than zero for all walk sizes. This criterion is less stringent compared with those implemented in the algorithm in Chapter 2, where the latter imposes an additional criterion that the *R* square must also be greater than 0.9 for an isarithm to be included. In sum, we used more walk sizes and less stringent criteria to derive the fractal dimensions of the cancer patterns so that we can examine the patterns and the behavior of the algorithm more thoroughly.

Fractal dimensions for each cancer pattern were computed using both row and column orientations. The dimension values derived from averaging all eligible isarithms for each orientation as well as from the mean isarithm of each surface are presented in Table 12.2. Figure 12.5 illustrates the relationship between the walk size and the number of boundary cells (the length of the line) in a double logarithmic form for the mean isarithms of the three cancers. Similar to previous studies (Goodchild, 1980; Shelberg et al., 1983), the regression plots are characterized by breaks of slopes, indicating the existence of distinct self-similarity ranges within each pattern.

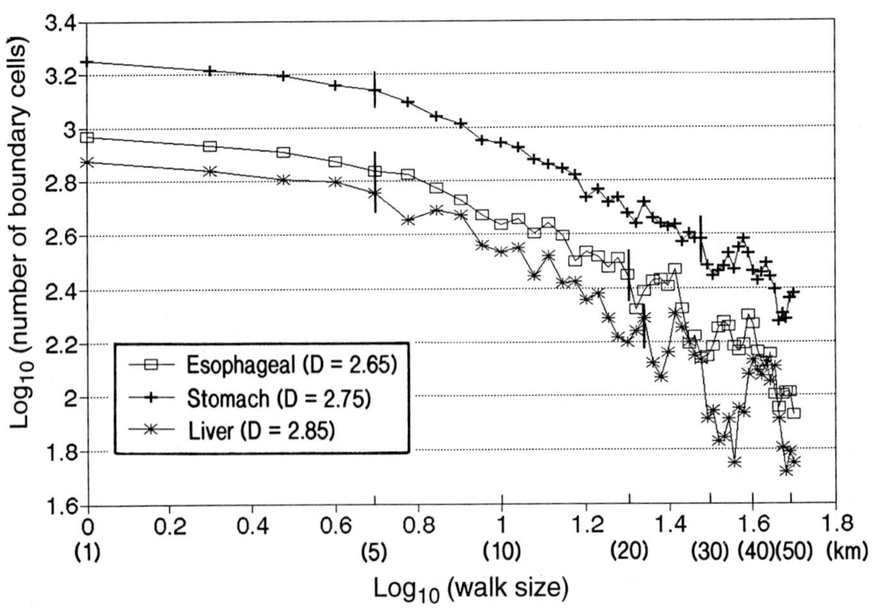

Figure 12.5 Scatterplots of the three mean isarithms (esophageal, 55; stomach, 50; liver, 40). The bars on each curve indicate the scale ranges over which fractal dimensions were recomputed.

TABLE 12.2 Comparison of Fractal Dimension Values Derived from the Isarithm and Variogram Methods

	Isarithm			
Cancer	Row	Column	Average	Variogram
Esophageal, all	2.68 (0.83)	2.57 (0.91)	2.63	2.71 (0.99)
Esophageal, mean (55)	2.69 (0.88)	2.61 (0.88)	2.65	[6–149 km]
5–20 km	2.65 (0.95)			
Stomach, all	2.62 (0.92)	2.61 (0.92)	2.61	2.78 (0.99)
Stomach, mean (50)	2.65 (0.92)	2.71 (0.93)	2.68	[6–98 km]
5–30 km	2.75 (0.98)			
Liver, all	2.62 (0.92)	2.64 (0.91)	2.63	2.86 (0.95)
Liver, mean (40)	2.76 (0.84)	2.70 (0.83)	2.73	[6–152 km]
5–22 km	2.85 (0.92)			

"all" denotes fractal dimensions and R squares estimated using all eligible isarithms; "mean" denotes values estimated using the mean isarithms only; "5–20 km" denotes values estimated from that scale range of the mean isarithms. R squares are in parentheses, and values in brackets denote the scale ranges in true distances for which the fractal dimension values are estimated (that is, self-similarity ranges) in the variogram method.

Variogram Method

The variogram method seems to be a more direct method for estimating the fractal dimension of a polygonal surface, although there have been some reservations with regard to its use in measuring multifractal phenomena (see Chapter 8). As discussed in Chapter 2, this method utilizes another important property of the surface, the variogram, to estimate the fractal dimension. Variograms are characteristic functions describing the relationship between variances in values among points and their distance. No polygon-to-grid conversion is required in this method. Instead, the centroids of the polygons are used to construct the variogram of the surface.

The algorithm first finds the distances between pairs of centroids and their squared differences in cancer mortality rates. It then divides the distances into 100 groups and computes the corresponding mean-squared difference in mortality rates (that is, variance) for each distance group. The variogram for each cancer pattern is illustrated by a plot of the distance values against their variances in a double logarithmic form (Figure 12.6). The fractal dimension can then be estimated by regressing the two variables linearly, and $D = 3.0 - (B/2)$, where B is the slope of the regression. The corresponding R square is also computed to indicate the regression fit.

Figure 12.6 indicates clearly that the variogram plots do not behave linearly for all distance groups. Breaks of slopes in the variograms are common, indicating the ranges at which the patterns show self-similarity and that the fractal dimension should be estimated using this range of points only. In this study, the exact breakpoint locations were identified by eye-balling the locations at which the greatest changes of direction occur, and the regressions were recomputed for these ranges. Table 12.2 lists the resultant D values and their corresponding self-similarity ranges from this method, along with the results from the isarithm method.

RESULTS AND DISCUSSION

As expected, the fractal dimensions of the cancer surfaces estimated in this study are much higher (from 2.61 to 2.86) than most real terrain surfaces (with a typical D of 2.2; see Mandelbrot, 1983). The three-dimensional displays of the cancer surfaces (Figure 12.4) illustrate a ruggedness that is uncommon in most topographic surfaces. It should be noted that the geographical patterns selected for this study are highly clustered. It would be of

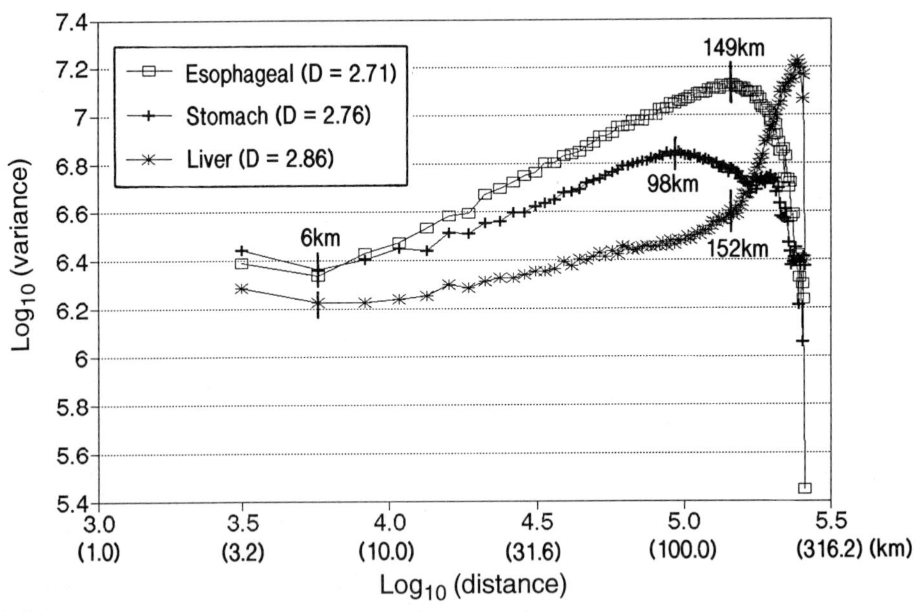

Figure 12.6 Variogram plots of esophageal, stomach, and liver cancer patterns for the Taihu Region, China. The bars on each curve indicate the scale ranges over which fractal dimensions were computed.

interest in the future to use some cancer surfaces that are basically random or regular in order to determine the bench values so that other cancer or socioeconomic surfaces can be tested against them. Similarly, testing the fractal method on some known patterns such as checkerboards or stripes and comparing the method with spatial autocorrelation measures on these known patterns and the existing cancer patterns would be of great use. The utility of the fractal method in analyzing socioeconomic patterns would not be fully recognized until the bench values are established.

Second, the *D* values computed for the three patterns are quite similar, with liver cancer yielding slightly higher dimensions than the other two in both methods. The difference in fractal dimension values among surfaces is considered minimal, despite the fact that these three patterns have distinct geographical patterns (Figures 12.3 and 12.4). We can conclude that, although the fractal dimension could be used to indicate the spatial complexity of the surface as a whole, the use of it in discriminating or reflecting the locational variations among these surfaces remains questionable.

Analysis of the results from the two measuring methods indicates that the dimension values derived from the variogram method are higher than those derived from the isarithm method (see the Average column in Table 12.2). Based on the variogram method, the liver cancer pattern was found to have the highest dimension ($D = 2.88$), with a scale range roughly between 6 and 152 km. The fractal dimensions of esophageal and stomach cancers are similar but are lower than that of liver cancer, with esophageal cancer having a larger self-similarity range of 6 to 149 km than that of stomach cancer (6 to 98 km). These self-similarity ranges suggest that, if the communes are aggregated to a distance greater than the limits, the patterns could look very different and would likely lead to a different interpretation of the underlying processes. It is also expected that whatever the underlying controlling factors (for example, climate, topography, or water resources), they are likely to operate in ranges of scales similar to that of the cancer surfaces, as determined here by the fractal method. Further research on how these patterns would respond to scale is much needed for a better understanding of the phenomena and better utilization of fractal analysis, provided multiple scale data are available. Beyond these scale ranges, the curves behave erratically such that dimension values derived from these parts of the variograms are not reliable (*D* will exceed 3.0). This problem has also been noted in Mark and Aronson (1984) when measuring topographic surfaces.

Interpretation of the results from the isarithm method is less direct, since the resultant dimension values vary considerably with the input parameters, including the maximum walk size, isarithmic interval, row or column orientation, and criteria used in determining which isarithms are eligible for deriving the final surface dimension. When the average of all included isarithms in both orientations is used, the resultant fractal dimensions of the three surfaces are very similar, with values from 2.61 to 2.63 (Table 12.2). But when only the mean isarithm of each surface is used, liver cancer stands out with a higher *D* of 2.73 (using the mean isarithm of 40.0). This phenomenon can easily be explained as follows. By excluding the isarithms whenever they have zero boundary cells at any step, the present algorithm tends to omit the isarithms of higher values (the peaks), since they are usually shorter and scattered. The final dimension is therefore heavily influenced by the isarithms of low values, resulting in a lower dimension than the one derived from the mean isarithm only. For example, in the case of liver cancer, only eight isarithms at the low end were included, with eight others at the high end excluded when deriving the final *D* value (see Figure 12.7 for part of an output from the method). Figure 12.7 also indicates that fractal dimension increases with isarithmic value, but with decreasing *R* square, suggesting more unstable measurement toward the high end of the surface.

```
$ run fract_lam

     FRACTAL SURFACE MEASUREMENT

     MODIFIED BY LAM FOR CANCER PATTERNS


     ENTER THE INPUT FILENAME ===> liverl.dat
     NUMBER OF COLUMNS ===> 264
     NUMBER OF ROWS ===> 222
     MAXIMUM Z =     87.040
     MINIMUM Z =      0.000

     USE ROWS OR COLUMNS ? ===> R
     ENTER ISARITHM INTERVAL ===> 5
     INPUT MAXIMUM CELL SIZE ===> 50

          CONTOUR  ELEVATION        B       D        RSQ

             1        5.00      -0.4151  2.4151  0.9450
             2       10.00      -0.4881  2.4881  0.9484
             3       15.00      -0.5409  2.5409  0.9401
             4       20.00      -0.6064  2.6064  0.9392
             5       25.00      -0.6624  2.6624  0.9287
             6       30.00      -0.7231  2.7231  0.9120
             7       35.00      -0.7783  2.7783  0.8785
             8       40.00      -0.7611  2.7611  0.8406


     NO. OF ISARITHM LINES INCLUDED =    8
     NO. OF ISARITHM LINES NOT INCLUDED =    8
     DIMENSION =   2.6219
     R-SQ = 0.916577
     ISARITHM INT. =    5.00
     MAX. CELL SIZE RANGE: 50
     ROWS =    1,222
     COLUMNS =    1,264


     NEED SUMMARY FOR ISARITHM LINES(?)==> n
 FORTRAN STOP
```

Figure 12.7 Part of an output from the isarithm method.

Another input parameter, the maximum walk size, also affects the resultant dimension values considerably. Increasing instability occurs when the value of the maximum walk size increases. The scatterplots for the mean isarithms of the three cancer surfaces demonstrate this scaling phenomenon effectively (Figure 12.5). Three distinct parts can be observed from the scatterplots. The first part consists of about five walk sizes (that is, five grids or 5 km), which corresponds roughly to the average diameter of the polygons. In other words, on average, it takes about five to six pixels to represent a polygon along one dimension, thus leading to an overrepresentation of the surfaces and biased estimates of fractal dimensions. When the walk size increases to a distance larger than the diameter of most polygons, changes in isarithm lengths (number of boundary cells) become more drastic, as represented by the second part of the curve from walk size 5 to walk sizes 20 to 30 (20 to 30 km). Beyond these ranges, the scatterplots become unstable, and regression results derived from these parts of the scatterplots are not reliable.

Based on these observations, additional regressions for the middle parts of the scatterplots were performed and the resultant fractal dimensions are surprisingly similar to those derived from the variogram method, with dimensions becoming 2.65, 2.75, and 2.85 for esophageal, stomach, and liver cancers, respectively (Table 12.2). Accordingly, we can conclude that all three cancer surfaces are spatially complex, with high variations across space, and that the liver cancer surface is slightly more complex than that for esophageal and stomach cancers. Nevertheless, more studies using geographical patterns that are different from the present ones, such as regular, random, checkerboard, stripes, and white-noise patterns, are needed to derive the bench values for socioeconomic data so that more in-depth interpretation can be achieved.

Our findings seem to indicate that the variogram method is more straightforward and easy to utilize for this type of application, since no polygon-to-grid conversion is required, higher R squares are obtained, and less ambiguity is involved. Factors influencing the results from the variogram method are polygon size and its size variation. If large polygons are used, the centroids become an inadequate representation of the polygons, and the distance values computed from the centroids would not be reliable. Another problem is that the variograms may have negative slopes after certain points, leading to dimension values exceeding 3.0, which is unexplainable. In addition, Clarke and Schweizer (1991) have warned that the variogram D measure is affected by the scale of the measurement units in use, although their mathematical proof and empirical tests do not really support the argument. The discrepancies in scale ranges detected by the two methods (the variogram method yields much larger scale ranges for all three cancers) suggest that the variogram method may be less sensitive to multifractals or multiple scale phenomena as argued in Chapter 8. Future studies on the performance of the measurement methods are urged.

On the other hand, the ambiguity, or rather flexibility, allowed in the isarithm method should be considered as an advantage. We can observe how different input parameters affect the resultant dimensions, and thus a good grasp of the behavior of the surface can be obtained. Moreover, multifractality, which in fact exists in most phenomena, is readily shown by the fractal dimension as well as the scale range of individual isarithms.

CONCLUSION

The utility of fractals will be curtailed if fractal dimensions and self-similarity ranges derived from different measuring methods vary considerably. Our study has shown that the fractal dimensions derived from both the variogram and the isarithm methods are very similar, but their scale ranges are markedly different. All three cancer surfaces have very high dimensions, much higher than most topographic surfaces on the earth. Liver cancer has the highest dimension (2.86 and 2.85), followed closely by stomach (2.78 and 2.75) and esophageal (2.71 and 2.65) cancers. The difference in fractal dimension values among these patterns is considered minimal. Although the fractal dimension can be used to quantify the complexity of spatial patterns as a whole, it does not differentiate adequately the locational variations among these patterns. In terms of scale ranges, the isarithm method yields ranges of approximately 5 to 30 km for all three cancer surfaces, whereas the variogram method yields much larger scale ranges of approximately 6 to 150 km. The discrepancy in scale ranges between the two methods indicates that the variogram method may be less sensitive to the multifractal nature of the cancer surfaces. These scale ranges could be used in the future to identify possible etiologic factors that operate at similar ranges.

We have also pointed out that derivation and interpretations of these results require some caution. It is important for researchers who utilize fractals to understand the principles underlying these methods, as well as the effects of the input parameters, before comparisons within and between studies can be made. Finally, we suggest that for fractal analysis to be truly useful in analyzing and linking patterns and processes, a comprehensive study on more real and controlled patterns, such as white noises, stripes, and checkboard patterns, is needed to derive bench values so that more accurate interpretation can be achieved. Comparisons with other spatial measures, such as spatial autocorrelation statistics and information statistics, on those controlled and real patterns should also be made. The utility of fractal analysis will not be fully recognized until these bench values are established and evaluations with other spatial measures are made.

ACKNOWLEDGMENTS

This research was partially funded by research grants from the National Program for Advanced Study and Research in China, administrated by the U.S. National Academy of Sciences, and from the Nanjing Institute of Geography of the Chinese Academy of Sciences. We thank David Mark and Lee De Cola for their helpful comments and Xie-Tau Hou and colleagues from the Nanjing Institute of Geography for their support.

REFERENCES

Clarke, K. C., and Schweizer, D. M. 1991. Measuring the fractal dimension of natural surfaces using a robust fractal estimator. *Cartography and Geographic Information Systems* 18(1):37–47.

Editorial Committee. 1979. *The Atlas of Cancer Mortality in the People's Republic of China.* Beijing, China: China Map Press.

Foster, D. 1989. Clues from the environment. *Environment* 31(3):12–39.

Goodchild, M. F. 1980. Fractals and the accuracy of geographical measures. *Mathematical Geology* 12: 85–98.

Howe, G. M., ed. 1986. *Global Geocancerology: A World Geography of Human Cancers.* Scotland: Churchill Livingstone.

Lam, N. S.-N. 1986. Geographical patterns of cancer mortality in China. *Social Science and Medicine* 23(3): 241–247.

———, et al. 1990. The role of geographical scale in analyzing cancer mortality patterns in China. Paper presented at the 1990 IGU Regional Conference on Asian Pacific Countries, Beijing, China.

Lavallée, D., et al. 1992. Nonlinear variability of landscape topography: Multifractal analysis and simulation. In *Fractals in Geography,* N. Lam and L. De Cola, eds. Englewood Cliffs, NJ: Prentice Hall.

Mandelbrot, B. B. 1983. *The Fractal Geometry of Nature.* San Francisco: W.H.Freeman & Co.

Mark, D. M., and Aronson, P. B. 1984. Scale-dependent fractal dimensions of topographic surfaces: An empirical investigation, with applications in geomorphology and computer mapping. *Mathematical Geology* 11:671–684.

Openshaw, S. 1974. The modifiable areal unit problem. *CATMOG Series,* No. 38. Norwich, England: Geo Books.

Shelberg, M. C., Lam, N. S.-N., and Moellering, H. 1983. Measuring the fractal dimensions of surfaces. In *Proceedings, Sixth International Symposium on Automated Cartography (Auto-Carto 6)* 2, Ottawa, Ontario, Canada, pp. 319–328.

Part IV

RESEARCH AND APPLICATIONS IN MAPPING SCIENCES

CHAPTER 13

ONE THOUSAND MOUNT EVERESTS?

Keith C. Clarke

Department of Geology and Geography, Hunter College and the Graduate School and University Center, City University of New York, NY

ABSTRACT

One of the most intriguing applications of fractal geometry in the geosciences has been to realistic simulations of terrain. Several fractal terrain-simulation models have been suggested and applied, including cascading, surface subdivision, and Fourier inversion. Some of these models have been shown to produce terrain with undesirable characteristics, such as seams, apparently incorrect fractal dimensions, and incorrect quantities of pits and peaks. This chapter reviews the use of fractal methods for cartographic simulation of terrain, covering the major methods published to date and giving examples of methods where available. A model of natural terrain, proposed earlier by the author, is summarized and its assumptions examined. This model describes topography as the summation of two major parts: a scale-dependent periodic spatial structure and a scale-independent fractal texture. The scale-dependent component of the model can be calibrated for any given landscape using Fourier analysis, as can the scale-independent component by measurement of the fractal dimension of a specific piece of terrain. The model is invertible, allowing simulation of terrain with specific surface characteristics. As a means of exploring the variation in the parameters associated with the model, a very large number of simulations were performed, loosely based on the terrain parameters of the Mount Everest summit area. A computer movie was generated showing variations in the parameters, and isolated frames from the movie were interspersed with views of the real terrain. A group of observers was then asked to select which images showed the real terrain. The distribution of correct and incorrect responses answers some of the questions on what a skilled terrain interpreter perceives as "correct" when viewing terrain and suggests how synthetic terrain can be tailor-made to appear realistic. Several color graphics, three-dimensional perspective views of the real and synthetic terrain are shown so that the test results can be viewed in context.

FRACTAL TERRAIN SIMULATION

Fractal geometry has taken geography and cartography by storm, and nowhere is the influence of the theory pioneered by Mandelbrot so visible as in the uses of fractals in simulations. Fractal geometry offered the opportunity to model and therefore simulate real-world objects which in the past had defied quantification. Early simulations were of two-dimensional objects, but recently three-dimensional objects, such as trees and clouds, and complex phenomena such as air turbulence and the weather have been successfully modeled and simulated using fractals. One of the most intriguing applications of fractal geometry in the geosciences has been to realistic simulations of terrain. Natural terrain has few regular, geometric features and highly complex detail. Terrain modeling and simulation have relied either on gross "geometric" representations of terrain or on massive real-world data bases containing actual renditions of objects. In both cases, the resultant terrain forgery is obviously synthetic, either too smooth or with blocky or spiky detail.

The modeling and simulation of terrain that is not obviously synthetic have provided a considerable challenge. In general, many methods are available to interpolate surfaces from gross generalizations, for example using Bezier and *B*-spline surfaces (Stepoway and Christiansen, 1989). Fractal geometry, however, has allowed the addition of texture (detail) at any level of scale or however much one's viewpoint zooms in toward the surface. The advantages of fractals for terrain representations are that only the required level of detail need be generated at any time, allowing fast, real-time simulation, and the use of the method for motion simulation. Computer graphics has enhanced such simulation methods by adding striking new forms of terrain representation, while the motion picture industry has added the dimension of animation (Fournier et al., 1982).

Several fractal terrain-simulation models have been suggested and applied, including cascading, surface subdivision, and Fourier inversion. The simplest of the methods is the recursive mid-point displacement described in Jeffery (1987), a surface subdivision technique. Jeffery attributed the method to Fournier et al. (1982) and included an algorithm and a computer program to implement the technique. The technique (Figure 13.1) works by taking successive neighborhoods of four grid cells and displacing the corners by an amount proportional to the scaling of the cell size. The displacement is therefore determined by the desired fractal dimension and by the step size. After one iteration, the step size is halved, and the process is repeated, leading to a nested set of displaced cells, getting smaller and smaller. Jeffery also presented a program using Voss's (1988) improved algorithm, which adds the displacements at all levels of the interpolation rather than only at the current. This technique eliminates the patchy effect of successive interpolations within larger squares.

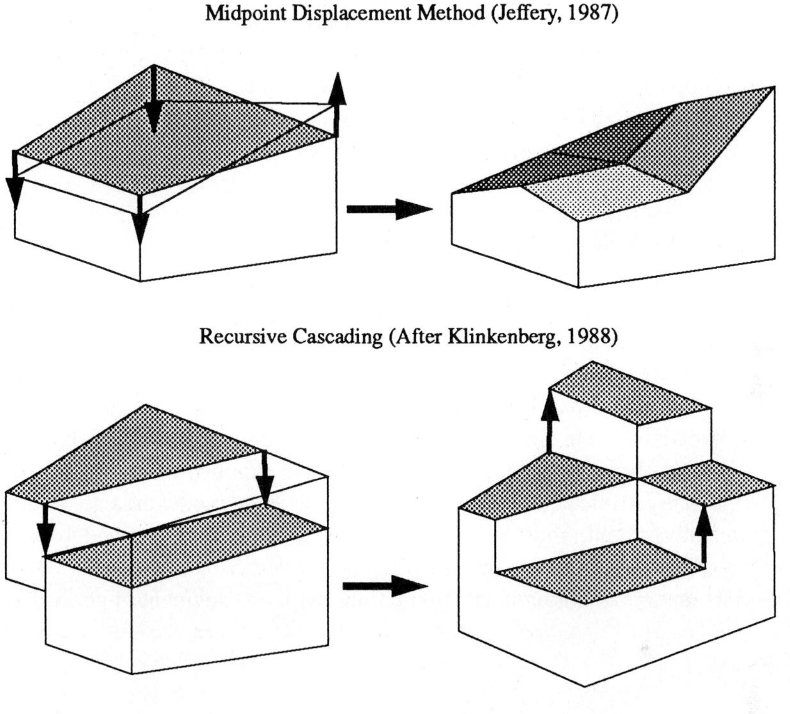

Figure 13.1 Surface subdivision and cascading methods for terrain simulation.

The cascading method (Figure 13.1) was first proposed by Mandelbrot (1977) and is described in a simulation context by Klinkenberg (1988). As a first step, a flat plane is generated across which a randomly oriented "fault" is placed randomly. The two half-planes that result are then displaced, one up and one down, to form a cliff. The amount of the displacement is computed as a function of the Hausdorff value H. When H is 0.5, the two halves are displaced equally. The process is then repeated until the effect of one cliff becomes negligible. Klinkenberg (1988) used 1000 cliffs for a grid with 256 rows and 256 columns. A problem noted by Klinkenberg is the fact that surfaces generated with a specific fractal dimension often yield different fractal dimensions when actually measured.

In many cases, the pure fractal model has been found to fit natural terrain only in a limited sense (Klinkenberg and Clarke, 1992; Mark and Aronson, 1984; Roy et al., 1987). For simulation purposes, some of these models have been shown to produce simulated terrain with undesirable characteristics, such as seams, apparently incorrect fractal dimensions, and incorrect quantities of pits and peaks. In many cases, simple adjustments can be made to eliminate these inadequacies. While the results are effective, the resulting terrain can still appear synthetic. What, then, does the observer use in a terrain representation to decide whether or not a piece of terrain is "realistic"? The hypothesis advanced here consists of the following. Natural terrain contains at least three components, all of which are anticipated when viewing terrain representations. These three components can be summarized as structure, texture, and process-texture. The structure represents the broad scale features and is periodic in nature. The texture represents the surface detail or roughness of the terrain, and the process-texture represents the modifications of the surface detail attributable to geomorphological processes at a detailed level.

A terrain-simulation model incorporating this approach was advanced in an earlier paper (Clarke, 1988a). A *scale-based* model of natural terrain was proposed, which describes topography as the summation of scale-dependent periodic spatial structure and scale-independent fractal noise. These two elements are the terrain structure and texture described above. This model can be calibrated using highly generalized digital representations of actual terrain and then inverted to simulate terrain. The goal of this approach is to produce simulated terrain that a cartographer would recognize as having familiar properties and to increase the realism and utility of synthetic terrain for cartographic uses.

The advantages of the scale-based model are fourfold: (1) The model uses mathematical background from two substantive areas of theory, spectral and fractal analysis, both of which have been applied to terrain for some time (Pike and Rozema, 1975; Goodchild, 1982; Pentland, 1984). (2) The model can be calibrated from highly generalized real terrain and is invertible across larger scales. (3) Since the model uses no iterative methods, it is fast and efficient, and produces terrain in digital elevation model format. (4) The inversion of the model allows parametric control over the appearance of the terrain structure and its texture, enabling many simulations from one segment of generalized terrain.

PROCESS-TEXTURE

Terrain structure and texture can be used effectively to generate synthetic terrain. The challenge is to include an additional terrain element, termed here process-texture. At larger map scales, the impact of geomorphological processes on terrain is more noticeable. Probably the biggest impact at these scales concerns the typical forms resulting from stream erosion and deposition, that is, the in-filling of depressions, the back-cutting of stream valleys, and the dendritic structure of valley and ridge arranged within a basin, where both structural and local forces determine form. For example, within a drainage basin, valleys and ridges may repeat at regular intervals as we make a circuit around the main drainage channel. Locally, however, numerous factors can divert streams and their valleys.

At the very large scale, down to the level where distances are in meters and tens of meters, individual boulders, mass-wasting, vegetation characteristics, human intervention, landslides, and a myriad of other local processes tend to dominate. These processes result in small features with a larger random component, but spatial structure nevertheless. The final part of the scale-based model has to do with these local processes, in particular mass-wasting. Since this process applies only at the very largest scale, the operator has a small neighborhood. As has been noted, the fractal terrain texture contains large numbers of local maxima and minima, which are not reflected in real terrain. In this case, the worst offender is the pits, which in real terrain are normally filled in by mass-wasting, except in a few special terrain types (for example, karst). In the original version of the scale-based model, a local operator was used to find pits and fill them by averaging their elevations with the next-lowest neighboring grid cell. The operator is invoked as many times as the user wishes, filling and smoothing more or less as required.

The initial tests of the model (Clarke, 1987; 1988a) used this simple pit-filling erosional operator, with no basis in geomorphology. As an extension of the original model and in an attempt to introduce the concept of process-texture, two additional erosion operators were developed. First, a fluvial erosion operator was developed. This operator was applied to an eight-cell neighborhood (the grid cells immediately contiguous to any given grid cell) and simulates the movement of water locally, moving material from the highest point to the lowest within the neighborhood. Without a channel-deepening effect, this operator would merely simulate soil creep or mass-wasting. Band (1986) and Douglas (1986) have done work using DEMs to simulate the direction of water flow and partitioning of the hill slopes between the stream segments.

The vastly simpler approach used here differs in that it attempts merely to simulate the forms resulting from erosion for cartographic purposes so that the surface process is seen as spatially continuous rather than partitioned. For each kernel, the elevation maximum, the secondary maximum, and the minimum are determined and located. Second, a mean elevation operator is computed over the eight-cell neighborhood. The operator then works in different ways for alternate rows and columns of the grid. For odd rows and columns, the kernel elevation is simply replaced with the eight-cell mean. This process simulates smoothing and may actually increase elevation at the kernel for concave neighborhoods (a likely outcome on real slopes, as material fills in the cavity). For even rows and columns, the operator replaces the kernel with the secondary low and replaces the secondary low with the low.

In this way, the "cut" of a leading terrain notch is propagated both upslope and downward. Elevations cannot go below the overall grid minimum, a sort of base level for the erosion. The alternation of the two operations is necessary first to combine the fluvial effects of cutting and deposition of material into a single pass and also to avoid the propagation effect of stream heads over the whole array in one pass. In the second erosion model, which may be more appropriate to extraterrestrial and desert environments, slopes are allowed to retreat in parallel. In this case, erosion takes place only in areas of high gradients, and material is allowed to disperse widely over the less steep areas. This form of erosion is suited to temperature extremes, large differences in elevation, and a general lack of water in the environment. In terms of the algorithm, a critical slope must be identified at which slopes become stable. The operator in this case simply computes the drop from the high to the low within the eight-cell neighborhood and replaces the kernel pixel with the mean over the eight-cell neighborhood if the gradient exceeds a value determined by the user. Thus, when gradients exceed the cutoff, regular smoothing takes place, tending toward regular linear slopes, until the critical gradient is reached, after which the slope becomes stable and stops eroding.

Figure 13.2 shows a diagram of a test of these two types of operator applied to a test landscape of a uniform plateau. As a starting condition, a flat plane was used with a single row of lows at one edge, a vertical cliff. In the center of the cliff, one extra grid cell was reduced to the elevation of the cliff bottom. While the cliff serves as the overall feature to be eliminated, the single grid cell serves as a seed to propagate local erosion effects, the trait desired when these operators were applied to fractal texture. Figure 13.2 shows the status of the terrain after multiple passes, while Figure 13.3 compares the contour map

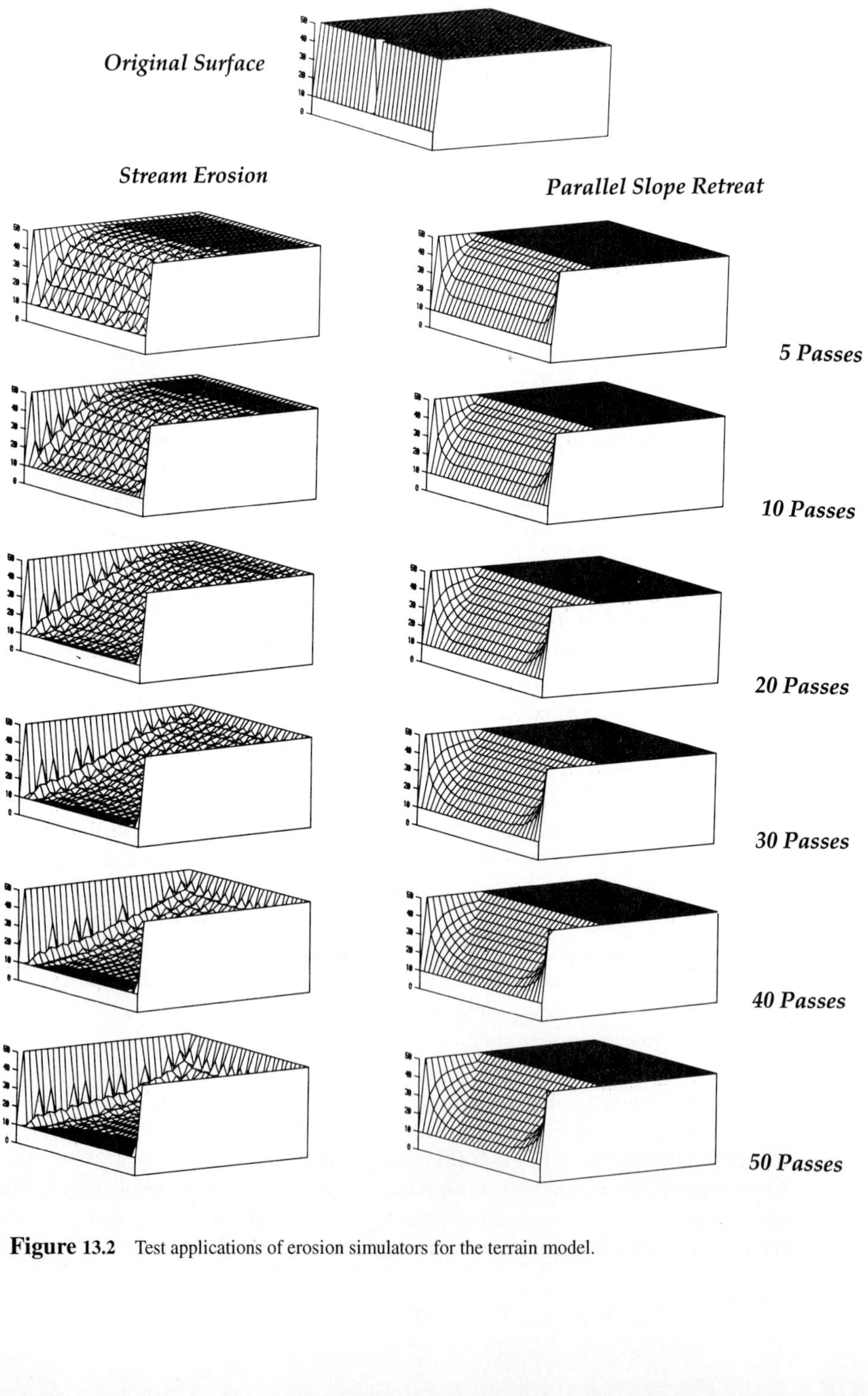

Figure 13.2 Test applications of erosion simulators for the terrain model.

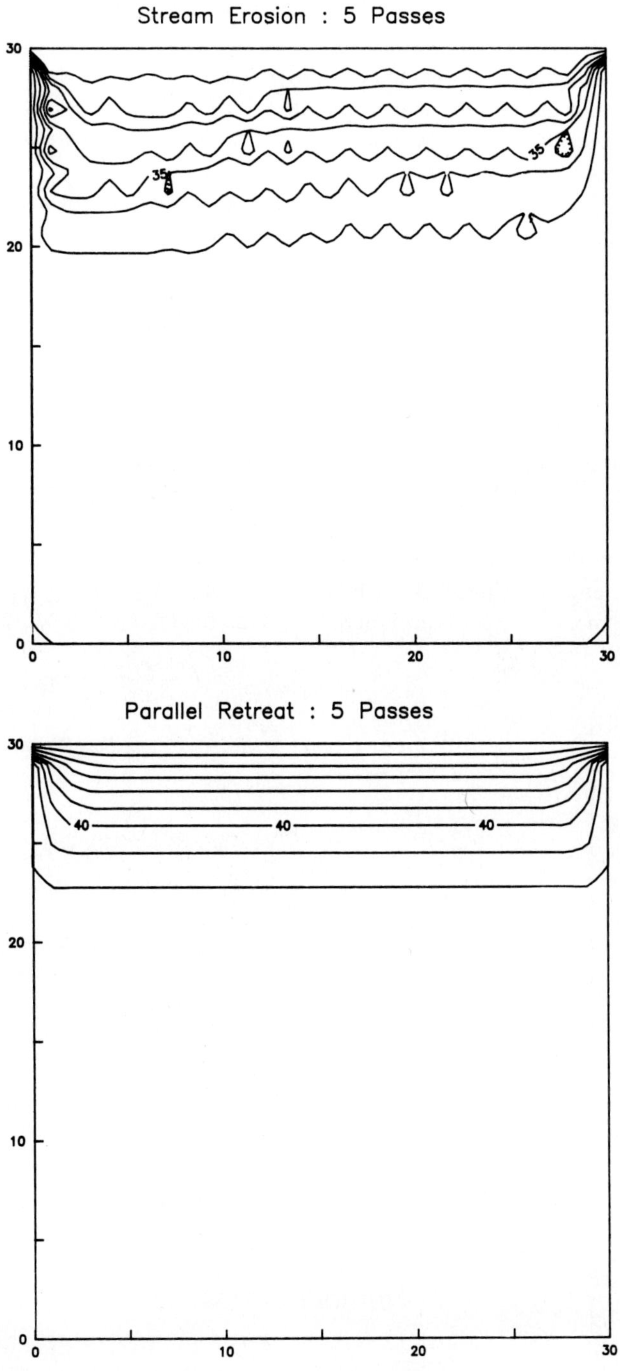

Figure 13.3 Test applications of erosion simulators, contour maps.

after five passes using the two methods. The obvious boundary effects are due to the fact that the operator can only work over an eight-cell neighborhood, thus avoiding changes in elevation in the first and last row and column. Figure 13.4 shows how the means and standard deviations of the elevations over the grid change with multiple passes of the operators. In both cases, the triangles represent the parallel retreat operator, while the circles represent stream erosion.

In the case of the parallel retreat operator, successive passes result in less material being removed, with a resulting increase and then stabilization of the elevation standard deviation, one indicator of surface roughness. On the contrary, the stream erosion continues to remove material, but results in smoother slopes after an initial increase in roughness. This is due to the existence of both an early peak rate of removal of material (Figure 13.5) and the change from convex to concave slopes (Figure 13.2). In addition, the stream erosion operator shows activity propagating from the seed grid cell, in addition to the overall erosion effect (Figure 13.3).

The two operators therefore show the following desirable characteristics. First, they erode toward a given base level for the terrain, in this case the overall low. Second, they operate systematically on the terrain structure in a predictable way and less systematically on the terrain texture, which tends to amplify, propagate, and divert the overall effect of the systematic erosion. Thus, process-texture simultaneously complicates both of the other two terrain elements, the periodic structure, and the fractal terrain texture.

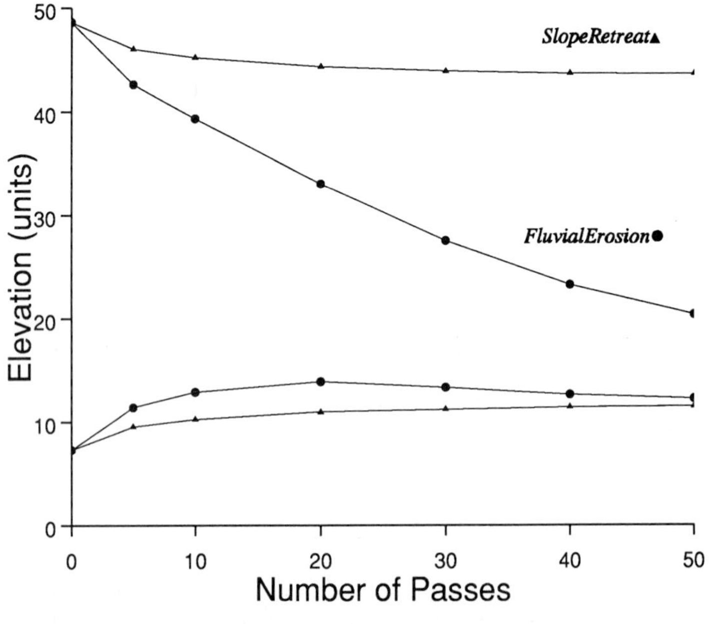

Figure 13.4 Changes in elevation means (the upper two curves) and standard deviations (the lower two curves) with multiple passes of the two operators.

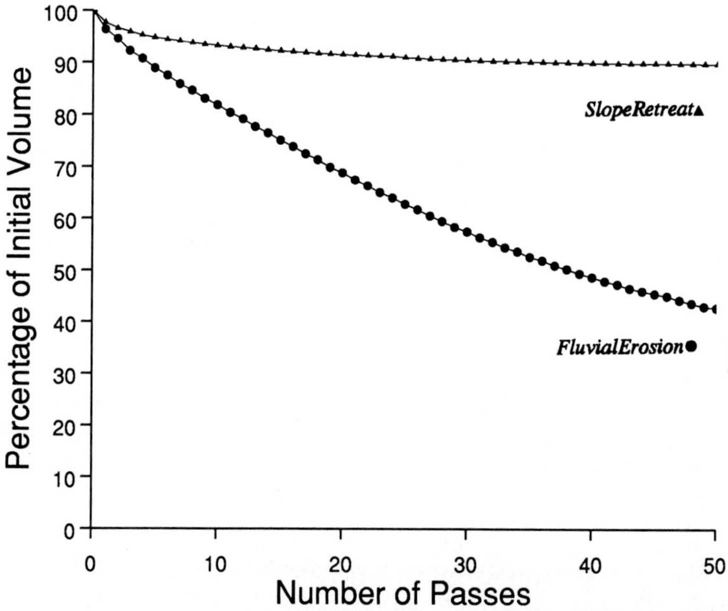

Figure 13.5 Relationship between percent of surface volume remaining and number of passes.

THE MODEL

The scale-based model takes the following form (Clarke, 1988a). Terrain is assumed to be represented as a set of discrete points in the form of a regular grid. Elevation z at implicit location (x, y) is given by

$$Z_{x,y} = T_{x,y} + cF_{x,y} + H_{x,y} + E_{x,y} \tag{13.1}$$

where

$$T_{x,y} = \beta_0 + \beta_1 x + \beta_2 y \tag{13.2}$$

defines a linear trend (very low frequency harmonic)

$$F_{x,y} = \sum_{i=1}^{i=k} \sum_{j=1}^{j=l} (A_{i,j}C_iC_j + B_{i,j}C_iS_j + C_{i,j}S_iC_j + D_{i,j}S_iS_j) \tag{13.3}$$

where the sine and cosine terms are given by

$$C_j = \frac{\cos(2j\Pi y)}{L_y} \tag{13.4}$$

$$C_i = \frac{\cos(2i\Pi x)}{L_x} \tag{13.5}$$

$$S_i = \frac{\sin(2i\Pi x)}{L_x} \tag{13.6}$$

$$S_j = \frac{\sin(2j\Pi y)}{L_y} \tag{13.7}$$

and where L is the length of the series in x and y as given by the subscripts. At high spatial frequencies, when the distances associated with harmonics approaches twice the sample resolution, Fourier analysis tends to highly overestimate the Fourier coefficients, a feature known as the Gibbs phenomenon. To avoid this problem, the maximum length of the series in the map in x and y, with harmonic numbers l and k above, are limited to one- third rather than one-half of the lengths of the map in x and y. The power spectrum for this two-dimensional Fourier series is an array P given by

$$P_{i,j} = A_{i,j}^2 + B_{i,j}^2 + C_{i,j}^2 + D_{i,j}^2 \tag{13.8}$$

The Fourier coefficients A, B, C, and D are given by

$$A_{i,j} = B_{i,j} = C_{i,j} = D_{i,j} = \sqrt{\frac{P_{i,j}}{4r^2}} \tag{13.9}$$

where r is 1 if i and j are zero, 2 if one but not both are zero, and 4 otherwise. The power associated with harmonic pair (i, j) is given by

$$\log P_{i,j} = \alpha + \beta \log d_{i,j} \tag{13.10}$$

where d is the ground resolution distance corresponding to harmonic pair (i, j). The slope of this linear equation, which can be computed by regression using the harmonic pairs as observations, gives the fractal dimension of the topography (Clarke, 1988a). The values of A, B, C, and D in Equation 13.9 can be either positive or negative and, if desired, could be assigned random values such that their sum fell within limits. In the implementation of the model here, they were assigned values derived from an empirical Fourier analysis used to yield the H component.

The equation for H is the same as Equation 13.3, with the constraint that A, B, C, and D are zero unless

$$\frac{100 P_{i,j}}{\sum_{i=1}^{i=k} \sum_{j=1}^{j=l} P_{i,j}} \geq \tau \tag{13.11}$$

where τ is a critical cutoff percentage of the total variance attributable to any harmonic, that is, the level defining significance. The value c in Equation 13.1 is a constant, given as a proportion of the topographic relief into which the self-similar texture is to be scaled.

Finally, E is the erosion effect, determined by iterating one of the two local operators and representing the addition of process-texture.

The model is therefore based on control of only four parameters, the scale factor of the fractal relief (c), the significance level of the Fourier inversion (τ), the type of erosion to be applied (stream, parallel, or none), and the number of cycles of erosion to be applied to the simulated terrain. These values control the smoothness, level of detail, and erosive character of the simulated terrain, respectively.

INVERTING THE MODEL

Since the model is based on parameterizing a Fourier transform, given a set of Fourier measurements the inverse Fourier transform can be used to generate the simulation. The model assumes the input of a small, highly generalized piece of terrain. This segment is used to compute the two-dimensional Fourier coefficients and the power spectrum and, from them, to estimate the fractal dimension and the log–log power-to-distance relationship. From these data, the Fourier inversion can be performed and the two major components of the landscape calculated. In the case of scale dependencies, the Fourier coefficients are zeroed out except in the cases when the harmonic is deemed to be significant; that is, the harmonic explains more than some cutoff level of the total surface variance. The Fourier transformation is then inverted, generating elevation values as summations of the Fourier expansion summed over significant harmonics. The resultant array is then split until it reaches a desired maximum size, the number of times being determined either by the scale at which the terrain is desired to be sampled or by the size of a given display device.

The texture is generated by the same inverse Fourier transform, except that the Fourier coefficients are given by values from the log–log regression of power and distance, equalized across the four values and using the same signs as the original Fourier coefficients. The texture is scaled by a percentage of the relief in the structure and is split by mirroring the grid onto itself so that each new quadrant takes the form of the original. Since the two-dimensional power spectrum has been computed and used in a log–log fit, the terrain's fractal dimension can be computed as a by-product. This dimension, however, applies only to the inverted texture and is changed by the scale-dependency effect of adding terrain structure. The final stage of the inversion involves passing the erosion operator over the terrain a number of times determined by the user. Successive passes of the operator result in more and more simulated erosion and a higher degree of process-texture, rather than purely fractal texture.

SYNTHETIC EVERESTS

To generate simulated surfaces using the model described above, a map of Mount Everest was digitized. The 5000 plus digitized elevation points, extracted from the source map for the 1988 1 : 50,000 Mount Everest summit area map described in Clarke (1988b), were assigned to approximately 280-m grid cells by averaging points within cells. Thus, a grid of 32 rows by 32 columns was generated to be used as a seed for the simulation model. Many simulations were performed, in each case generating a 248 row by 248 column grid of simulated elevations. Sequences of simulations were also performed, in each case holding the model parameters constant, but varying a single value. First, the cutoff in harmonics for the Fourier inversion (τ in Equation 13.11) was held as 5.0%, and the proportion of the relief attributed to texture (c in Equation 13.1) was increased from 0.0 to 0.1. This generated a sequence of 11 images. Each was colored by color level slicing the elevations and was hill-shaded using ERDAS. Perspective views were then generated of the "east face" using the same viewpoint in all images and using a standardized range of elevations from high to low.

Next, similar sets of perspective views were generated using texture as 0.03 of relief, but increasing the number of harmonics in the simulated structure by decreasing the cutoff value. This value (τ) is the proportion of the total power related to a particular harmonic pair and, as such, is similar to a proportion of variance explained by a given cosine wave in both x and y in the data. The cutoff determined which harmonic pairs are used in the Fourier inversion to create the structure.

The cutoff value was varied from 0% of the total variance (all harmonics computed are used, making the terrain more detailed) to 10% (only a few are used, smoothing the terrain). Again, colored perspective views were generated. Finally, the two erosion operators were applied to the data, using a standard simulation with 3% texture and a cutoff of 1%.

Simulations were performed using from zero to nine passes for each of the two erosion operators. Again, the simulated surfaces were colored and projected into perspective views. Some of the images generated are shown in Figure 13.6 (see color insert). Each of the four major parameters of the model was therefore given 10 values, resulting in 40 simulations. The views were pieced together using video projectors with fade mechanisms onto videotape and can be viewed as slow sequences showing change in the model parameters. Just as easily, the model parameters could each be given 250 different values to generate the 1000 Mount Everests alluded to in the title of this chapter, resulting in a smoother animation. However, since each perspective view involved about 2.4 hours of CPU time on a COMPAQ 386/20, to generate the 1000 views would require about three months of processing time. Each simulated Mount Everest can be thought of as a possible Mount Everest. The next stage was to determine which are probable Mount Everests.

To determine what qualities made the terrain images produced by the simulations appear realistic or artificial, a preliminary test was undertaken to quantify these subjective parameters. As a casual, quasi-scientific test, color photographs were taken of the perspective views. Each image showed the same amount of relief, since elevations were given a standardized range, used the same hypsometric colors, and were illuminated from the same sun angle and elevation. Thus, the images were comparable with each other. Four sets of three images (Figure 13.6) were selected to vary the model parameters. In the first set, the three images had a harmonic cutoff of 1%, used fractal texture as 3% of relief, but had different types of erosion. One image had no erosion, one had five passes of stream erosion, and one had five passes of parallel retreat erosion. In the second set of images, the erosion was eliminated, texture remained at 3% of relief, and the number of the harmonic was varied, with the cutoff set at 0%, 6%, and 8%. In the third set, the actual terrain data for Mount Everest were included with two of the most convincing simulations, cutoffs of 0% and 5%, with no erosion and 3% texture. Finally, with no erosion and a fixed 5% harmonic cutoff, textures of 1%, 5%, and 9% were used.

The ordering of pictures within each set was randomized, as was the order of sets, and the images were pasted to pages of a notebook with numbered pages and three lettered images on each page. A total of 30 subjects, selected not at random but including mostly geography, geology, and computer science faculty and students at Hunter College, was then shown the images and instructed to annotate an answer sheet with which image appeared most "realistic" and most "artificial" in their own opinion. Instructions were worded such that the one nine-year–old included in the study could follow them with ease. Almost all the subjects were quickly able to grasp the real versus artificial nature of the study. No subject was informed of the purpose of the task, other than the fact that the images showed computer pictures of mountainous terrain and that the study was designed to "help in selecting which methods for simulating natural terrain produce the best simulations." The booklet used in the test assured that only the three images to be compared at any one time were visible together.

The raw test scores were converted to measures of realism. This was done by assuming that an image judged "most realistic" was worth three realism points, one judged "most artificial" was worth one realism point, and the unannotated third selection was worth two points. Total numbers of responses for the 30 people were then averaged, and realism scores were standardized to range from a 3 (most realistic) to a 1 (most artificial).

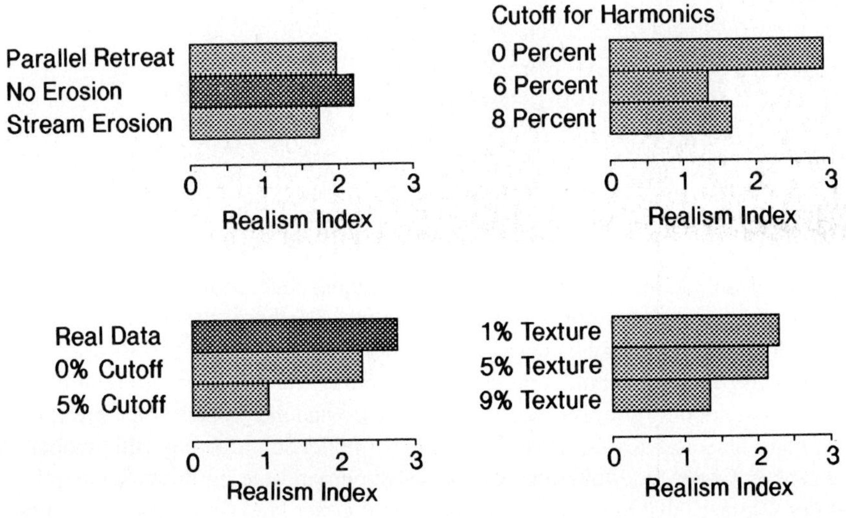

Figure 13.7 Realism values from the test results.

Table 13.1 shows the standardized and raw scores. The orders of the images have been reorganized to show continuous variation along the variable under test. Figure 13.7 shows the realism statistics from the test.

TABLE 13.1 Realism Values of Simulated Terrain

Image			Response (% of 30 subjects)			
Cutoff (t)	Texture (*c*)	Erosion (no. of passes)	High	Medium	Low	Realism Index
1.0	3.0	Parallel (5)	40	20	40	2.00
1.0	3.0	None	47	26	27	2.20
1.0	3.0	Stream (5)	13	54	33	1.80
0.0	3.0	None	90	10	63	2.90
6.0	3.0	None	0	37	63	1.37
8.0	3.0	None	10	53	37	1.73
Real	Real	None	67	30	3	2.64
0.0	3.0	None	33	64	3	2.30
5.0	3.0	None	0	6	94	1.06
5.0	1.0	None	54	29	17	2.37
5.0	5.0	None	33	57	10	2.23
5.0	9.0	None	13	14	73	1.40

DISCUSSION

The terrain model formulated and tested assumes that terrain consists of the three properties of structure, texture, and process-texture. Of these, the terrain model uses Fourier methods, fractal methods derived from the power spectrum, and local erosion operators, respectively, in the simulation process. A goal of the model was to produce simulated terrain that not only simply "looks good" but also had the property of subjective realism. The preliminary test results have some direct implications for how this model continues to evolve. Given the limitations of the test group and the no doubt improvable test conditions, images that have undergone erosion of either kind are perceived as no more realistic than images without erosion. Clearly, if process-texture is to become a significant component in terrain simulations, either many passes of the erosion operator are required or different erosion simulators must be developed. Since the erosion stage of the model uses the most computer time, dispensing with this component altogether has some attractions. Second, in terms of the Fourier structural component of the model, the more components that are used in the inverse Fourier transform, the more realistic the image appears. The Fourier inversion with most harmonics, in fact, scored as the most realistic of all measured images. Unfortunately, this was tested with a constant fractal texture. Clearly, some subjects preferred smooth, untextured images over all else. This may be attributable to the type of image (perspective view), and further testing with contour maps, hill shading, and other terrain representations may prove valuable.

Third, the subjects were easily able to detect the realism of the real versus synthetic data. While advances may have been made toward answering the question of what makes terrain look real, clearly the human map reader can still very quickly detect the real thing, even when represented in the same way as the best forgery. Finally, the amount of fractal texture used can very soon obscure the realism of the terrain, appearing as too rough. Fractal texture reaches the artificial side of the realism index at about 5% of relief. Many terrain simulations obscure this effect by transforming elevations to their logarithms before plotting so that only a few elevation extremes appear as "fractal mountains" and the remainder becomes low relief texture. Clearly, a more sophisticated method is called for in this case.

As a final stage of this investigation, the aggregate surface statistics were computed for the simulated terrain surfaces. Of most interest was the surface standard deviation of elevations, computed simply as the standard deviation of all the 248×248 elevations used in each simulation. The real Mount Everest data had a low of 5148 m, a high of 8848 m (Mount Everest summit), a mean of 6730 m, and a standard deviation of 807.47 m. In the simulations the highs and lows were largely retained, as also were the means. The standard deviations of elevation, a crude measure of surface roughness or texture, varied considerably. Figure 13.8 shows the effects of changing the cutoff, texture, and type and amount of erosion on this statistic. As harmonics are removed from the simulation, there is little effect on the standard deviation until the cutoff level is reached where only a few harmonics remain, resulting in a gross smoothing of the simulated surface. This occurs around the 7% level, but is determined by the actual generalized seed data used to drive the simulation. On the other hand, as the amount of texture as a percentage of relief is increased, the surface standard deviation increases smoothly. For erosion, parallel retreat gradually removed surface roughness, while stream erosion reduces the load suddenly, and then at a lesser rate, but still faster than parallel retreat.

In summary, a terrain model based on both periodic scale-dependent and fractal scale-independent properties has been tested, both statistically and subjectively, in terms of its ability to generate realistic looking terrain. The results allow some modifications and refinements to the technique. The model has some distinct advantages over the alternative models that have been used to date, since it is based on two bodies of theory and seems more closely linked to reality. Fractal terrain simulations have much to offer theory, in addition to providing interesting images. Perhaps, through continued pursuit of the perfect cartographic simulation, we will discover a Mount Everest as important and exciting to geography as the real thing.

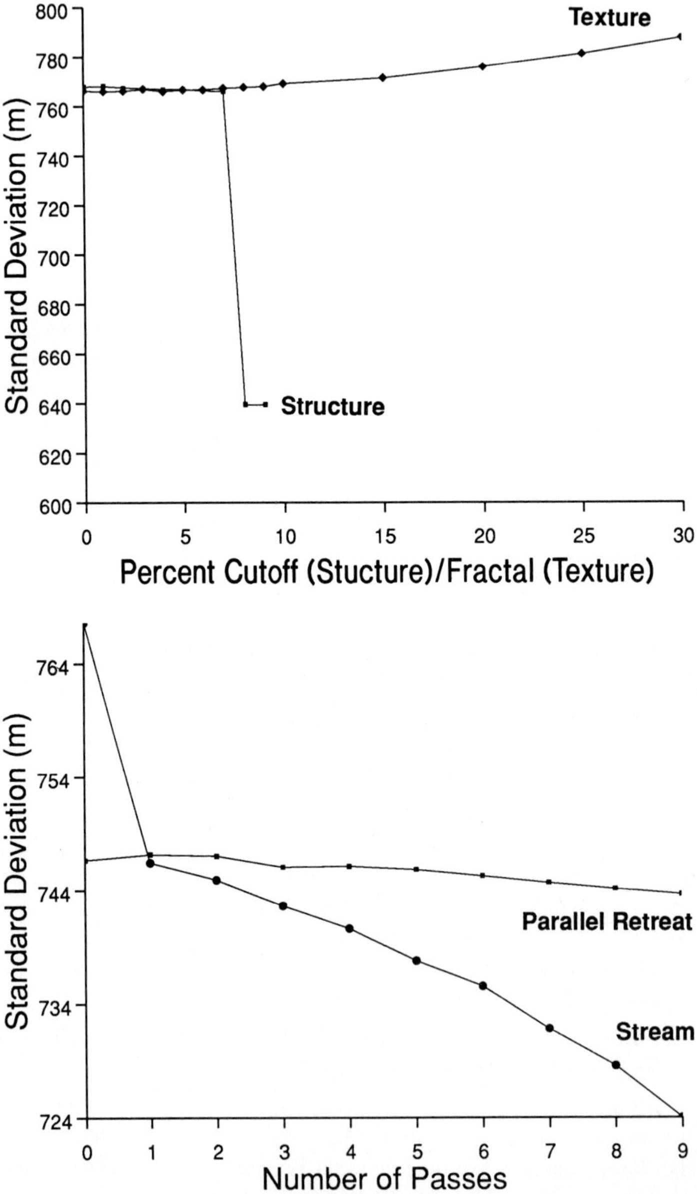

Figure 13.8 Elevation statistics for the simulated terrain.

REFERENCES

Band, L. E. 1986. Topographic partition of watersheds with digital elevation models. *Water Resources Research* 22(1):15–24.

Clarke, K. C. 1987. Scale-based simulation of topography. In *Proceedings, Eighth International Symposium on Computer-assisted Cartography (Auto-Carto 8)*, Baltimore, pp. 680–688.

———. 1988a. Scale-based simulation of topographic relief. *The American Cartographer* 15:173–181.

———. 1988b. Everest, a new perspective. *Explorers Journal* 66(4):183.

Douglas, D. H. 1986. Experiments to locate ridges and channels to create a new type of digital elevation model. *Cartographica* 23(4):29–61.

Fournier, A., Fussell, D. and Carpenter, L. 1982. Computer rendering of stochastic models. *Communications, Association for Computing Machinery* 25:371–384.

Goodchild, M. F. 1982. The fractional Brownian process as a terrain simulation model. *Modeling and Simulation* 3:1133–1137. Proceedings, 13th Annual Pittsburgh Conference on Modeling and Simulation.

Jeffery, T. 1987. Mimicking mountains. *Byte* 12:337–344.

Klinkenberg, B. 1988. *Tests of a fractal model of topography.* Ph.D. Dissertation. London, Ontario: Department of Geography, University of Western Ontario.

———., and Clarke, K. C. 1992. Exploring the fractal mountains. In *Automated Pattern Analysis in Petroleum Exploration*, I. Palaz and S. K. Sengupta, eds. New York: Springer-Verlag.

Mandelbrot, B. B. 1977. *Fractals: Form, Chance and Dimension.* San Francisco: W.H. Freeman & Co.

Mark, D. M., and Aronson, P. B. 1984. Scale-dependent fractal dimensions of topographic surfaces: An empirical investigation, with applications in geomorphology and computer mapping. *Mathematical Geology* 16(7):671–683.

Pentland, A. P. 1984. Fractal-based description of natural scenes. *IEEE Transactions on Pattern Analysis and Machine Intelligence* PAMI-6(6):661–674.

Pike, R. J., and Rozema, W. J. 1975. Spectral analysis of landforms. *Annals of the Association of American Geographers* 65:499–516.

Roy, A. G., Gravel, G., and Gauthier, C. 1987. Measuring the dimension of surfaces: A review and appraisal of different methods. In *Proceedings, Eighth International Symposium on Computer-assisted Cartography (Auto-Carto 8)*,. Baltimore, pp. 68–77.

Stepoway, S. L., and Christiansen, M. G. 1989. Object-oriented fractal modeling on a shared-memory MIMD machine. *Journal of Object-oriented Programming* 2(1):20–25.

Voss, R. F. 1988. Fractals in nature: From characterization to simulation. In *The Science of Fractal Images*, H. Peitgen and D. Saupe, eds. New York: Springer-Verlag.

CHAPTER 14

MULTIFRACTALS IN IMAGE PROCESSING AND PROCESS IMAGING[1]

Lee De Cola
U.S. Geological Survey, Reston, VA

ABSTRACT

Image data may be analyzed for a range of scale levels by generalizing high resolution measurements into an image pyramid whose statistical description is used to estimate a fractal dimension D_t for each value t of the image histogram. This multifractal analysis is demonstrated for a satellite image of Reston, Virginia, which is compared to data simulated by Gaussian and random-walk processes. The behavior of the multifractal dimension is used to characterize simulated and empirical data and to detect low- and high-dimension features in the image.

INTRODUCTION

Casual observation, as well as a great deal of empirical research, reveals the following:

- Spatial *phenomena* are complicated and irregular,
- Spatial *patterns* tend to be self-similar among scales,

[1]Any use of trade, product, or firm names in this publication is for descriptive purposes only and does not imply endorsement by the U.S. government.

- Spatial *data* cannot be completely characterized using Euclidean notions of points, lines, planes, and volumes.

Fractal sets, which are irregular, self-similar, and non-Euclidean, can be used to analyze data for many kinds of spatial phenomena. This paper presents results from experiments using fractal concepts in the analysis of images. Four kinds of sets, topological, ideal, stochastic, and empirical, can in part be characterized by multiple fractal dimensions *D*. These measures are applied to empirical data, and the results are then compared to data simulated to mimic spectral characteristics of the real images. This comparison suggests a hierarchy of spatial processes that give empirical data their richness and complexity. Although the objective of extracting features from images is demonstrated, the exercise also provides a broader understanding of spatial process.

Let us make a distinction among these four kinds of sets, which are not always self-similar across scales or space, and not all of which correspond to a single (monofractal) dimension (De Cola, 1991):

Fractal set	Irregular	Self-similar	Fractal Dimension
Topological	No	Yes	Single (integer)
Ideal	Yes	Yes	Single (real)
Stochastic	Yes	Yes	Multiple
Empirical	Yes	Limited	Multiple

Topological spaces establish the Euclidean framework, usually $\mathbf{R}^3$, within which data are gathered, and may be used to define locations of observations, usually bounded by $\mathbf{R}^1$ lines in $\mathbf{R}^2$. Such topological fractal sets are self-similar and have unique integral non-negative dimensions $D \in \mathrm{N}$.

Ideal fractal sets illustrate the fundamental ideas of fractal geometry, the irregular self-similarity of single-dimension (monofractal) sets. The infinite Koch "snowflake" construction is an example of an ideal fractal set with dimension $D = \log 4/\log 3 \sim 1.26$ identically at all scales, uniformly at all locations, and consistently at all times. Much like their Euclidean counterparts, the point, line, or plane, ideal fractals are perfect and unchanging.

Stochastic fractal sets require that we drop the monofractal assumption and measure many dimensions in order to represent the varying form of a phenomenon as a function of local variables across space. These sets, which have a self-similar structure compounded with a random component, are more complex and are particularly useful in the simulation of realistic data. But, unlike ideal fractal sets, stochastic data are multifractals whose dimensions change with intensity, location, time, and even across scales. Figure 14.1, for example, is a rendering of a cellular random walk in two dimensions, where the vertical axis is the number of times a given cell is visited (Kaye, 1989). This simple cellular model generates an irregular self-similar pattern—small pieces look like the whole—yet the pattern can only be characterized as having more than a single (monofractal) dimension (Barnsley, 1988; Falconer, 1990).

Finally, *empirical* fractal data, which represent a decidedly nonfractal world, require that we abandon self-similarity except within a range of scales. Because real spatial phenomena are the products of scale-bound processes, their descriptions vary depending on how, where, and when they are measured. Far from being a liability, this situation means that a measurement such as fractal dimension is a function of local variables. Multifractal measurement can be used as a description of how spatial patterns, as well as hypothetical underlying processes, vary in irregularity at different intensities from place to place and time to time and even among scales. This paper describes how the multifractal model can be used to explore spatial structure and to make inferences about processes in the real world.

The world is therefore not "fractal" in the ideal sense, for the processes underlying real phenomena are too complex to be characterized by any single model (Nicolis and Prigogine, 1989). But empirical data are increasingly amenable to models embodying the fractal paradigm of self-similar irregularity represented over a wide range of scales by nonintegral dimension. It should also be recognized that fractal analysis is just one approach to the measurement of spatial variation; two other multiscale approaches are Fourier and variogram analysis (Roy, et al., 1986; Burrough, 1989). Although there is not enough space here to develop this argument, we can demonstrate that the fractal model is quite robust, parsimonious, and therefore at least as powerful a characterization of spatial data as these alternative techniques.

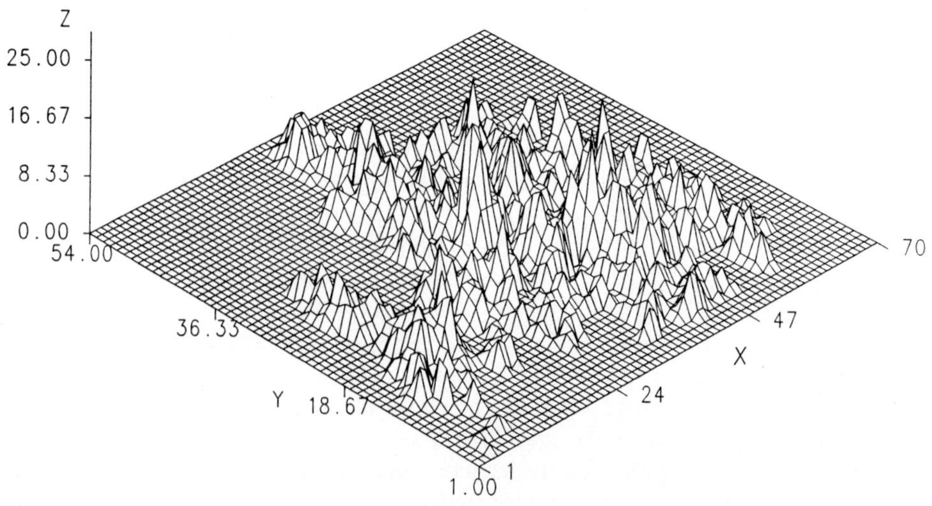

Figure 14.1 Three-dimensional rendering of an orthogonal, unit-step random walk in two dimensions.

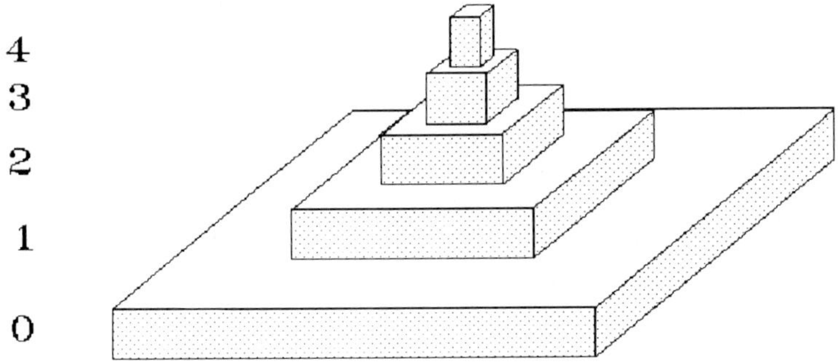

Figure 14.2 An $L = 4$ data pyramid in which each layer is an aggregation of the one below.

MULTISCALE ANALYSIS

Before we turn to the application of these ideas to simulated and empirical data, we must establish an algorithmic foundation for the multiscale analysis of spatial form. This foundation is particularly valuable because several approaches to the measurement of fractal dimension exist. We begin with topological specifications that provide data about a phenomenon within some maximum spatial extent and for a smallest resolution element. Let $L > 0$ be an integer with $n = 2^L$ and let X_0 be an $n \times n$ array (raster) of integers representing the value of some variable x_0. One way to generalize this 0-level array is by sampling every other cell (pixel) in every other row, yielding an $(n/2) \times (n/2)$ array (Justice et al., 1989). The technique used here, however, is to take the mean value of each 2×2 subarray, which generalization may be repeated to yield a stack of $L + 1$ layers X_l, $l = 0,..., L$, as shown in Figure 14.2, for $L = 4$. Each layer in the stack is a generalization of the layer below (Uhr, 1987).

Three things should be noted. First, it is possible to use subarrays other than the 2×2 chosen, but this is a natural choice that also allows rapid bit arithmetic. Second, while the size of the original $l = 0$ layer is $(2^L)^2$, the size of the complete image pyramid is only $4/3(2^L)^2$. Third, although the size of each layer is one-fourth that of the one below it, the amount of information (as measured, for example, by some qualitative assessment of image "clarity") certainly declines more slowly than this.

The fractal analysis of empirical data is based on the examination of statistics computed at each level. At level $l = 0$, let a subset of X_0 be $X_{t0} = \{x_0\colon x_0 \geq t\}$, where t is a threshold value of the array (threshold brightness in the case of an image), and let e_{t0} be the length of the boundary of X_{t0}, that is, a count of the cell edges separating X_{t0} from its complement $X_0 \backslash X_{t0}$ (Culling, 1989). It is possible to generalize this measurement to the set $X_{tl} = \{x_l\colon x_l \geq t\}$, which yields e_{tl}, the cell boundary of the generalized set X_{tl}. Fractal dimension is a measure of the scaling relationship between a region's boundary and its scale level $l{:}e_{tl} = a_t(2^l)\text{-}b_t$ which, taking logs, yields

$$\log_2(e_{tl}) = \log_2(a_t) - b_t l + \varepsilon_t \qquad (14.1)$$

where ε_t is an ordinary least-squares error term. Because the data X exist in the topological three-dimensional space $[\mathbf{R}^2 \times \mathrm{I}]$ of location × value, $D_t = 3 - b_t$ (Falconer, 1990, p. 42). The regression also yields an estimate of a_t, a measure of size, which will not be discussed here (De Cola, 1989a, b; see Whalley and Orford, 1989, for other approaches).

This multiscale analysis provides not merely a single *mono*fractal D, but also a *multi*fractal function D_t (Lovejoy and Schertzer, 1988) that varies with the intensity of a phenomenon (in this case t, the brightness values of an image). The expectation is that $0 \le D_t \le 2$; the irregularity of the phenomenon measured by the data X may range between that of a point singularity, for which $D = 0$, and that of a Euclidean set (such as a white disk on a black field) for which $D = 2$. D_t is therefore a tool whereby we can measure the form of a phenomenon across its range and, by selecting different windows, throughout the space it occupies.

A simple way to illustrate these ideas is with a remotely sensed image. The SPOT high-resolution visible (HRV) panchromatic sensor detects energy in the range of 0.45 to 0.80 μm for (10 m)^2 pixels. Figure 14.3 shows the location of a SPOT scene chosen for analysis, a 10-level (10.24 km)^2 image centered on the U.S.G.S. National Center in Reston, Virginia. (Permission has been given to use this image courtesy of SPOT Image Corporation, Reston, Virginia © 1990 CNES.) The image (hereinafter called Image to distinguish it from simulated data) is shown as a pyramid in Figure 14.4, in which frame 0 is at the highest resolution [1024 × 1024 pixels of (10 m)^2] while frame 1 shows the data aggregated into 512 × 512 pixels each of 20 × 20 m^2, and so forth. Note how each scale level, while similar to its neighbors in the pyramid, has its own characteristics. The higher-resolution images ($l = 0, 1$) are best for the identification of individual features such as lots and roads; middle resolutions ($l = 4, 5$) are suitable for the detection of large, complex areas; and low resolutions ($l = 8, 9$) highlight broad areas of high and low brightness. A (10.24 km)^2 single-pixel, level-10 image could also be regarded as part of an even larger

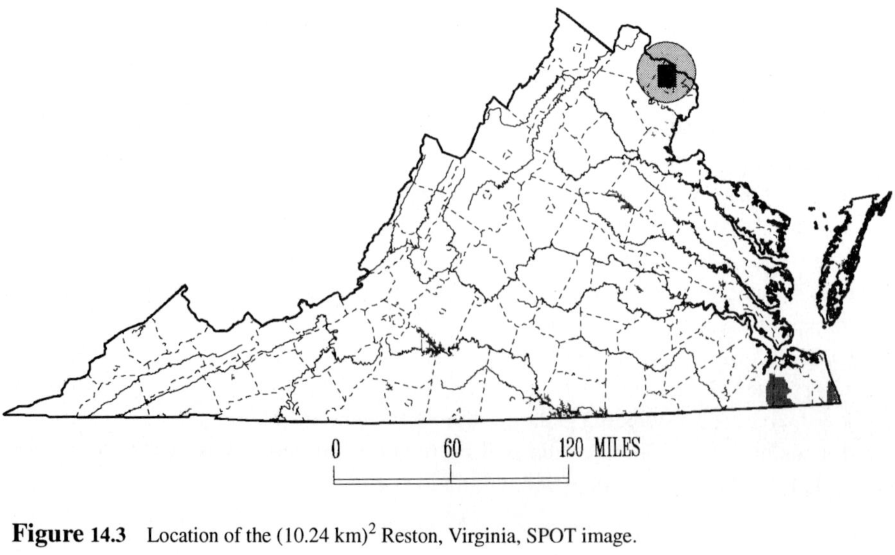

Figure 14.3 Location of the (10.24 km)^2 Reston, Virginia, SPOT image.

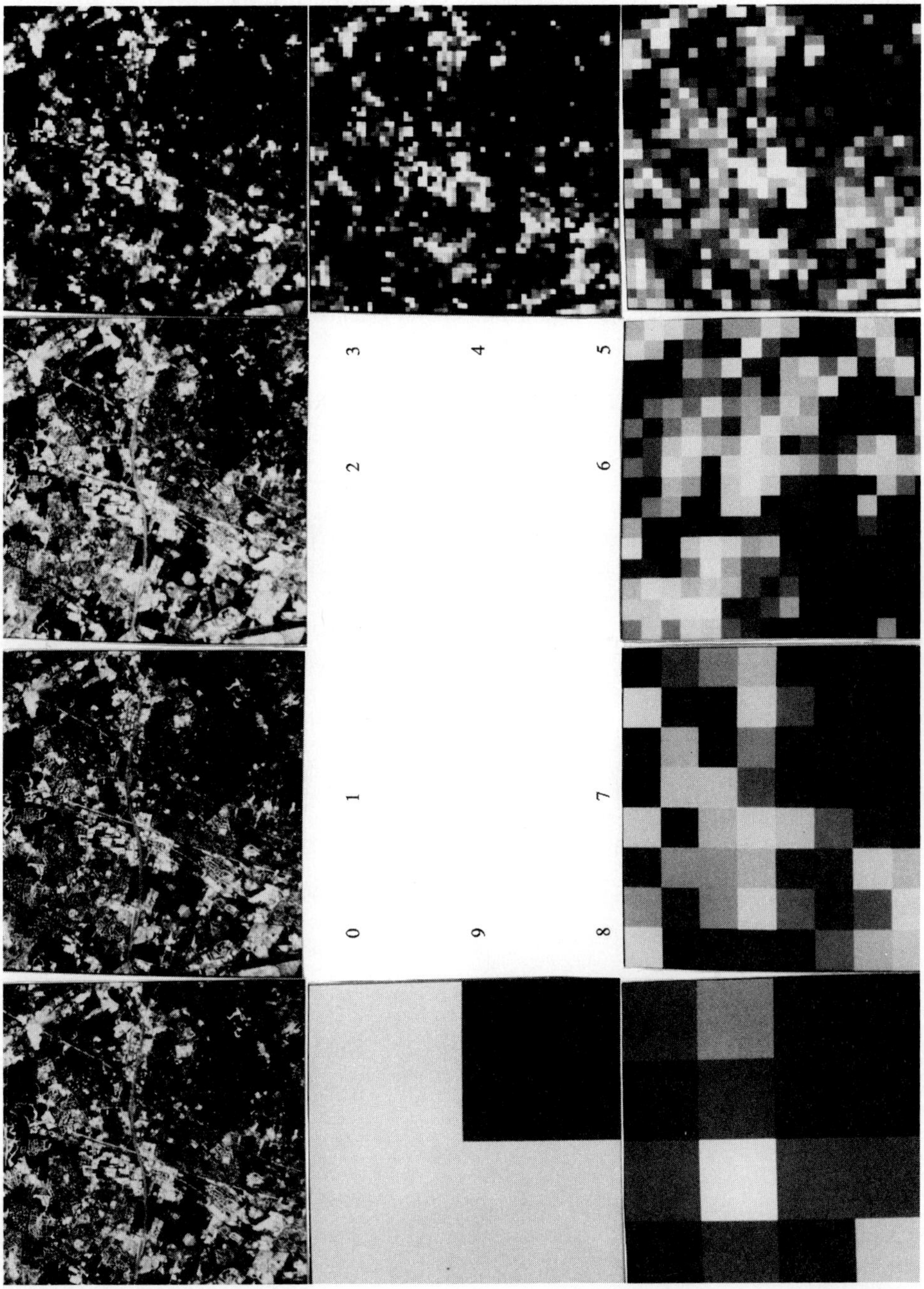

Figure 14.4 An $L = 10$ data pyramid for the Reston, Virginia, image. Frame 0 is the original 1024^2 10-m image, frame 1 is a 512^2 20-m aggregated image, and so on. Each frame is 10.24 km wide.

array, in keeping with the idea of movement up an open-ended scale pyramid. [In fact, a number of the frames are similar in resolution, but not necessarily in spectral sensitivity, to such sensors as SPOT multispectral (frame 1), Landsat Thematic Mapper (frame 2), Landsat multispectral (frame 3), Coastal Zone Color Scanner (frame 6), and Advanced Very High Resolution Radiometer (frame 7).]

Table 14.1 shows the characteristics of this pyramid. Because each layer is an aggregation of the one below it, $\bar{X}_l = 46.1$ for $l = 0, \ldots, 10$, but the standard deviation of the data s_l declines with scale. Such a data structure, like the related quadtree scheme, allows us to move within the $[0, L]$ continuum of a scale space (Samet, 1990). Upward movement is toward aggregation, large features, and generalization, while downward movement is toward disaggregation, greater detail, and the segmentation of space (Goodchild and Mark, 1987; Hummel, 1987).

TABLE 14.1 Statistics of the Data Pyramid for the Image of Reston, Virginia

	Level (l)	Edge (2^l)	Cell size (meters)	Number of cells (4^{L-l})	Standard deviation of data (s_l)
$L =$	10	1,024	10,240	1	—
	9	512	5,120	4	2.17
	8	256	2,560	16	2.30
	7	128	1,280	64	3.47
	6	64	640	256	4.38
	5	32	320	1,024	5.65
	4	16	160	4,096	6.85
	3	8	80	16,384	7.93
	2	4	40	65,536	9.09
	1	2	20	262,144	9.68
	0	1	10	1,048,576	10.21

Figure 14.5 shows a plot for the Image of the values of $\{D_t: t \in [28, 81]\}$, each estimated by using Equation 14.1. (The minimum R^2 for these 54 estimates is 0.95.) Clearly, the data are not monofractal: $D_t = 0.402 \pm 0.515$ and these dimensions have a typically multifractal parabolic shape (Falconer, 1990). For low values of t, the phenomenon is essentially pointlike, with few edges separating the sparse set from its background complement. Low values of D_t are also found for high values of t, but here the roles of figure

and background (set and complement) are reversed. And the maximum $D_{46} = 1.21$ occurs at $\overline{X}$. Shown on the figure is a quadratic fit to the data,

$$\hat{D}_t = -2.67 + 0.144t - 0.000177t^2, \qquad R^2 = 0.70 \tag{14.2}$$

(where the caret signifies prediction) that gives a predicted maximum of $\hat{D}_{48} = 0.87$. Moreover, there is significant autocorrelation among the values of D_t; the Durbin–Watson statistic = 0.611 ($DW_{0.01} = 1.36$), which does not support the hypothesis of nonautocorrelated disturbances (Theil, 1971). Not only is D a parabolic multifractal, each value is correlated with its neighbors, and this should be considered when modeling the behavior of D_t.

SIMULATION: PROCESS IMAGING

To better understand what D_t tells us about the Image, it will be useful to explore how D_t behaves for data over whose parameters we have some control. While the objective of image *processing* is often simply to improve the appearance of an image in some way, the more ambitious task of digital image *analysis* is to make inferences about the real world

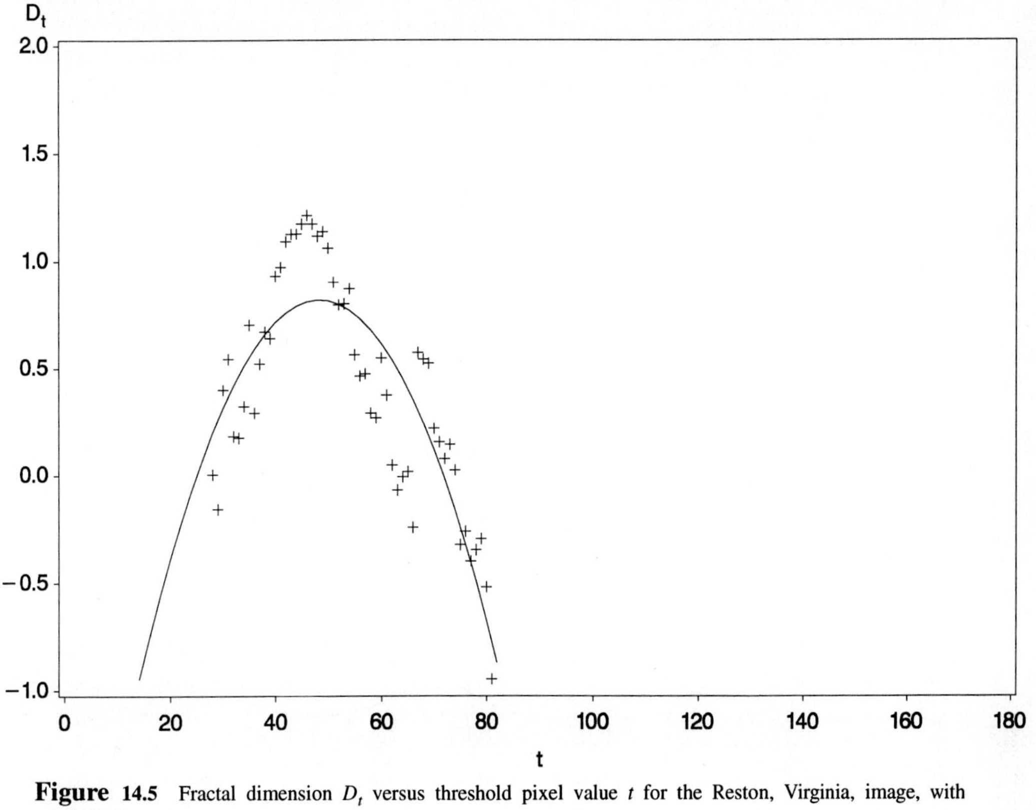

Figure 14.5 Fractal dimension D_t versus threshold pixel value t for the Reston, Virginia, image, with quadratic fit.

based on the information from raster data (Muller, 1988). For example, the 10^7 bits of the X_0 Image are reduced to the three coefficients of Equation 14.2 relating D_t to t — a mere 10^2 bits, a good example of data reduction. Of course, a large amount of data are lost in this process, but we gain information for inferences about the processes that might have given rise to the image.

To consider what fractal analysis has revealed about the Image data, we need to explore alternative spatial processes that might give rise to such an array of numbers. This approach to spatial process modeling may be called *process imaging*. This technique, which is a form of visualization but more explicitly deductive, reverses the direction of image processing by producing large data sets from relatively brief descriptions of spatial process consisting of a few instructions. For example, the 10^2 bits that specify the first two moments of the Image data will be used to simulate an image of 10^7 bits, a good example of data expansion (Frenkel, 1988).

Consider, therefore, what processes might simulate various aspects of the Image data whose mean $\overline{X} = 46.1$ and standard deviation $s = 10.2$. Figure 14.6 shows one way of organizing a class of cellular spatial models according to whether or not they manifest three characteristics:

1. Are the values of a cell *quantities* (numerical measurements) as opposed to nominal classes, as in a raster GIS? An example of a quantitative process would be the Poisson model of the completely random location of points among cells (Schachter and Ahuja, 1979; De Cola, 1991).

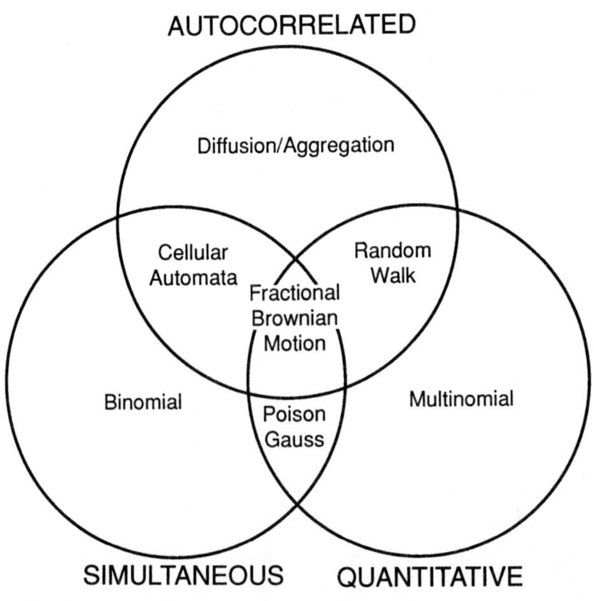

Figure 14.6 A typology of cellular processes according to whether they have three characteristics.

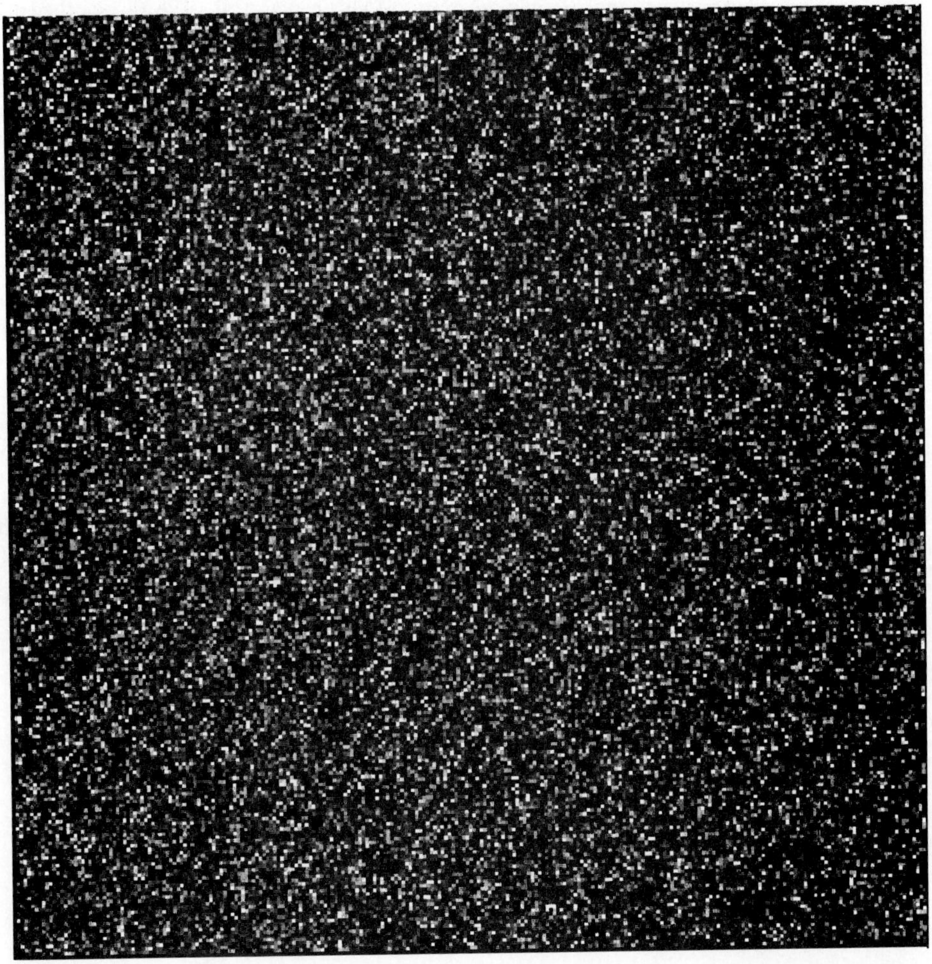

Figure 14.7 Gauss simulation on a 256^2 array using the same mean and standard deviation as the Reston, Virginia, image.

2. Are all values determined *simultaneously* or sequentially over some period? Cellular automata are examples of processes in which array values are computed across space at the same time (Wolfram, 1984; Phipps, 1988).
3. Is the value of one cell *autocorrelated* with (affected by the value of) any other cell? Diffusion-limited aggregation is an example of a process that sequentially assigns a nominal binary value to a cell based on the values of its neighbor (Batty et al., 1989).

These characteristics may be combined in various ways to provide a wide range of hybrid models for the study of a given real-world phenomenon. Perhaps the most useful such hybrid is fractional Brownian motion (Musgrave et al., 1989), but there are many other examples.

To reproduce certain characteristics of the Image data, we have used two relatively simple processes, called Gauss and Randomwalk, to simulate images of 256^2 pixels each. The simplest simulation of the image is a "heap" of pixels whose spectrum is normally distributed with the same mean and standard deviation as the Image. (Think of taking the pixels from Image, scrambling them, and then withdrawing a smaller set of 65,536 pixels.) Figure 14.7 shows one such simulation, and although Gauss and Image share the first two moments of their histograms, their appearance is very different. The complete entropy of the Gauss reveals virtually no spatial information. In fact, this image is one representation of what an absolutely fractal world might look like: stochastically identical at all scales and no cues about the size or location of anything.

The Gauss model is the most parsimonious way to simulate raster data spectrally similar to a given image. The problem is that this process manifests no spatial autocorrelation because the value of each cell is unrelated to that of any other. This totally random behavior is indicated in Figure 14.8 by the fact that the estimation of D_t is confined to a very narrow range of the image values, $38 \leq t \leq 55$, centered on the mean of the image histogram, although $x_0 \in [4, 86]$. Moreover, $D_t < 0$ for all but three values of t. This behavior appears to reflect the extreme spatial dispersion of the Gauss process, where spatial auto-

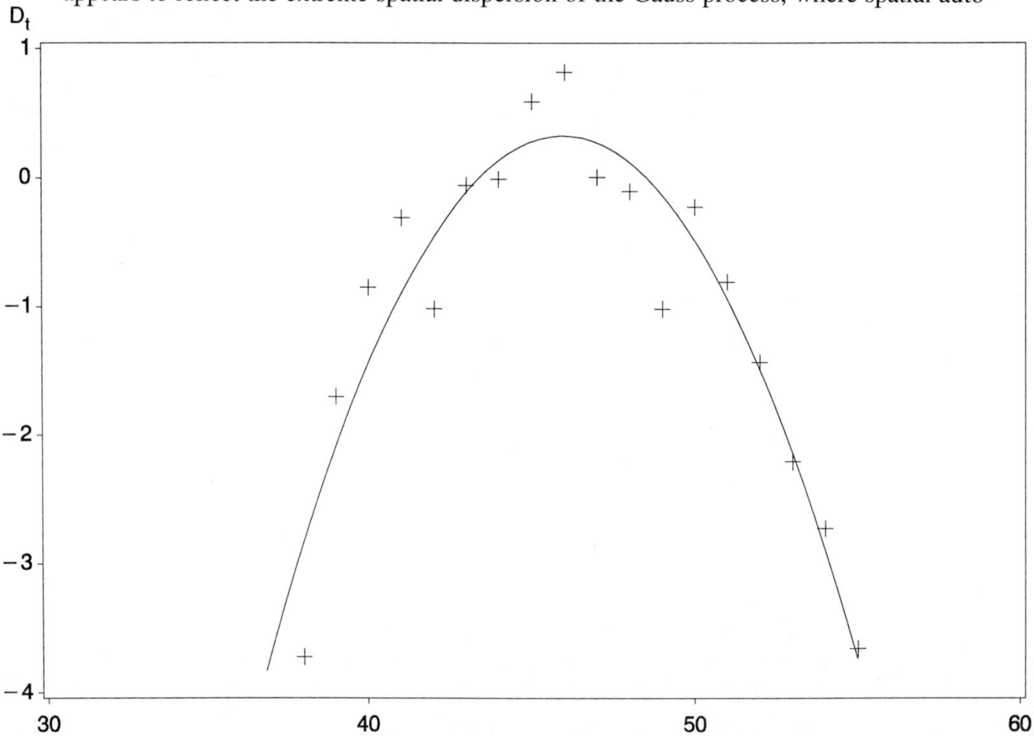

Figure 14.8 Fractal dimensions for a Gauss simulation.

correlation is absent, clusters of homogeneous pixels are rare, and therefore the dimensionality of the process is low (the possibility of negative multifractal dimensions is discussed in Mandelbrot, 1989). Certainly, this process corresponds neither in appearance nor in process to the real world.

It is illuminating to compare the Gauss simulations to a spatial process shown in the typology of Figure 14.6 that *is* clearly autocorrelated. We simulated 16 random walks that begin with a square lattice window of 256^2 empty cells, that is $x_0 = 0 \ \forall \ x$, starting at the central cell. The simulation moves randomly according to the uniform distribution of $Pr(\uparrow) = Pr(\downarrow) = Pr(\rightarrow) = Pr(\leftarrow) = 1/4$. Each time a cell is visited, the value of x_0 is incremented and the process stops when $\Sigma x_0 = \overline{X} \times 256^2 = 46.1 \times 65{,}536 = 3{,}020{,}554$ steps. If the process leaves one edge of the array it returns at the opposite edge, so a unit-step orthogonal random walk on a torus is simulated (De Cola, 1991). An example of this simulation is shown in Figure 14.9.

The Randomwalk simulation corresponds to the Image data if we imagine that the brightness values for each pixel are numbers representing events at $(10 \text{ m})^2$ locations; indeed, the radiance values are linear and homogeneous transformations of the energy received within the frequency limits of the sensor (SPOT, 1988). Although the simulation shown in Figure 14.9 might be an adequate representation of cloud data (Joseph, 1985) or perhaps chaotic terrain (Goodchild, 1982; Jones et al., 1989), it bears only weak resemblance to the study image but is certainly closer than Gauss. The fractal plot of this simulation, shown in Figure 14.10, is also different from that of the Gauss process. The process yields a characteristic parabola, but the curve spans a wider domain and range than Gauss. Even for low values of brightness the image gives a wide range of dimensions, generally in the range $0 \le D_t \le 1.3$. When fitted to a quadratic polynomial, the simulation yields

$$\hat{D}_t = -0.328 + 0.0519t - 0.000481t^2, \qquad R^2 = 0.65$$

which has a predicted maximum $\hat{D} = 1.07$. Although these results share the parabolic shape of the Gauss process, much more acceptable dimensions are produced and for a much larger domain of t values (compare Figures 14.5 and 14.8).

So far the analysis has been largely visual and qualitative. Two stochastic processes (Gauss and Randomwalk) and one empirical data set (Image) have similar histograms, but give very different images. While the behavior of D_t for each of the three has been shown to have the parabolic form typical of multifractal sets, the domain of dimension for Gauss is too restricted and its range too frequently negative to be an appropriate model for the empirical image. We may therefore abandon Gauss as a model and pursue Randomwalk.

To examine the general behavior of Image over space and to provide data with a sufficient sample size, we divided the full 1024 × 1024-pixel image into sixteen 256 × 256-pixel subwindows and computed D_t for each value of t for each subwindow. Figure 14.11 represents a fractal plot of the results. D_t spans almost all its expected range, and although this scatter has no particular mathematical form, a cubic polynomial has been plotted to suggest that the data may have an underlying parabolic shape significantly distorted by a third-degree term at the positive extreme.

Using the mean of the Image histogram, 16 Randomwalk simulations were also run. Their fractal plot (Figure 14.12) indicates that the cubic polynomial gives a better fit.

Table 14.2 displays the results of fitting third-order polynomials to the relationship between D_t and t for the two sets of 16 images. Terms were added to the model if they were statistically $\neq 0$ at the 10% level of significance. The table, as well as a comparison of Figures 14.11 and 14.12, reveals similarities among and differences between the 16 Randomwalk simulations and the 16 subwindows of Image. On the one hand, both roughly correspond to a cubic polynomial, the estimates of whose terms (except for the intercept) differ by no more than 10%. On the other hand, the two data sets differ in the following ways: (1) Randomwalk corresponds to a cubic relationship in all but two of the simulations, while the results of Image subwindows are less consistent; (2) the distributions of all the Image coefficients are highly and inconsistently skewed; and (3) the mean R^2 for the Randomwalk regressions is over 50% higher than that for Image. (A curve such

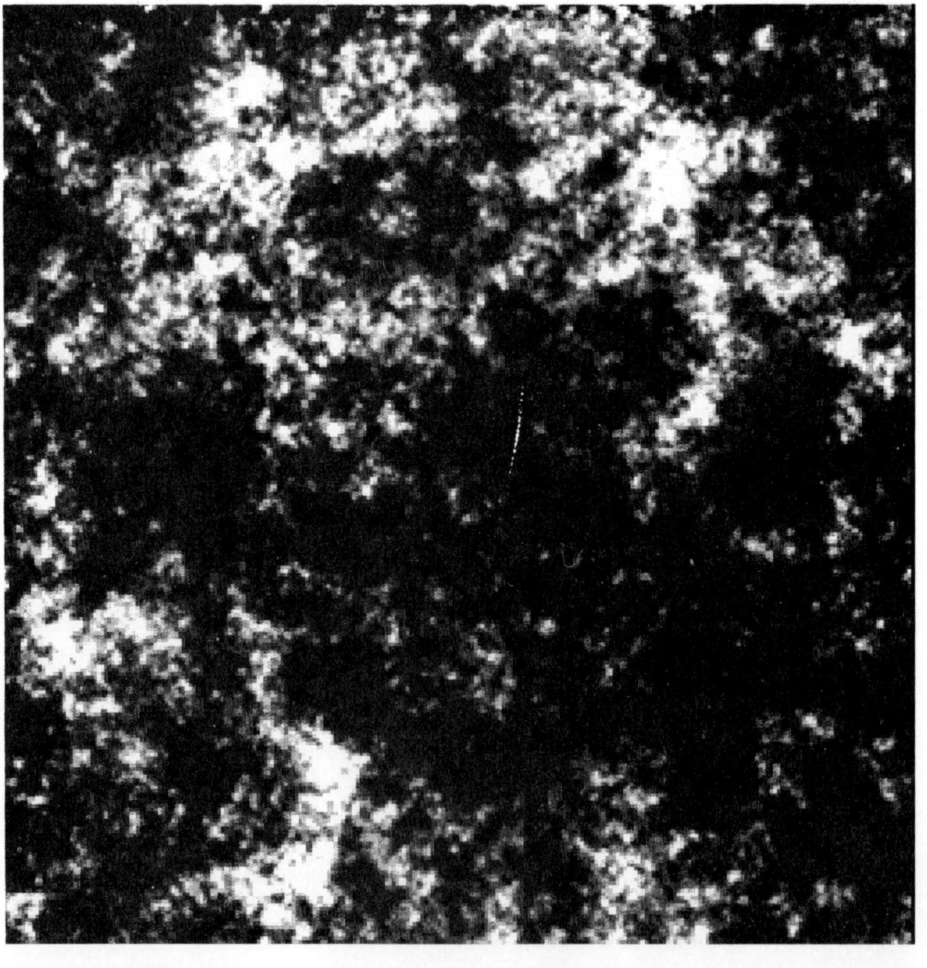

Figure 14.9 Randomwalk simulation on a 256^2 lattice, with the same mean as the Reston, Virginia, image.

as $D_t = \alpha t^{\beta} e^{\gamma t}$, with a maximum at $t = -\beta/\gamma$, might better fit the Image results. A similar curve appears in Mandelbrot, 1989.)

TABLE 14.2 Result of Fitting Third-degree Polynomials to the 16 Randomwalk Simulations and 16 Image Subwindows

	16 Randomwalk Simulations				16 Image Subwindows			
Term	*N*	Mean	Standard Deviation	Skewness	*N*	Mean	Standard Deviation	Skewness
0	16	-0.749	0.579	-0.661	16	-0.515	2.41	-1.275
1	16	0.0915	0.0336	0.871	13	0.0930	0.128	1.230
2	16	-0.00138	0.000626	-0.654	14	-0.00127	0.00206	-1.502
3	14	0.00000665	0.00000315	0.561	12	0.00000633	0.0000120	1.288
R^2	16	0.774	0.0802	-0.468	16	0.507	0.206	-0.289

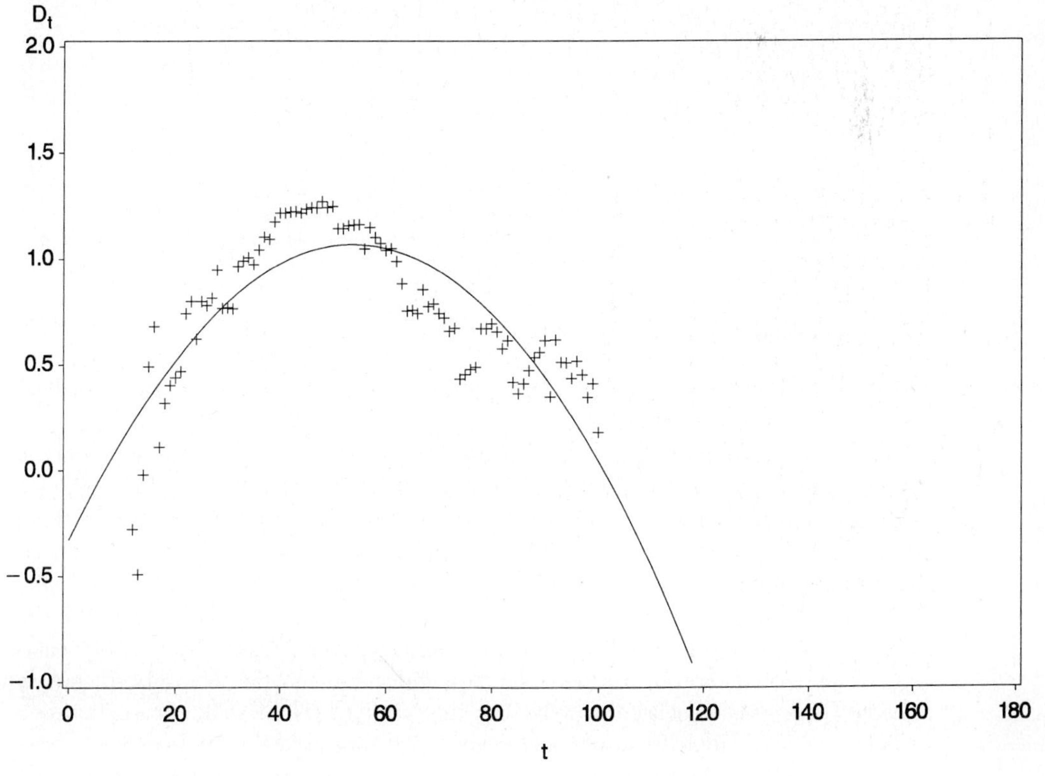

Figure 14.10 Fractal dimensions for a single Randomwalk simulation.

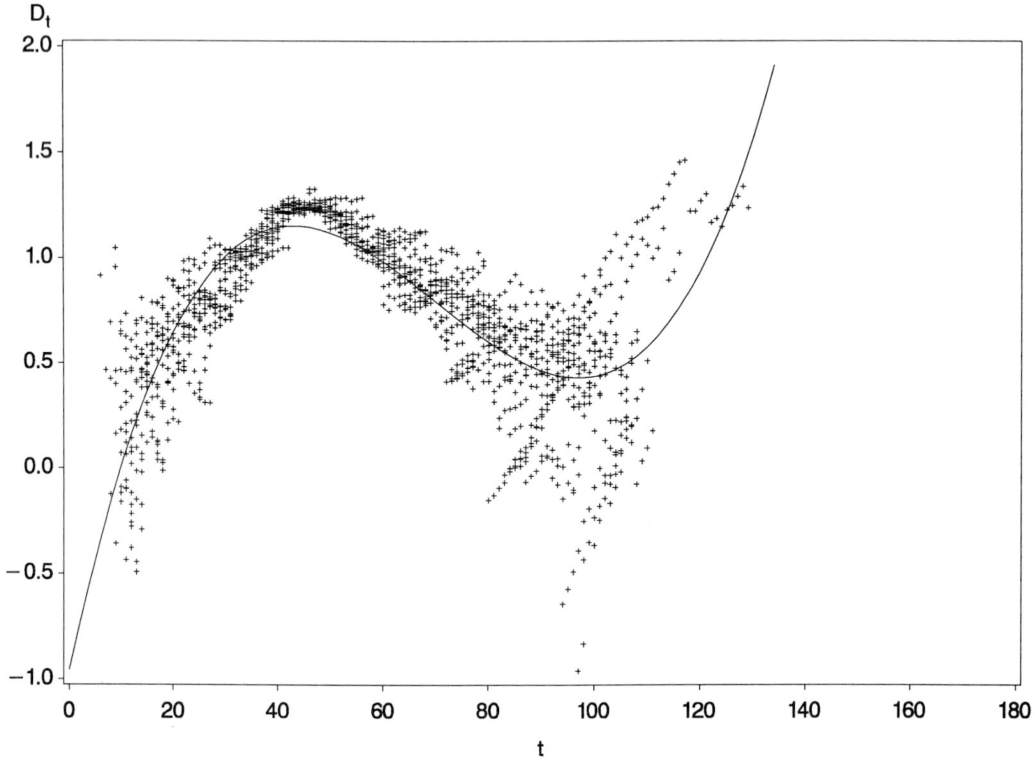

Figure 14.11 Fractal dimensions for sixteen 256 × 256-pixel subwindows of the Reston, Virginia, image, with cubic fit.

FROM IMAGE TO MAP

These experiments suggest that multifractal dimensions reflect different characteristics of spatial process. In general, $\partial^2 D_t / \partial t^2 < 0$, and D_t may have complex behavior for autocorrelated processes. We have observed that, although the mean values of the three processes (Gauss, Randomwalk, and Image) are the same and their histograms are all roughly normal, their spatial appearances (and consequently their fractal behavior) are quite different. Gauss is a diffuse multifractal whose lack of spatial autocorrelation results in largely negative values of D_t. Randomwalk is a spatially coherent stochastic multifractal for which $D_t = 0.76 \pm 0.38$, varying with intensity. The subimages of the Image are real multifractals for which $D_t = 0.98 \pm 0.33$, varying not only with intensity but also strongly with location; that is, the spatial form of a given brightness value is a function not only of (possibly inherent) *processes* associated with the data but also of *where* the value occurs. In keeping with the above notion of multifractals, it therefore appears that combinations or hierarchies of models will be necessary to explain real spatial patterns (Getis and Boots, 1978). Certainly we have had to abandon the monodimensional restriction of ideal fractal sets.

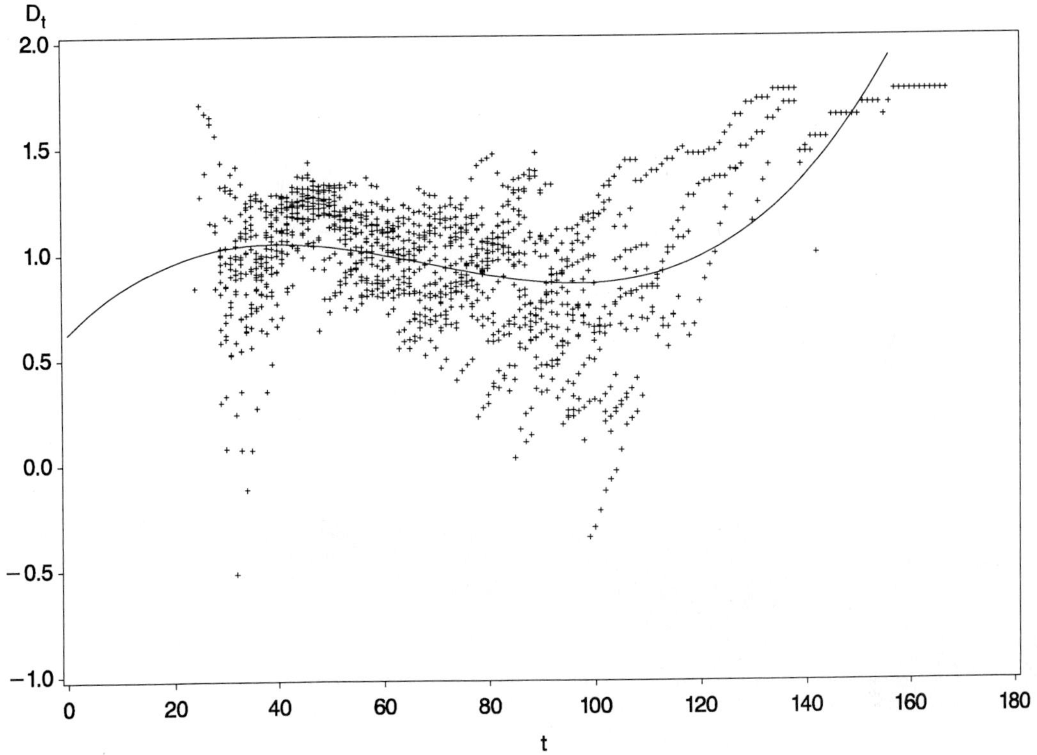

Figure 14.12 Fractal dimensions for 16 Randomwalk simulations on a 256^2 lattice.

In fact, the spatial variation of the Image process may be seen by examining a particular layer of the data pyramid. At $l = 5$, for example, there are $[2^{(10-5)}]^2 = 1024$ subwindows, each of which has an ensemble of associated multifractal dimensions. Figure 14.13 maps the values of $D_{43} = 1.28 \pm 0.313$. We can see exactly how D_t varies over space, and, in fact, this picture could be viewed as a fractal of a fractal, or a *metafractal.* Whether or not this is a suitable neologism, the figure is a clear illustration of how far this real-world phenomenon deviates from the isotropism of ideal fractality. The process appears to be stationary, however, as fitting the data to a plane reveals no significant north–south or east–west trend.

Any automated technique designed to extract features from images must be sensitive to dependencies on intensity and location in spatial pattern. The above exercise illustrates moreover that considerable intelligence, natural as well as artificial, is necessary to interpret spatial and spectral information (Uhr, 1987). Because fractal concepts provide a key to the multiscale measurement of self-similar spatial patterns, it is possible to use scale information to distinguish among patterns in real-world images. Dimension can be used to spatially identify features in a way that suggests how automated image processing based on multiscale analysis might work.

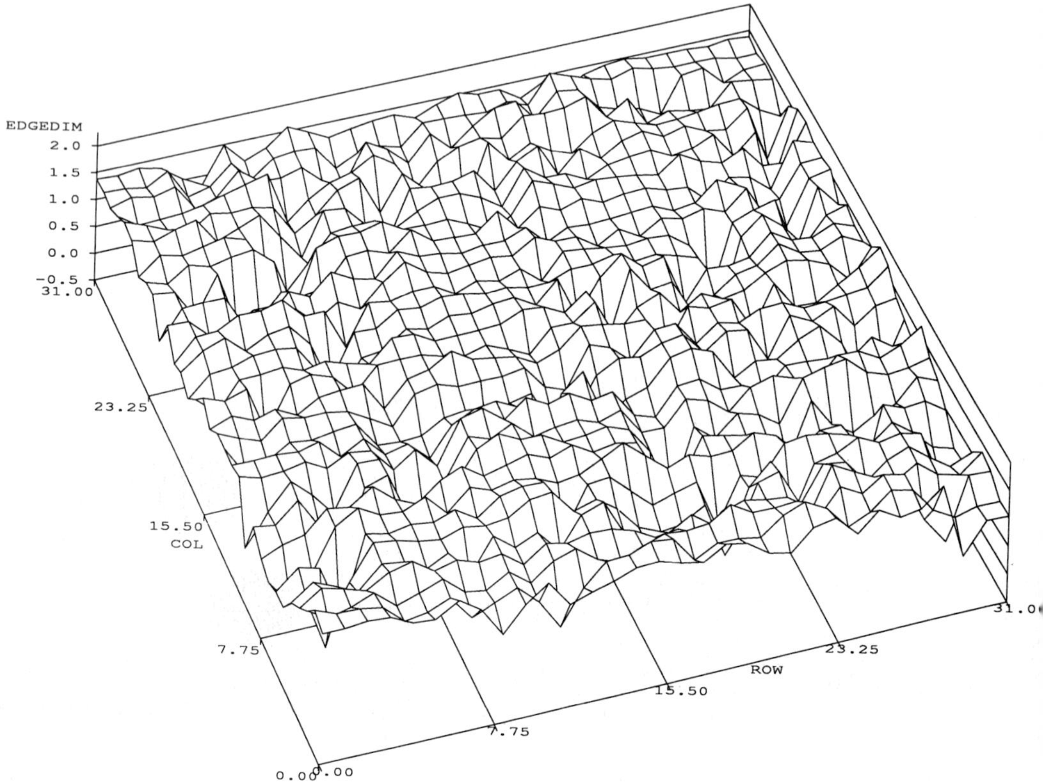

Figure 14.13 Three-dimensional plot of the values of D_{43} for 1024 (= 32 × 32) subwindows of the Reston, Virginia, image.

We have used the technique developed here to do some image processing on a subwindow of the Image, specifically a 256 × 256-pixel subwindow around the South Lakes region of the scene (Figure 14.14). A parabolic fit to the multifractal dimensions of this image gives the equation

$$\hat{D}_t = 1.107 + 0.00887t - 0.000130t^2 \qquad R^2 = 0.72$$

Figure 14.15 shows the residuals $\hat{\varepsilon}_t = D_t - \hat{D}_t$ from this regression. Consider first the maximum residual $\hat{\varepsilon}_{27} = 0.37$, implying that the values of the image for $t \geq 27$ have a more regular boundary than do other thresholds. When the values of the image $x_0 \leq 27$ are colored black, as in Figure 14.16, we see that they are associated with the lakes of the region. Another interesting interval is defined by $47 \leq x_0 \leq 57$, the longest run of negative residuals (that is, relatively low dimensional features whose perimeters quickly disappear at higher scales). When the pixels associated with this interval are colored gray, we see in Figure 14.16 that they may be associated with some of the roads in the subwindow. We therefore arrive at a crude thematic map of the region, created by using a semiautomated technique that can be used to extract other features as well.

The above distinction among fractal processes consequently applies to extracting information—and ultimately making maps—from images. Although the relatively scale independent structure of the Randomwalk model can be used to characterize the fractal behavior of the Image data, some of the interesting natural and cultural features we wish to highlight in a map are revealed by the extent to which they deviate from simulated processes. It is useful to speculate on other processes that might have given rise to the patterns in which we are interested to see why the exceptions that make the world so interesting actually occur, particularly as a result of human activity. Indeed, the detection and depiction of these exceptions are two of the main functions of the mapping process.

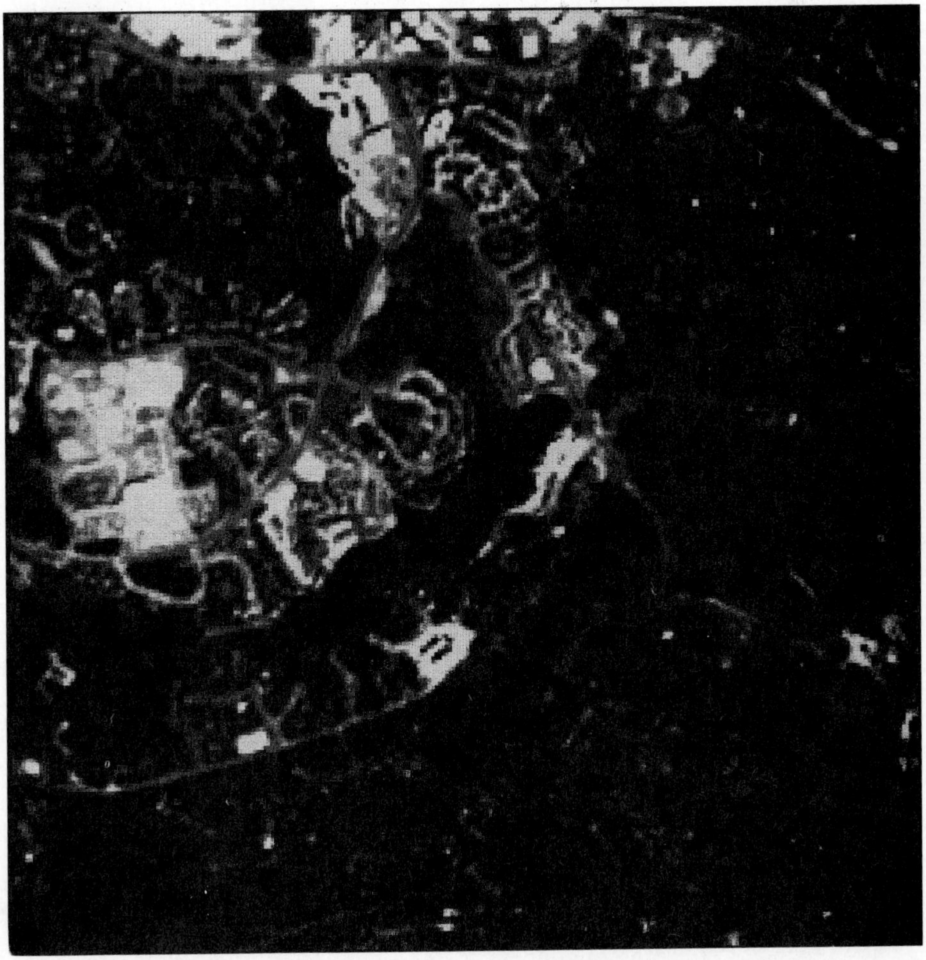

Figure 14.14 Panchromatic image of the South Lakes subwindow of the Reston Image.

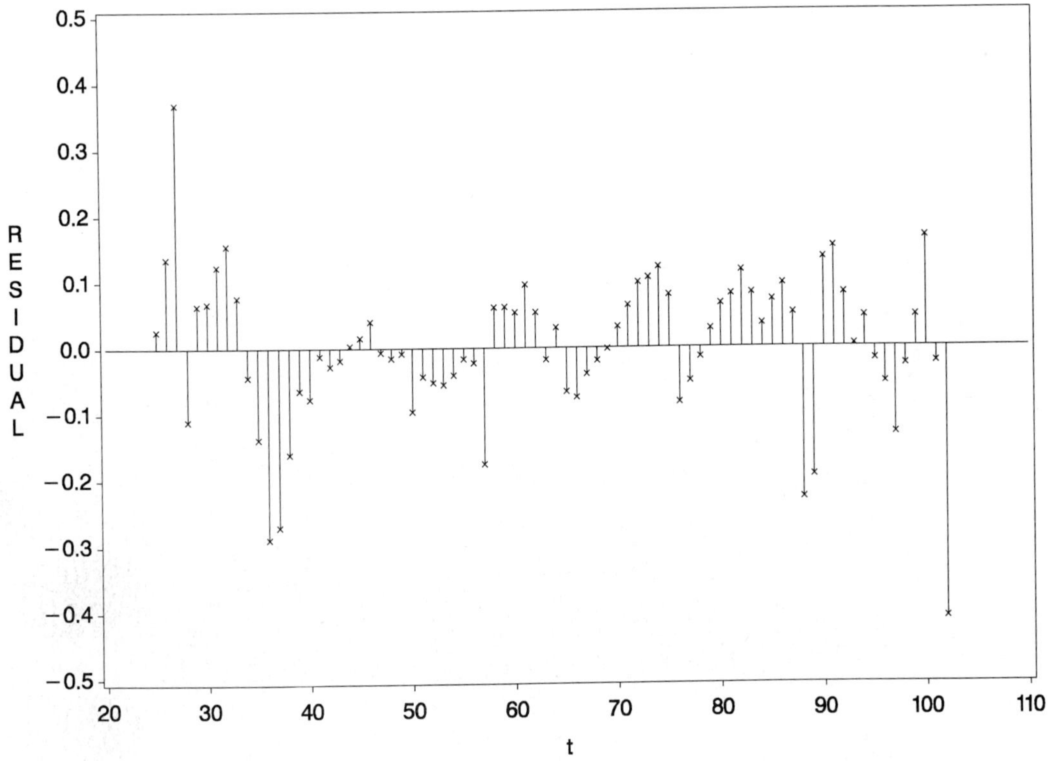

Figure 14.15 Residuals from quadratic fit of D_t for South Lakes subwindow.

CONCLUSION

On the basis of the above results, we may therefore postulate that four kinds of models coexist in the real world:

1. True *noise*, like that of Gauss, that is often inherent in instruments and is completely scale independent,
2. More complex *physical* processes that can often be visualized by stochastic models (like that of Randomwalk), which in the real world look like mountains and clouds and result from generally chaotic processes that are relatively scale independent,
3. *Natural* processes, such as vegetation and other land-cover evolution, that are the result of complex multifractal systems working within relatively narrow scales,
4. *Cultural* or human activities that are frequently erratic, difficult to predict, and often strikingly scale dependent.

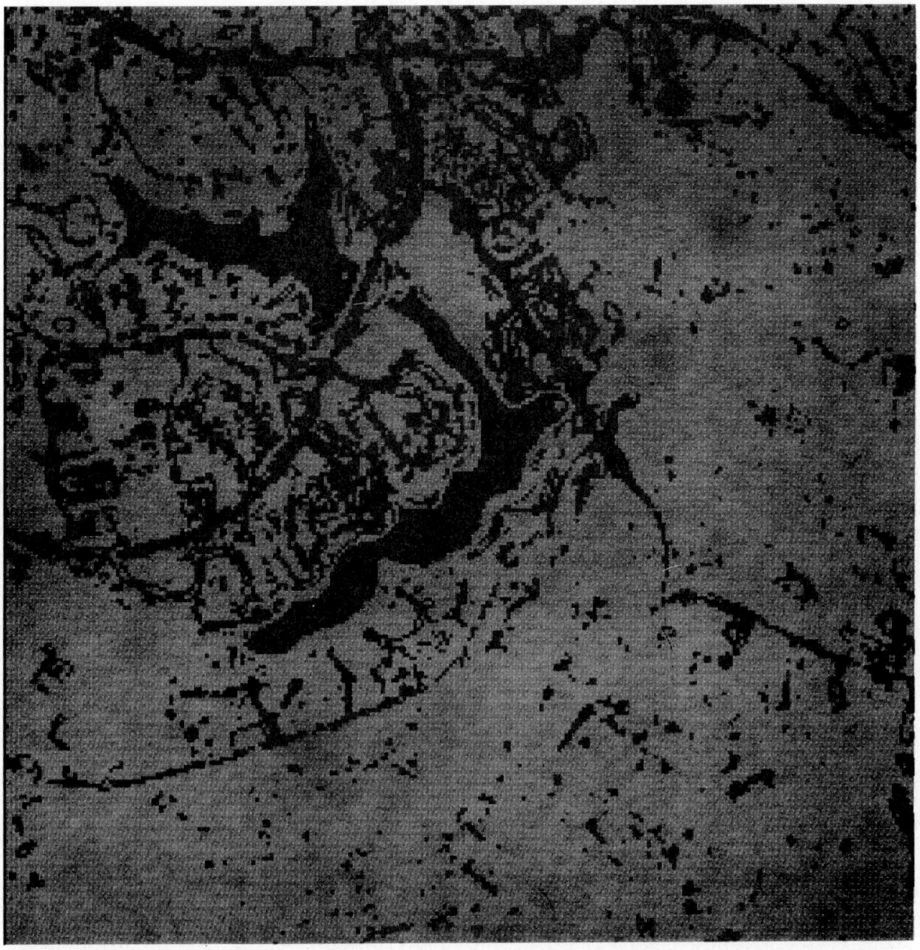

Figure 14.16 The South Lakes subwindow classified into high-dimension "lakes" and low-dimension "roads."

The analysis has also illustrated how movement within a scale space allows a scientist to aggregate or subdivide space in order to focus at will on low-level measurement, high-level description, and feature detection. Those phenomena that are more scale independent (closer to the classical fractal ideal) require fewer data than others whose scale and location dependence reveal exceptions. This approach is obviously of value for the management of data and extends the notion of the quadtree by asserting the pertinence of *all* scales of the image pyramid to the understanding of process (Samet, 1990). No level is universally appropriate for examination; each reveals important scale dependencies, and, taken together, all may be used both to summarize the irregularity of pattern and to make inferences about the complexity of process. While it is obviously true that the level-0 image contains all the data that exist, higher-level images are more suitable for more generalized analysis and, from a data management perspective, they are considerably smaller.

The analysis also suggests the value of simulation and visualization in image analysis. The notion of *process modeling* includes and extends specific techniques of image processing and land-use classification to encompass broad multidisciplinary studies that analyze varieties of data to make complex inferences about real-world processes (Holdridge et al., 1971; Szeliski and Terzopoulos, 1989; Schlesinger et al., 1990). Spatial scientists are beginning to merge sophisticated multivariate analytical techniques with the manipulation of image data, and we need to acquire skills in this area that are commensurate with our integral-dimension (point–arc–polygon) capabilities. We are also witnessing a convergence between image processing, which ideally reduces large amounts of data to smaller amounts of information, and process imaging, which produces large amounts of data from the information contained within small models. A dialectic between these techniques is developing.

Finally, this exercise has demonstrated how multifractal analysis of spatial data can be used to make inferences about the processes that give rise to spatial patterns. The techniques of remote sensing and geographic information system analysis are being extended by a wide range of disciplines (Holden, 1988; Chrisman, 1990). Such fields as geophysics, geomorphology, hydrology, geology, and economic geography provide a rich set of spatial models with predictions of spatial pattern that have been associated not merely with specific fractal dimensions, but with multifractal behavior. It is clear that we have only just begun to make use of a powerful idea that enables us to reach beyond the concepts and techniques of any one discipline.

ACKNOWLEDGMENTS

I thank Norman Bliss, Pat S. Chavez, Jr., and Nicholas C. Matalas of the U.S. Geological Survey and Nina Lam for their helpful comments and the staff of the U.S. Geological Survey's EROS Data Center in Sioux Falls, South Dakota, for providing imagery. This research was conducted under the auspices of a National Research Council Research Associateship at the U.S. Geological Survey National Center in Reston, Virginia.

REFERENCES

Barnsley, M. F. 1988. *Fractals Everywhere.* San Diego, CA: Academic Press.

Batty, M., Fotheringham, A. S., and Longley, P. 1989. Urban growth and form: Scaling, fractal geometry, and diffusion-limited aggregation. *Environment and Planning A* 21:1447–1472.

Burrough, P. A. 1989. Fractals and geochemistry. In *The fractal approach to Heterogeneous Chemistry: Surfaces, Colloids, Polymers*, D. Avnir, ed. New York: John Wiley & Sons.

Chrisman, N. R. 1990. Review of building databases for global science. *Annals of Association of American Geographers* 80(1):149–151.

Culling, W. E. H. 1989. The characterization of regular/irregular surfaces in the soil-covered landscape by Gaussian random fields. *Computers & Geosciences* 15(2):219–226.

De Cola, L. 1989a. Fractal analysis of a classified Landsat scene. *Photogrammetric Engineering and Remote Sensing* 55(5):601–610.

———. 1989b. Pareto and fractal description of regions from a binomial lattice. *Geographical Analysis* 21(1):74–81.

———. 1991. Fractal analysis of multiscale spatial autocorrelation among point data. *Environment and Planning A* 23:545–556.

Falconer, K. J. 1990. *Fractal Geometry: Mathematical Foundations and Applications*. New York: John Wiley & Sons.

Frenkel, K. A. 1988. The art and science of visualizing data. *Communications of the Association for Computing Machinery* 31(2):111–121.

Getis, A., and Boots, B. 1978. *Models of Spatial Process*. New York: Cambridge University Press.

Goodchild, M. F. 1982. The fractional Brownian process as a terrain simulation model. *Modeling and Simulation* 13:1133–1137.

———. and Mark, D. M. 1987. The fractal nature of geographic phenomena. *Annals of the Association of American Geographers* 77:265–278.

Holden, C. 1988. The ecosystem and human behavior. *Science* 242:663.

Holdridge, L. R., et al. 1971. *Forest Environments in Tropical Life Zones: A Pilot Study*. Elmsford, NY: Pergamon Press.

Hummel, R. 1987. The scale-space formulation of pyramid data structures. In *Parallel Computer Vision*, L. Uhr, ed., pp. 107–124. Orlando, FL: Academic Press.

Jones, J. G., Thomas, R. W., and Earwicher, P. G. 1989. Fractal properties of computer-generated and natural geophysical data. *Computers & Geosciences* 15:227–235.

Joseph, J. H. 1985. Morphology of fair weather cumulus cloud fields as remotely sensed from satellites, and some applications. *Advances in Space Research* 5(6):213–216.

Justice, C. O., et. al. 1989. Spatial degradation of satellite data. *International Journal of Remote Sensing* 10(9):1539–1561.

Kaye, B. 1989. *A Random Walk through Fractal Dimensions*. New York: VCH Publications.

Lovejoy, S., and Schertzer, D. 1988. Extreme variability, scaling and fractals in remote sensing: Analysis and simulation. In *Digital Image Processing*, J. P. Muller, ed. pp. 177–212. Philadelphia: Taylor & Francis.

Mandelbrot, B. B. 1989. Negative fractal dimensions and multifractals. *Physica A* 163:306–315.

Muller, J. P., ed. 1988. *Digital Image Processing*. Philadelphia: Taylor & Francis.

Musgrave, F. K., Kolm, C. E., and Mace, R. S. 1989. The synthesis and rendering of eroded fractal terrains. *Computer Graphics* 23(3):41–50.

Nicolis, G., and Prigogine, I. 1989. *Exploring Complexity*. San Francisco:W.H.Freeman & Co.

Roy, A., Gravel, G., and Gauthier, C. 1987. Measuring the dimension of surfaces: A review and appraisal of difficult methods. *Proceedings, Eighth International Symposium on Computer-Assisted Cartography (Auto-Carto 8)*, Baltimore, pp. 68–77.

Phipps, M. 1988. Dynamical behavior of cellular automata under the constraint of neighborhood coherence. *Geographical Analysis* 21(3):197–215.

Samet, H. 1990. The design and analysis of spatial data structures. Reading, Massachusetts: Addison-Wesley.

Schachter, B., and Ahuja, N. 1979. Random pattern generation processes. *Computer Graphics and Image Processing* 10:95–114.

Schlesinger, W. H., et. al. 1990. Biological feedbacks in global desertification. *Science* 247:104–1048.

SPOT Image Corporation 1988. *SPOT User's Handbook.* Reston, VA: SPOT Image Corp.

Szeliski, R., and Terzopoulos, D. 1989. From splines to fractals. *Computer Graphics* 23(3):51–60.

Theil, H. 1971. *Principles of econometrics.* New York: John Wiley & Sons.

Uhr, L., ed. 1987. *Parallel computer vision.* Orlando, Florida: Academic Press.

Whalley, W. B., and Orford, J. D. 1989. The use of fractals and pseudofractals in the analysis of two-dimensional outlines: Review and further exploration. *Computers & Geosciences* 15(2):185–197.

Wolfram, S. 1984. Cellular automata as models of complexity. *Nature* 31(4): 419–424.

INDEX